CP OBGYN Clinical Perspectives in Obstetrics and Gynecology

Series Editor:

The late
Herbert J. Buchsbaum, M.D.

CP OBGYN Clinical Perspectives in Obstetrics and Gynecology

---

perspective *noun:* . . . the capacity to view subjects in their true relations or relative importance.

Each volume in Clinical Perspectives in Obstetrics and Gynecology will cover in depth a major clinical area in the health care of women. The objective is to present to the reader the pathophysiologic and biochemical basis of the condition under discussion and to provide a scientific basis for clinical management. These volumes are not intended as "how to" books, but as a ready reference by authorities in the field.

Though the obstetrician and gynecologist may be the primary provider of health care for the female, this role is shared with family practitioners, pediatricians, medical and surgical specialists, and geriatricians. It is to all these physicians that the series is addressed.

Series Editor: The late Herbert J. Buchsbaum, M.D.

Published Volumes:

Buchsbaum (ed.): *The Menopause*
Aiman (ed.): *Infertility*
Futterweit: *Polycystic Ovarian Disease*
Lavery and Sanfilippo (eds.): *Pediatric and Adolescent Obstetrics and Gynecology*
Galask and Larsen (eds.): *Infectious Diseases in the Female Patient*
Buchsbaum and Walton (eds.): *Strategies in Gynecologic Surgery*
Szulman and Buchsbaum (eds.): *Gestational Trophoblastic Disease*
Cibils (ed.): *Surgical Diseases in Pregnancy*

Forthcoming Volumes:

Collins (ed.): *Ovulation Induction*
Altchek and Deligdisch (eds.): *The Uterus*

Surgical Diseases in Pregnancy

Luis A. Cibils
Editor

With 28 Illustrations

Springer Science+Business Media, LLC

Editor:
Luis A. Cibils, M.D.
Mary Campau Ryerson Professor of Obstetrics and Gynecology
University of Chicago, The Chicago Lying-in Hospital
Chicago, Illinois, USA

Series Editor:
The late
Herbert J. Buchsbaum, M.D.

Library of Congress Cataloging-in-Publication Data
Surgical diseases in pregnancy / edited by Luis A. Cibils.
p. cm. — (Clinical perspectives in obstetrics and gynecology)
Includes bibliographical references.

ISBN 978-1-4613-8981-1 ISBN 978-1-4613-8979-8 (eBook)
DOI 10.1007/978-1-4613-8979-8

1. Pregnant women—Surgery. 2. Pregnancy, Complications of.
I. Cibils, Luis A., 1927– . II. Series.
[DNLM: 1. Pregnancy Complications. 2. Surgery—in pregnancy. WQ 240 S9611]
RD144.S87 1990
618.3′059—dc20
DNLM/DLC
for Library of Congress 90-9401

Printed on acid-free paper.

Typeset by David E. Seham Associates, Inc., Metuchen, New Jersey.

9 8 7 6 5 4 3 2 1

Preface

The management of some clinical conditions occurring during pregnancy pose distinctive, difficult problems because of the physiologic changes brought about by pregnancy as well as by the need to preserve fetal well-being in the majority of cases. There are many excellent books dedicated to so-called "medical" complications of pregnancy but there is a paucity of those focusing on surgical problems, that is to say, complications requiring surgical intervention, or pregnancy following important surgical procedures.

This volume has been conceived with the idea of covering some of the more common syndromes (or situations) confronting the obstetrician-gynecologist. By the nature of the subjects discussed, it should be most useful to colleagues practicing Emergency Medicine, Surgical consultants, and, of course, general Obstetrician-Gynecologists, who may be the first to see the patient or to be called in consultation.

The authors invited to write the various chapters have had extensive experience in the management of each condition, and it is with that background that they discuss the pertinent literature and offer what they consider the best approach at this time.

The first four chapters deal with complications of the pregnancy condition, very common in daily practice, but which contribute a disproportionate number of maternal deaths in this country and the rest of the world. The permanent infertility of many who survive these syndromes attests to the difficulties in appropriate management in spite of great advances in medical technology.

The following six chapters deal with acute problems of various organ systems requiring immediate intervention. These complications may not be as prevalent as the previous four but they are as lethal when treatment is delayed.

The diagnosis of carcinoma (or a "tumor") of genital organs during gestation is, if possible, more shocking than usual because it implies a management decision directly affecting the life of the conceptus as well as the long-term prognosis of the young mother. The more common types are discussed in the subsequent three chapters.

Finally, the last two deal with situations created by the great advances in surgery in the last few decades. The possibility of a successful pregnancy following the implantation of artificial heart valves or an allograft kidney is a reality but a fragile one because the medications taken to prevent severe complications may affect the fetus. On occasion, emergency interventions may be needed to save the mother.

The authors hope that the concepts elaborated in this book, and their recommendations for treatment, will help their colleagues to choose the correct one when confronted with some of the serious problems described.

I wish to express my thanks to the authors who so kindly agreed to spend time summarizing their experience and to abide by the guidelines suggested by the publishers. To them, and to Springer-Verlag of New York, my thanks for cooperation and understanding.

Luis A. Cibils

Contents

Contributors

ALBERTO ALVARADO-DURAN, M.D.
Professor of Obstetrics and Gynecology, Universidad Nacional Autónoma de México; Medical Director, Gynecology-Obstetric Hospital, Centro Médico Nacional, Mexican Institute of Social Security, México, DF

EMILY BAKER, M.D.
Chief Resident, Department of Obstetrics and Gynecology, The Chicago Lying-in Hospital, Chicago, Illinois, USA

JOHN BARRETT, M.D.
Director, Trauma Unit, Cook County Hospital, Chicago, Illinois; Associate Professor of Surgery, University of Illinois at Chicago, Chicago, Illinois, USA

DAVID CHARLES, M.D.
Former Professor and Chairman, Department of Obstetrics and Gynecology, Marshall University School of Medicine, Huntington, West Virginia, USA

LUIS A. CIBILS, M.D.
Mary Campau Ryerson Professor of Obstetrics and Gynecology, University of Chicago, The Chicago Lying-in Hospital, Chicago, Illinois, USA

FREDRIC L. COE, M.D.
Professor of Medicine and Physiology; Director of Nephrology, University of Chicago, Chicago, Illinois, USA

JAMES R. DINGFELDER, M.D.
Associate Clinical Professor, Obstetrics and Gynecology, University of North Carolina School of Medicine, Chapel Hill, North Carolina, USA

Donald J. Ferguson, M.D., Ph.D.
Professor Emeritus of Surgery, University of Chicago, Chicago, Illinois, USA

Charles S. Field, M.D.
Head, Section of Obstetrics; Consultant, Section of Medical Gynecology and of Reproductive Endocrinology, Mayo Clinic and Mayo Foundation; Assistant Professor of Obstetrics and Gynecology, Mayo Medical School, Rochester, Minnesota, USA

Douglas D. Glover, M.D.
Professor, Department of Obstetrics and Gynecology, West Virginia University School of Medicine, Morgantown, West Virginia, USA

Carlos Ibarra-Perez, M.D.
Professor of Cardiology, Universidad Nacional Autónoma de México; Director, Department of Research and Education, Cardiology Hospital, Centro Médico Nacional, Mexican Institute of Social Security, México, DF

Mahmoud A. Ismail, M.D.
Associate Professor of Obstetrics and Gynecology, University of Chicago, The Chicago Lying-in Hospital, Chicago, Illinois, USA

Mark H. Kadowaki, M.D.
Resident in Surgery, University of Chicago, Pritzker School of Medicine, Chicago, Illinois, USA

Edwin L. Kaplan, M.D.
Professor of Surgery, University of Chicago, Pritzker School of Medicine, Chicago, Illinois, USA

Marshall D. Lindheimer, M.D.
Professor of Medicine and Obstetrics and Gynecology, University of Chicago, Chicago, Illinois, USA

Patsy Maikranz, M.D.
Former Fellow, Department of Internal Medicine, University of Chicago, Section of Nephrology, Chicago, Illinois, USA; Current Physician, Indiana Nephrology and Internal Medicine PC, Indianapolis, Indiana, USA

Fabrizio Michelassi, M.D.
Associate Professor, Department of Surgery, University of Chicago, Pritzker School of Medicine, Chicago, Illinois, USA

A.H. Moawad, M.D.
Professor of Obstetrics and Gynecology, University of Chicago, The Chicago Lying-in Hospital, Chicago, Illinois, USA

JOSE NIZ-RAMOS, M.D.
Professor of Obstetrics and Gynecology, Universidad Nacional Autónoma de México; Director of Perinatology, Gynecology-Obstetric Hospital, Centro Médico Nacional, Mexican Institute of Social Security, México, DF

ANDREA PIETRABISSA, M.D.
Istituto di Patologia Chirurgica, University of Pisa, Pisa, Italy

KARL C. PODRATZ, M.D., Ph.D.
Chairman, Department of Obstetrics and Gynecology and Consultant, Section of Gynecologic Surgery, Mayo Clinic and Mayo Foundation; Professor of Obstetrics and Gynecology, Mayo Medical School, Rochester, Minnesota, USA

WOLFGANG H. SCHRAUT, M.D., Ph.D.
Professor of Surgery, Department of Surgery, University of Pittsburgh School of Medicine, Pittsburgh, Pennsylvania, USA

ELIZABETH K. SENEKJIAN, M.D.
Resident in Psychiatry, Boston University School of Medicine, Boston, Massachusetts; Former Associate Professor, Department of Obstetrics and Gynecology, University of Chicago, Chicago, Illinois, USA

FRANK P. STUART, M.D.
Professor, Department of Surgery; Director of Organ Transplantation, University of Chicago, Chicago, Illinois, USA

JOAN K. STUART, R.N., B.S., M.A.
Clinical Research Associate, Department of Surgery, University of Chicago, Chicago, Illinois, USA

RICHARD L. SWEET, M.D.
Professor, Obstetrics, Gynecology, and Reproductive Sciences, University of California San Francisco, San Francisco, California, USA

1
Incompetent Cervical Os

DAVID CHARLES and DOUGLAS D. GLOVER

Deferral of childbearing until later in life and the anticipation of a small, planned family mean that the emotional impact of recurrent pregnancy failures is now far greater than it used to be. When a planned pregnancy is interrupted for any reason, the parents expect competent, detailed counseling.[1] Because of changing clinical social attitudes, a greater responsibility is placed on the physician who has to evaluate the etiology of the unsuccessful pregnancy. Likewise, physicians consulted by such patients must realize that they have a pivotal role in the success of any subsequent pregnancy.

Much interest has been generated in cervical incompetence as a cause of recurrent second trimester abortion. The term *incompetent cervix* is attributed to Lash and Lash,[2] although in 1865 Gream[3] pointed out that a damaged cervix affects subsequent pregnancy outcome. Little attention was paid to Gream's report probably because the scourge of puerperal infection overshadowed all other obstetric complications. He lived in the era of high forceps and traumatic vaginal delivery when the more venturesome physician employed forcible dilatation of the cervix either manually or instrumentally *(accouchement force)* to expedite delivery of the fetus. This procedure was one of the most formidable obstetric operations at the end of the last century, but such trauma to the cervix was not implicated as a major factor in pregnancy wastage.

Early in this century it was suggested that cervical lacerations might result in abortion of a subsequent gestation, and as a result repairs of such lacerations were considered a method of avoiding such pregnancy loss. Various forms of trauma have been considered to result in structural defects of the cervix and thereby cause cervical insufficiency.

The first in-depth investigations on the histopathology of the cervix were reported by Aschoff[4] in 1905. He described in some detail the morphologic features of the uterine cervix but failed to provide a structural explanation of the role of the cervix in retaining a conceptus in the uterus.

Stieve,[5] describing the histology of the cervix, stated that it consisted largely of smooth arranged in a circular manner. In addition, he identified a poorly defined layer of longitudinal muscle medial to the circular muscle and a well marked longitudinal layer of smooth muscle on the outside. He noted that the muscular component was markedly reduced in the lower portion of the cervix and was replaced by connective tissue. Stieve was, however, unable to demonstrate the existence of a cervical sphincter. It is important to recognize the means by which the cervix acts as a retaining sphincter and how it adapts to the progressively expanding gestational sac during pregnancy. The internal cervical os is a ridge of the upper end of the cervical canal that recedes into the uterine cavity. To consider the internal cervical os merely as a fixed point gives too static a view, as the internal cervical os and the adjacent supravaginal cervix are, from a physiologic point of view, part of the uterine corpus and thereby are part of the wall of the gestational sac.

The junction between the corpus and the cervix was defined by Danforth[6] as a fibromuscular ring. The composition of the cervix below this point is predominantly fibrous tissue. On the other hand, the upper part including the anatomic internal os is predominantly muscular and comprises the isthmus of Aschoff. It is now accepted that the cervix is composed predominantly of connective tissue in which collagen provides the mechanical strength that is necessary to retain the conceptus in the uterus. Various studies have demonstrated that the connective tissue component of the cervix undergoes important changes during pregnancy and parturition. Effacement and dilatation of the cervix probably involve partial breakdown of collagen, which requires the specific action of collagenase. Studies by Leppert and associates[7,8] indicated that elastin plays a role in the mechanism of cervical dilatation. A decreased elastin has been observed in incompetent cervices, which may be a significant factor in cervical insufficiency.

In the nongravid state the cervix is rigid and fibrous, but during pregnancy it softens. This change is believed to be the result of the activity of estrogens, progesterone, prostaglandins, and relaxin.

Cervical incompetence is basically a defect in the sphincteric mechanism of the internal os, which is usually an acquired state, although rarely it may be congenital. When a critical weight is reached during gestation the contents of the gravid uterus begin to extrude because of this defect. Herniation and rupture of the fetal membranes then occur, and there is no doubt that this entity during the middle trimester of pregnancy is related to incompetence of the supravaginal cervix. Dilatation of he cervix and bulging of the fetal membranes were assumed in the past to occur only in response to uterine action in such a manner that abortion was inevitable. This concept is no longer valid, and the acceptance of a mechanical defect thus introduces the possibility of surgical correction. Palmer and Lacomme[9] indicated a possible relation between isthmic cervical os weakness and the occurrence of abortion during the second trimester of a pregnancy. Consequently, such observations have allowed one to enunciate the definition of an incompetent internal cervical os as one that is incapable of maintaining a gestation until such a time that a viable fetus can be delivered.

Incidence

It is difficult to assess the incidence of cervical incompetence in a given community. Direct comparison of the incidence of this entity in diverse obstetric populations is compounded because of selective referral of high risk patients to certain institutions and the haste with which some women with recurrent abortions are evaluated and treated empirically by surgical procedures. The frequency of cervical incompetence in relation to the number of live births has not been accurately determined. The incidence of cervical cerclage has been variously reported to range from 1 per 54 to 1 per 1000 deliveries (Table 1-1). Such a marked difference may be partly explained by the variable criteria used for the diagnosis. There is no doubt that many individuals are treated for cervical incompetence when it does not exist. In fact, in some institutions cervical incompetence is probably one of the most overdiagnosed and overtreated conditions.

Stromme and Haywa[15] considered that

TABLE 1-1. Frequency of cervical cerclage procedures per deliveries.

Authors	Frequency
Toaff et al.[10]	1/54
Kuhn & Pepperell[11]	1/182
Barter et al.[12]	1/850
Ritter[13]	1/350
Lipshitz[14]	1/1000

cervical incompetence may account for 16% of second trimester abortions. Mann and associates[16] were of the opinion that one in every five cases of spontaneous midtrimester abortion is the result of an incompetent cervix. If so, it must be concluded that if more than 20% of cases of midtrimester abortion are diagnosed as due to cervical incompetence in any institution it is far too frequently diagnosed. It is well to remember that second trimester abortion can result from causes other than cervical incompetence, and the patients who have had a midtrimester abortion must not all be considered to have an incompetent cervix.

Etiology

Cervical incompetence can be either congenital or acquired, but in most patients it is an acquired condition caused by structural damage to the internal cervical os. Most authorities acknowledge the existence of the congenital form of cervical incompetence as a cause of pregnancy loss, albeit infrequent; and in fact Lash and Lash[2] did not consider it a cause of pregnancy wastage.

The diagnosis of congenital cervical incompetence is entertained when the cause of the initial pregnancy loss is consistent with incompetence of the cervix and there is no history of cervical trauma. Many have postulated the causes of cervical incompetence in a primigravid patient, and the terms congenital, dysfunctional, and physiologic have been used. Sherman[17] believed that abnormal myometrial activity in the lower uterine segment and cervix, in the absence of true labor, may result in cervical dilatation and preterm delivery. Congenital cervical incompetence may be associated with müllerian duct anomalies, but hysterographic studies rarely reveal any abnormality of the upper genital tract.

Palmer,[18] from radiologic investigations, concluded that cervical incompetence is infrequently associated with the more marked degrees of müllerian abnormalities such as uterus didelphys but is more commonly encountered in conjunction with the lesser degrees of bicornuate uterus. Heinonen and associates[19] evaluated 265 pregnancies in 126 women with uterine anomalies and noted that 17% of the patients with a partial bicornuate uterus had improved fetal salvage after a cerclage procedure for cervical incompetence.

Congenital malformations of the uterus, such as uterus didelphys and septate uterus, are known to be associated with a significant increase in the incidence of preterm labor, which in some reports is as high as 30%. The concept of a congenital defect of the cervix as a cause of pregnancy failure has been advanced by Jennings,[20] who reported cervical incompetence in four pairs of sisters. Peters and associates[21] also reported cervical incompetence in twin sisters. These investigators reported that 31% of their patients had congenital weakness of the cervix resulting in cervical insufficiency during their initial pregnancy.

Peters et al.[21] also reported that after completion of their study they evaluated two primigravid patients with an incompetent cervix who had been exposed in utero to diethylstilbestrol (DES). Singer and Hochman[22] reported the first case of cervical incompetence in a patient who had been expose to DES. That same year Goldstein[23] reported five patients who had had exposure to DES; four had had successful pregnancy outcomes following a cerclage procedure as soon as cervical dilatation was observed. This investigator suggested that when gross anatomic abnormalities of the cervix are noted a hysterosalpingogram should be performed; and if a cervical abnormality is apparent, a prophylactic or early second trimester cerclage is done. Nunley and Kitchin[24] likewise reported a successful outcome of pregnancy following cerclage in a DES-exposed patient who had an incompetent cervix. Kaufman and associates[25] reported gross anatomic changes of the upper genital tract of females exposed to DES in utero. Their hysterosalpingographic studies demonstrated that the endometrial cavity was T-shaped and the uterus hyperplastic in a high percentage of these women. Haney and associates,[26] in a study of women exposed to DES, observed a markedly narrow endocervical canal in the patients who had been exposed to this agent. They considered that functional impairment of the internal cervical os could be present in such patients despite the absence of demonstrable anatomic changes.

Cousins[27] compared the reproductive histories of 71 women who had been exposed to DES with those of 69 demographically matched women who had had no such exposure. This study failed to reveal either a significant difference in the number of pregnancies and frequency of infertility problems between the two groups or any difference in the incidence of spontaneous abortion. Cousins stated that the incidence of preterm labor and perinatal mortality significantly increased in women who had been expose to DES. Furthermore, it was noted that preterm delivery is three times as common among women who exhibited such features of DES exposure as vaginal adenosis, extensive ectropion, and deformities of the anterior cervical lip. Prematurity was positively related in this study to the presence of cervical incompetence.

Barnes and associates[28] made no mention of cervical incompetence as a pregnancy complication in their report of 289 individuals who had been exposed to DES. Sandberg and associates,[29] in a report pertaining to 167 parous women who had been exposed to DES in utero, stated that seven of the women had been diagnosed as having cervical incompetence during pregnancy. These investigators, however, reported that it was impossible to determine the appropriateness of the diagnosis because it was made in other institutions. In addition, the histories of the individuals implied that the clinical management of most of them was dictated by the premise that cervical incompetence was a proved pregnancy complication of any woman exposed to DES during intrauterine life. Haney and associates[26] considered that a functional impairment of the integrity of the internal cervical os was probable in patients exposed to DES despite the absence of any typical anatomic changes, and that this condition may manifest when the individual becomes gravid. In their opinion, attempts at operative correction of these defects should await the same documentation as in the patient without DES exposure, i.e., proved second trimester abortion.

The extensive report of uterine and tubal anomalies due to intrauterine DES exposure by Kaufman and associates[25] suggests that most uterine and cervical changes are observed when exposure to DES had occurred prior to the 18th week of gestation; and, furthermore, that the incidence of uterine anomalies is 91% when anomalies of the cervix are present. Although habitual abortions, premature deliveries, and premature rupture of the membranes may be attributed to upper genital tract anomalies in women who had been exposed to DES during intrauterine life, it must be emphasized that an increased awareness of the possibility of in utero DES exposure must be entertained in women who have a history of repeated pregnancy loss. However, one should not go to the extent of performing a cerclage procedure during the first pregnancy in any DES-exposed patient when uterine anomalies are demonstrated, as is recommended by Ben-Baruch and associates.[30] Prophylactic cerclage procedures are not justified in DES-exposed individuals who become pregnant. If such a patient has had a previous premature delivery, she is instructed to seek prenatal care at an early date during any subsequent pregnancy. It is important for physicians to realize that, although reproductive performance is DES-exposed women has been associated with an increased risk of an unfavorable outcome, more than 80% of such women have successfully delivered one or more live-born infants.

It had been suggested that cervical lacerations result in abortion during a subsequent pregnancy. Consequently, repairs of such lacerations were considered to be a method of preventing future pregnancy loss. Various forms of trauma have been considered to result in structural defects of the cervix and therefore cause cervical incompetence. There is no unanimity, however, on whether cervical conization is associated with an increased incidence of abortion or incompetence of the cervical os. It is agreed that the more extensive the conization procedure the greater is the risk of cervical incompetence and midtrimester abortion during a subsequent pregnancy. Cervical conization has been used with increased frequency as both a diagnostic and a therapeutic measure for individuals with cervical carcinoma in situ. Consequently, an increasing number of women of reproductive age are subjected to this procedure, and various reports have indicated that it affects fertility.

Weber and Obel,[31] from their review of the literature, found no support for the concept that there was an increased spontaneous abortion rate in women following cervical conization. They did, however, note a slightly increased risk of prematurity. These authors indicated that virtually all the studies that had been published at that time had been uncontrolled and thus difficult to interpret appropriately.

Buller and Jones[32] evaluated the role of cervical cone biopsy on fertility, pregnancy, labor, and delivery among a group of patients who served as their own controls, as they had had pregnancies prior to the conization. They concluded that conization did not influence the rates of spontaneous abortion, preterm delivery, or cesarean section in their patients. The only change they noted from the conization procedure was an alteration in the duration of the first stage of labor.

Larsson and associates,[33] in a review of 635 pregnancies in 197 women before and after conization, noted that there was a significant increase in premature deliveries in young women in the age group 21 to 25 after conization. In addition, they observed that the highest risk for preterm labors was among nulliparous patients. They reported that prophylactic cerclage did not reduce the incidence of premature labor or late spontaneous abortion in those patients.

The conflicting findings reported in the literature on the effects of cervical conization on future pregnancies might be explained by differences in operative technique. Leiman and associates[34] reported that the risk of premature delivery and late spontaneous abortion increased in direct proportion to the size of the cone. These investigators found a correlation between small cones and cervical stenosis. Cold conization has now been replaced in many instances by colposcopically guided laser conization, which probably will reduce the proportion of unnecessary large cones and thereby minimize adverse sequelae of conization. Most reports indicate that unless an unduly large cone is performed the risks of cervical insufficiency during a subsequent pregnancy are no greater than in women who have not undergone such a procedure. Moinian and Andersch[35] failed to find any difference in the incidence of spontaneous abortion following conization of the cervix.

Lash and Lash[2] ascribed the etiology of cervical incompetence to trauma of the cervix. In fact, most authors consider trauma to the cervix at the time of delivery or vaginal termination of pregnancy as the most frequent cause of cervical incompetence. Kantor,[36] discussing cervical incompetence, stated: "It is interesting to speculate that cervical dilatation for voluntary interruption of pregnancy, especially in the primigravida, may sufficiently disrupt the musculature of the cervix to cause incompetence in a later pregnancy." Ratten and Beischer,[37] in a series of 520 patients who had previously undergone termination of a first trimester pregnancy, noted that 8.1% experienced midtrimester abortion during their subsequent pregnancy and 8.6% had premature onset of labor, compared to 2.4% and 4.4%, respectively, in the controls. These authors concluded that patients undergoing termination of pregnancy must be told of the risk of cervical incompetence and that their future obstetrician be informed of the abortion. This study supports the concept that therapeutic termination of pregnancy during the first trimester is associated with a statistically significant increase in the incidence of second trimester abortion and premature labor.

There is some evidence that cervical dilatation attained by the use of Hegar dilators carries a greater risk of subsequent cervical incompetence when compared with dilatation of the cervix by laminaria tents. The latter have the advantage of being the only available modality that acts directly on the cervix without necessarily stimulating uterine contractions. They act mechanically by swelling and slowly dilating the cervix; however, it is of interest that they contain a considerable amount of arachidonic acid, which is a prostaglandin precursor. During dilatation of the cervix, arachidonic acid or its metabolites may stimulate collagen breakdown on the cervix. It is also possible that laminaria tents induce softening of the cervix by altering the ground substance. Follow-up studies have shown a low incidence of cervical incompetence during subsequent pregnancies when laminaria tents have been employed.

Prostaglandins E_2 and $F_{2\alpha}$ are used for dilating the cervix prior to vacuum aspiration, second trimester abortion, ripening of the cervix, and labor induction. It is generally agreed that the frequency of complications with vacuum aspiration increases with advancing gestational age. Some of these complications, e.g., cervical laceration, are directly related to the mechanical dilatation necessary for the procedure, especially during the last weeks of the first trimester or the initial weeks of the second trimester of pregnancy. Although most first trimester abortions are done surgically and complications are generally few, potential risks from injury can be associated with rapid mechanical dilatation of the cervix. Prostaglandins, specifically the analogs were found by Rath and associates[38] to reduce the collagen content and ripen the cervix thereby diminishing the risk of cervical incompetence during a subsequent pregnancy.

Occasionally difficulties arise when dilating the cervix for termination of a pregnancy. If cervical trauma occurs at the time of pregnancy termination, it often goes unnoticed; and consequently any predisposition to subsequent cervical incompetence cannot be assessed.

When cervical incompetence is present, the exact cervical pathology is not clearly defined. Some patients may have suffered either obstetric or surgical trauma to the cervix, and the absence of such a history suggests some form of congenital defect. In other instances an abnormal cervical response to the hormonal changes of pregnancy has been invoked as a cause of cervical incompetence. Asplund[39] reported hysterographic studies that demonstrated the internal cervical os to be narrower during the secretory phase than the proliferative phase of the menstrual cycle. Borell and Fernstrom[40] observed cyclic alterations in the diameter of the isthmus and cervical canal that can be reproduced by the administration of estrogens and progesterone. Their work was confirmed by studies conducted by Youssef.[41] Estrogens may directly affect the ground substance of the cervix and alter the water content, collagen synthesis, and metabolism. As previously mentioned, it is well established that prostaglandins are necessary for normal cervical ripening during pregnancy. Premature cervical effacement and dilatation may result because of inappropriate production of prostaglandins. Prostaglandins are known to be blocked by progesterone, which may explain the efficacy of progesterone in the management of some cases of cervical incompetence.

Diagnosis

There is no irrefutable method for diagnosing the presence of cervical incompetence. Classically, the reproductive history associated with incompetence of the cervix is one of recurrent second trimester abortion or premature labor with painless dilatation unaccompanied by bleeding. As stated above, one of the causes of morbidity during the second trimester of pregnancy is an incompetent cervix. The reproductive history of an individual with cervical incompetence presents as painless dilatation of the cervix, rupture of the fetal membranes during the second trimester without prior contractions, delivery of the fetus after a short, relatively painless labor, and the loss of one or more pregnancies during the second trimester associated with these symptoms. As indicated, establishment of a diagnosis of cervical incompetence in most instances is based on the clinical history. It is often difficult to investigate all the possible etiologies in many of the patients with a poor reproductive history. Likewise, patients who have had a successful pregnancy following a cerclage operation undergo an identical procedure during subsequent pregnancies without a firm diagnosis of this entity. Other causes of midtrimester fetal loss, such as amnionitis, abnormal placentation, submucous leiomyomas, and uterine abnormalities, must always be excluded.

Many patients furnish a history of previous cervical trauma due to abortion, dilatation of the cervix, previous traumatic obstetric deliveries, or cervical surgical procedures. As pointed out by McDonald,[42] however, approximately 2% of patients have no apparent etiology, and their form of incompetence is often believed to be congenital.

Various methods have been advocated to help establish a diagnosis of cervical incompetence during interconceptional care. The ready passage of a No. 8 Hegar dilator into the cervical canal, without resistance, is suggestive of cervical incompetence. Although some authors consider the obstetric history as the most important factor when formulating a diagnosis of cervical incompetence, others disagree. It is apparent that the clinical presentation is not uniform in all patients whose gestations are interrupted by this complication. Undoubtedly, the incidence would be higher and the results of therapy better in any series of patients where significant numbers were treated on the basis of history alone. It is important to carefully evaluate the cervix at weekly intervals in patients with a history of recurrent second trimester abortion. Such as assessment of the cervix is important, as the condition of the cervix in the nongravid state does not reflect its ability to retain a conceptus. Therefore, as indicated above, the mere presence of a patulous cervix and the ready passage of a No. 8 Hegar dilator may not indicate that a patient will have cervical incompetence in a future pregnancy.

Hysterography

Hysterosalpingography, which was initially advocated by Jeffcoate and Wilson[43] for the diagnosis of cervical incompetence, has been used by numerous investigators who consider that abnormal funneling of the cervical isthmus is diagnostic of this entity. Palmer[18] had previously reported that the region of the cervical isthmus may be abnormally wide at hysterography in individuals with a history of recurrent abortion. From his use of hysterosalpingographic studies, Asplund[39] reported an average isthmus width of 2.63 mm with an increased diameter to 6.09 mm in patients with proved cervical incompetence. He concluded that in cases of recurrent abortion in which no other anomaly was discernible the internal os was often abnormally wide.

Hysteroscopy can confirm cervical incompetence by the ready introduction of the instrument into the uterus. It frequently also reveals loss of the circular shape of the internal cervical os.

Ultrasonography

Ultrasonic visualization of the cervix during pregnancy has been shown to be of value in the diagnosis of this condition. The technologic advances in ultrasonography have resulted in improved understanding of normal and abnormal anatomy. Sarti and associates[44] suggested ultrasound evaluation of patients with a history of second trimester abortions for management as well as diagnostic purposes. These investigators recommended that the studies, in patients with suspected cervical incompetence, be performed with the urinary bladder completely distended and then repeated after partial emptying.

Brook and associates[45] confirmed the efficacy of ultrasonography for the diagnosis of cervical incompetence. They evaluated 24 patients between the 11th and 18th weeks of pregnancy because of a history of second trimester abortion. A group of 19 patients at a comparable stage of gestation, but with no such history, were used as controls. The width of the internal os was measured for both groups and revealed that the mean width of the internal os of the patients with cervical incompetence was 2.75 ± 0.36 cm compared with 1.67 ± 0.23 cm in the control group. Brook et al. concluded that an internal cervical os that was 1.9 cm or more in width was highly suggestive of cervical incompetence. Large randomly selected groups of patients, however, must be evaluated by the technique before their conclusion can be accepted.

Bernstine and associates[46] conducted prospective studies in patients who were at risk of cervical incompetence during pregnancy and paid particular attention to the diameter of the internal os and length of the cervix. The internal os is not visible or palpable clinically, yet ultrasonography can provide accurate information about its dilatation and shape. These authors demonstrated that the cervical length during pregnancy varies from 3 cm to about 6 to 7 cm. Their report was confirmed by Zemlyn,[47] who emphasized that patients at risk may have a progressively shortened cervix and consequently should benefit from serial ultrasonographic evaluation.

The routine use of ultrasonographic screening of all gravid patients is controversial, but there is agreement that ultrasonographic assessment of the cervix may define a specific group of patients who require some type of operative procedure. Such an evaluation may be helpful in patients with a history of cervical trauma, conization of the cervix, or instrumental dilatation of the cervix, as well as individuals who have been exposed to DES in utero.

Costantini and associates[48] reported that by ultrasound studies the internal uterine os can be measured with electronic calipers after identification of the cervical canal. There is no doubt that by paying due attention to detail and understanding the technical problems associated with ultrasonographic identification of the cervix and cervical canal, competent and reliable data can be obtained. It is a safe technique that can be used in gravid patients to assess all forms of cervical dysfunction that cannot be detected by studies performed on nongravid patients. Ultrasonography cannot always be considered the first test, but it is certainly a reliable test for the diagnosis of cervical incompetence. Jackson and associates[49] stated that ultrasonic monitoring can be used to choose the optimum time for operative interference and to decide when to perform and to determine the most appropriate point for placement of the cerclage. Furthermore, ultrasonic follow-up studies can be used to assess the results of surgical therapy. For such reasons ultrasonography should be performed during early pregnancy on a woman with a history suggestive of cervical incompetence, as it allows the clinician to establish an accurate diagnosis and perform the cerclage procedure at the optimum time.

Treatment

As Kipling wrote: "There are nine and sixty ways of writing tribal lays — and — every — single — one — of — them — is — right!" Many treatment modalities have been proposed for cervical incompetence, but some form of cerclage procedure has received the greatest acceptance.

Pessary

Numerous authors have implied that simple bed rest or the use of a Hodge pessary is as efficacious as any operative procedure. Vitsky[50] was the first author to describe the use of a Hodge pessary and concluded that its effectiveness was the result of posterior displacement of the cervix. Over the ensuing years several articles appeared indicating that it was an acceptable approach to the problem, albeit without confirmation until the late 1970s, when several papers from Europe appeared confirming the usefulness of this procedure in averting second trimester abortion in women with a history of cervical incompetence.[51,52] Although the pessary seems to be effective, it has received little support from contemporary obstetricians.

Hormone Therapy

Another therapeutic approach that has been tried but never used extensively is progestin therapy. The premise for the administration of such agents is the proposed blocking effect of progesterone on uterine contractility. Sherman[17] administered 17α-hydroxyprogesterone in a weekly dose of 500 to 1000 mg to women with cervical incompetence. His 74 gravid patients were considered, from their history, to have cervical incompetence. These women had a total of 114 previous pregnancies associated with a 25.1% survival rate. They received the progestin for the duration of their pregnancies, and 68 of the 74 (91.9%) were delivered of healthy infants. The use of high doses of progestational agents has not been accepted because of possible teratogenic effects of progestins, although a report by Wilson and Brent[53] suggested that nongenital malformations are rare.

There is no doubt that the mere support of an understanding physician and bed rest can do a great deal for women who have been afflicted with recurrent abortions. Habitual abortion has the attached stigma of progesterone deficiency leading to inadequate decidua formation. The success of progestins could merely be due to such ancillary measures as bed rest and psychological support. Critical observers are not convinced of any benefit from progestin therapy. Some physicians advocate medical therapy, which includes bed rest, progestational agents, tocolytics, and antibiotics. Most clinicians concur that bed rest, which is included in all perioperative measures, is an important part of the therapy. There are criticisms of bed rest, however—specifically because of the increased possibility of phlebitis and phlebothrombosis. Staying in bed for any length of time is virtually impossible for a woman who has dependent children.

Surgical Treatment

For the past several decades, surgical therapy in the form of cerclage has been the most popular approach to cervical incompetence.

Nongravid Patients

In the nongravid patient who has a history of recurrent pregnancy loss as well as clinical and radiologic evidence of cervical incompetence, the raison d'etre of the surgical approach is to enhance the resistance of the cervix by repairing the anatomic defect.

Palmer and Lacomme[9] advocated excising a wedge of tissue and reconstructing the cervical canal in the nongravid state. Lash and Lash[2] also believed that operative correction is appropriate in the nongravid state; they recommended excising a diamond-shaped area of tissue at the level of the internal os and repairing the defect with two layers of chromic catgut suture. They made a deliberate attempt, by such measures, to reshape the cervix on the assumption that the defect is localized and usually located anteriorly in the midline. It is estimated that in their procedure at least one-third of the circumference of the canal must be removed to produce effective narrowing. In a few cases, however, a localized defect is discernible on palpation or by examination of the tissue. The procedures conducted in the nongravid patient have only limited applicability, as there is concern about their effect on subsequent fertility. By modern standards such procedures have proved less than adequate and therefore have been abandoned by most physicians.

Gravid Patients

Most operations for an incompetent cervix that are conducted in pregnant patients attempt to produce total or partial occlusion of the cervical canal by passing an encircling band or ligature at the level of the internal os. A number of methods have been proposed, but currently the Shirodkar[54] and McDonald[55] techniques are the principal operations used for correcting cervical incompetence. Shirodkar,[54] with whose name the repair of the incompetent cervix is inseparably associated, in 1955 described a technique of encircling the cervix at the level of the internal os with a band of fascia lata. In a subsequent communication in 1960 he recommended nonirritative synthetic tapes instead of the original autogenous strip of fascia lata.[56]

Shirodkar Operation

The Shirodkar procedure (Fig. 1-1) is conducted under general anesthesia, with the patient in the lithotomy position and a slight Trendelenburg tilt. The cervix is gently grasped with sponge forceps, one applied to the anterior lip and one to the posterior lip. It is preferable to use nonserrated tissue forceps to grasp the cervix, thereby obviating undue traction and ensuring minimal cervical trauma. A small transverse incision is then made on the anterior fornix at the level of the cervicovaginal junction. Such an incision allows mobilization of the bladder and its reflection above the level of the internal os. A small vertical incision is made in the midline on the posterior cervicovaginal junction; these incisions allow passage of an aneurysm needle around the cervix carrying a cerclage suture. The suture material is available on swaged atraumatic needles and can be readily tied so that the diameter of the endocervical canal is markedly reduced. It is important to avoid placing the cerclage tape too deeply because of the risk of rupturing the fetal membranes; it must also not be placed too superficially because it may slip off and thus defeat the purpose of the operative procedure. Anchoring sutures can be used to ensure stabilization of the en-

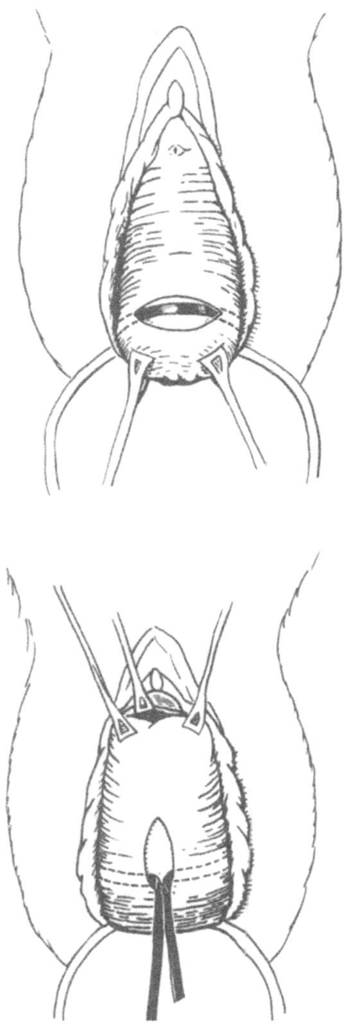

A

B

FIGURE 1-1. Shirodkar method of cervical cerclage. (A) The anterior surface of the cervix is exposed and the vaginal epithelium incised at the bladder reflection. The bladder is elevated to expose the supravaginal cervix and the mersilene band inserted. (B) The posterior aspect of the cervix is exposed through a sagittal incision and the band passed around the cervix. The ends of the cerclage material are ready to be tied, after which the vaginal incisions are repaired.

circlement material to the cervical stroma. The operation is then concluded by closing the vaginal incisions with interrupted chromic or vicryl sutures. Many modifications of the original Shirodkar procedure have been introduced, and few today perform the operation as originally described because of the wide dissection required. The site of the internal os is just above the level of the uterosacral ligaments, and therefore it is necessary to insert the cerclage above these ligaments and through the base of the cardinal ligaments if a satisfactory correction of the cervical incompetence is to be attained at the correct level. The operation and its variations are used extensively today, and presumably all operators are convinced that they improve fetal salvage. Objective evidence of the value of the procedure, however, is more difficult to obtain.

McDonald Procedure

Success has been obtained with many methods, but simplicity has its advantages. McDonald[55] treated 70 patients with cervical incompetence using a simple pursestring suture inserted by successive bites at the level of the cervicovaginal reflexion. The McDonald technique (Fig. 1-2) does not, however, make the cervix mechanically competent, although it is used extensively because of its simplicity and the fact that it involves no dissection of the cervicovaginal anatomy. The procedure described by McDonald in 1978[57] utilizes a simple pursestring suture of No. 4 mersilene. After the bladder is empty the suture is inserted on an atraumatic needle, and four or five bites are made circumferentially in the cervix. The suture is then tied tight enough to reduce significantly the diameter of the internal os. This pursestring suture is adequate as a primary measure in most cases, as it produces less trauma than the cerclage of the original Shirodkar operation. McDonald[58] reported that of 148 patients treated by pursestring suture, 86.5% had live-born infants. However, on reviewing the literature, it is apparent that both the Shirodkar and McDonald techniques are efficacious methods of treating cervical incompetence in the gravid patient.

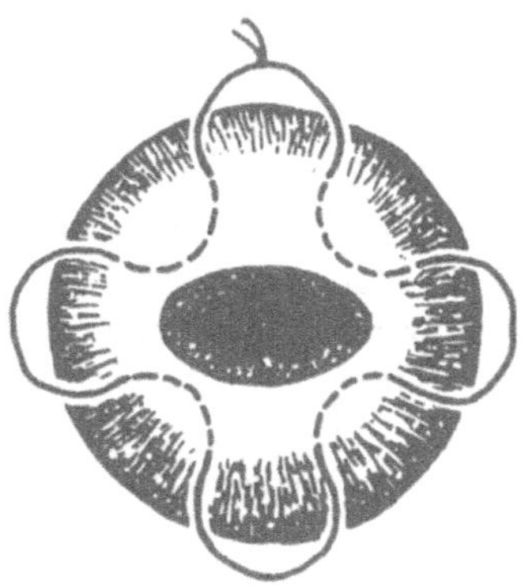

FIGURE 1-2. McDonald method of cervical cerclage. The suture is inserted deep into the cervical tissue at four sites in such a manner that when tied anteriorly it almost completely closes the internal os.

McDonald was of the opinion that the cerclage procedure must be performed about the 14th week of gestation, as this timing allows for early and natural abortion from chromosomal anomalies and avoids potential problems caused by bulging fetal membranes. The choice of anesthesia is also important because no drug that is likely to stimulate intrauterine contractions should be administered. Most anesthesiologists use thiopental induction followed by inhalation anesthesia, such as nitrous oxide and oxygen, with an intravenous relaxant. Such a combination is safe to mother and fetus and does not stimulate uterine contractions.

The pursestring suture is adequate as a primary measure in most instances and produces less trauma than the Shirodkar method. It permits vaginal delivery by simply removing the suture at or near term.

Abdominal Procedures

Sometimes it is not feasible to perform cervical cerclage by means of a vaginal approach. It is particularly true when a high amputation of the cervix has been performed in the Manchester Fothergill operation or when the cervix has been shortened following unsuccessful cerclage procedures. Under such circumstances a transabdominal approach is indicated. Hunter and associates[59] advocated the abdominal approach for cervical cerclage in the presence of vaginal stenosis. Benson and Durfee[60] emphasized that careful case selection and meticulous attention to detail are mandatory if a successful cerclage is to be attained by a transabdominal approach. They stated that transabdominal cervicoisthmic cerclage is ideally suited for patients whose cervical incompetence is complicated by a deformed cervix.

The transabdominal procedure may be performed by entering the peritoneal cavity by either a vertical or a transverse incision and then dividing the uterovesical fold of peritoneum so that the bladder can be reflected from the uterus. After the uterine artery and vein on either side have been identified, the posterior leaf of the broad ligament is perforated in an avascular area on each side so that a 15-cm segment of mersilene ribbon, 0.5 cm in width, can be passed circumferentially, medial to the vessels, around the cervical isthmus. The mersilene ribbon is grasped and brought forward on the other side so that the free ends lie anterior to the internal os. No attempt is made to bury the posterior part of the cerclage material, but the free ends are tied anteriorly by a single reef knot. Benson and Durfee[60] stated that anchoring sutures to hold the mersilene ribbon in place are not necessary because as the gestation progresses the enlarging uterus above the cerclage and the uterosacral and cardinal ligaments inferiorly keep it in place. After placement of the mersilene ribbon, the peritoneum and abdomen are closed according to standard procedures. This form of cervical cerclage is permanent, and all subsequent pregnancies must be delivered by cesarean section.

Watkins[61] reported that the transabdominal procedure is highly efficacious when cervical incompetence is the result of high amputation of the cervix. This author emphasized that resort to such an operative procedure is necessary when it is technically not feasible to insert a cervical suture by the vaginal route. He recommended that this procedure be performed only after the 12th week of pregnancy because of the need for laparotomy to divide the band should abortion occur. Mahran[62] considered that the transabdominal approach avoids the risk of dissecting through contaminated tissues, as is sometimes necessary with the Shirodkar procedure. He modified the operative technique by transfixing the uterine isthmus as well as the transacral and cardinal ligaments by the cerclage material. As a result, when the cerclage is properly tied, it is unlikely to be displaced. He recommended that the transabdominal approach be used whenever the transvaginal procedure is fraught with the risk of failure, specifically if there are large forniceal lacerations from the cervix.

In a review of the literature pertaining to cervical cerclage, Cousins[27] reported an impressive improvement in overall pregnancy outcome by both the vaginal and transabdominal procedures. In the case of the latter surgical procedure, he recorded that a viable pregnancy was achieved in 91 of 107 cases (85%) compared to 34 of 185 cases (18.4%) prior to operative intervention.

Novy[63] indicated that the transabdominal approach is essential for adequate therapy of cervical incompetence in a select group of patients—specifically those who have had unsuccessful pregnancies after a transvaginal approach or individuals in whom a satisfactory transvaginal cerclage is not technically feasible. The major disadvantage of the transabdominal approach is that a further abdominal operation is necessary to deliver the child. During a 15-year period there were 26,730 deliveries at Novy's center, and cervical cerclage (by transvaginal or transabdominal route) was performed in 103 pregnancies during that time. That is, some form of cerclage procedure was conducted on one of every 260 pregnancies (0.38%). Novy reported that the vaginal approach was used six times as often as the transabdominal method, which means that the latter had an incidence of 1 in every 1672 deliveries. Such figures coming from the same university center where Benson and Durfee[60] performed their operative procedure imply that transabdominal cervicoisthmic cerclage is not required in many patients, and proper patient selection is mandatory if the reported success rate is to be maintained. All authors have noted that the transabdominal cerclage procedure is conducted in a highly vascular area and on occasion is associated with undue blood loss.

Prospective Trials

In any evaluation of cerclage procedures, one must be cognizant of the fact that populations differ between centers in the same locality and that patients vary in parity and in past obstetric management. Although the results of cerclage appear to be good in most of the reports in the literature, some physicians remain skeptical and require more objective evidence of the efficacy of cerclage procedures during pregnancy. A sequential crossover trial cannot be conducted, so a randomized trial must be considered, but it too would be difficult to organize when the treatment is already widely practiced. Sufficient doubt persists regarding the value of cerclage procedures, however, that a properly randomized study of its use is ethically justifiable and of considerable clinical importance. Controlled prospective studies on the therapy of cervical incompetence are not available. During the past few years randomized trials of cerclage procedures have been reported, and they justify doubts about the established practice of many physicians, particularly the obstetricians who perform prophylactic cerclage without establishing a diagnosis of cervical incompetence. A mere history of repeated fetal losses during the second trimester of pregnancy may lead one to suspect the presence of cervical incompetence, but this information is usually subjective and far too often incomplete for a definitive diagnosis.

Assuming an ability to accurately diagnose this entity, the problem of management remains. Tocolytic agents are not expected to prevent a complication that appears to be a primary problem of the cervical tissues. Simple bed rest, preferably in hospital, may have more to offer. There is little convincing evidence that cerclage is of great benefit. A major factor in its use is the temptation to do *something* rather than nothing for these patients. Loss of the pregnancy may result despite cerclage or, in some instances, perhaps because of it. We have all seen abortions resulting from proved chronic infection in the area of the cervical suture. It is a paradox that the less well the patients are selected for this therapy and the more the unsuitable are the cases that are included, the better the results appear. It is only by comparing like with like that the supporters of cervical cerclage will convince the skeptics who maintain with W.B. Yeats that "our stitching and unstitching has been naught."

Retrospective Studies

Cole[64] collated hospital data for a 12-month period in Scotland pertaining to patients who had undergone a cerclage procedure during pregnancy. She pointed out that comparison data using the entire obstetric population of Scotland over 1 year could eliminate bias due to population groups. The report showed that differing management philosophies and conditions, other than cervical incompetence, may have played a causative role in recurrent pregnancy failures. Cole indicated that a large number of cases can be identified in the country in a single year, which avoids some of the problems of unusual catchment populations and of individual developments in clinical practice. By using the entire pregnant population as the background to the experience of the cervical cerclage group, additional data on other complications become less important. She noted that the patients who were selected for cervical suture were more highly gravid and had a worse obstetric history than other pregnant women during the year even when the effects of high gravidity were taken into consideration. However, she did note that after cervical suture the rate of both premature delivery and perinatal mortality were still higher than in the general pregnant population. The perinatal mortality rate in the women who had had a cerclage procedure showed a greater improvement after suture than was seen to occur between previous and current pregnancies of the whole population. However, the improvement in pregnancy outcome that occurred after cervical suture could not definitely be credited to the suture. The results of her analysis suggest that there is an improvement in the perinatal mortality rate after cervical suture in a group with an extremely poor obstetric history. Some of the improvement, however, may be due to the phenomenon of regression to the mean, whereby a variable that is extreme when it is first measured tends to be closer to the mean of distribution in subsequent observations. Improvement observed in this retrospective study still leaves the cervical suture group with a worse outcome than might be expected in comparison with the outcome in a pregnant population with similar gravidity.

Any study that assesses which alternative clinical approach is preferable must attempt to minimize both bias and imprecision in the comparison. Collection bias is controlled by random allocation to alternative clinical approaches; and as the randomization process is such a crucial element in the study design, it is important to ensure that there is no conscious or unconscious violation. Lazar and associates[65] emphasized that most published studies have simply compared pregnancy in the same woman, which they deemed a wholly inadequate basis for inferring causality. These investigators, through the collaboration of four obstetric teams, recruited a total of 506 women at moderate risk for preterm delivery who were randomly allocated to either cervical cerclage or control group. In the final analysis, 268 were allocated to cerclage and 238 to nonintervention. There were no significant differences in maternal age, parity, or history of previous preterm births. Women allocated to the cerclage group, however, were significantly more likely to have had previous abortions.

The difference was largely a reflection of a difference between the experimental and control groups in one of the four centers. Lazar et al.[65] found that more women in the group allocated to a cerclage procedure were admitted to hospital for reasons other than the operation per se, and more of these individuals received oral tocolytic agents. They noted that more cesarean sections were performed and more preterm deliveries occurred in the women who had a cerclage operation, although the differences between the two groups were small and not significant. They stated that, contrary to their expectations, the results of the randomized controlled studies did not support the hypothesis that cerclage improves the ultimate outcome of pregnancy by lowering the rate of preterm birth. They also noted that cerclage procedures may be associated with higher rates of various clinical interventions such as the use of β-mimetic agents, hospitalization, and cesarean section. They do not know why this result occurred, but they suspected that the woman with a cerclage was identified as being at higher risk than the comparable woman who did not have such an operation. They implied that after a cerclage obstetricans are more likely to prescribe oral β-mimetics and that cervical cerclage per se induces uterine contractions. Furthermore, they indicated that in view of the doubts engendered by this study it would be appropriate to conduct a randomized trial of cervical cerclage in women at higher risk.

Rush and associates[66] conducted a similar study. They attempted to determine whether placement of a cervical suture in patients with a history of late abortion or preterm delivery prolonged the gestation. They drew up specific guidelines for the study and included in their study 194 patients who were randomly allocated either to have a cervical cerclage procedure or to be managed without suture. There were 96 patients who had a cervical cerclage and 98 patients who had no suture (controls). Women allocated to the cerclage and control groups were comparable in respect to their age, body build, and stage of pregnancy at which randomization occurred. Although the frequency of two or more previous second trimester abortions or preterm deliveries were somewhat greater in women allocated to cerclage, this difference was not significant. They noted that women who underwent cerclage procedure spent significantly longer time in hospital, and this difference remained so after discounting the hospital admission for cerclage. Although more patients in the cerclage group received tocolytic agents, the difference between the two groups was not significant. There was no difference between the groups with respect to cesarean section.

The study of Rush et al.[66] also examined the effect of an elective McDonald suture on gestational length of women at risk of delivery prior to the 37th week of gestation. The results suggested no beneficial effect on prolongation of pregnancy or on perinatal survival. Indeed, the investigators concluded that cerclage may have had a deleterious effect on the duration of pregnancy because of an increased tendency for the amniotic sac to rupture spontaneously in women who had it. They emphasized that cervical incompetence was inferred on the basis of the reproductive history in these patients. Although it was possible that some selected women did not have true cervical incompetence, those entered were certainly at exceptionally high risk for a preterm delivery. The results of this investigation did not exclude the possibility that such patients may benefit from cervical cerclage.

Postoperative Care

If the cervical cerclage procedure is performed by the vaginal route, the patient must remain in the hospital for 24 to 48 hours. Early ambulation is encouraged and analgesics liberally prescribed for any discomfort. Following transabdominal cervicoisthmic cerclage, the patient remains in hospital for 4 to 5 days.

All patients who have had a cerclage procedure are monitored for evidence of uterine irritability, which may be due to the rise in serum prostaglandin that follows the operation. Bibby and associates[67] demonstrated that there is a rise in the serum 13,14-dihydro-15-keto prostaglandin F (PGF) level following cervical cerclage but not after a vaginal examination. Ellwood and co-workers[68] demonstrated that during the first trimester of pregnancy cervical tissues produce various prostaglandins, specifically PGE and PGF, as well as the major circulating metabolites of $PGF_{2\alpha}$, 13,14-dihydro-15-keto-PGF, and 6-oxo-$PGF_{1\alpha}$, but little thromboxane. Green and associates[69] recorded that the prostaglandin activity was located predominantly in the cervical mucosa. Although definitive studies are not available, it is probable that prostaglandins have a direct action on biochemical events in the cervix that are separate from the oxytocic action on the uterus. Levin and co-workers[70] observed a marked increase in uterine size during the initial 24 hours subsequent to cervical cerclage. They measured the distance between the symphysis pubis and upper border of the uterus before and after the operation in 71 patients who had cerclage performed between the 12th and 17th weeks of pregnancy. They recorded a mean increase of uterine size of 1.97 cm (SD $\pm$ 0.22) and found no correlation between the postcerclage enlargement and gestational age or parity. However, ultrasonographic studies failed to demonstrate any increase in size of the uterine cavity or gestational sac. They considered that the observed changes could be explained by congestion of the uterine wall due to increased production of $PGF_{2\alpha}$, which is known to influence capillary permeability.

As prostaglandins are potent stimulators of myometrial activity, it is not surprising that tocolytic agents have been prescribed to patients following cerclage operations. There are no available controlled studies that demonstrate the efficacy of β-adrenergic agonists in the postoperative management of these patients. The oral administration of these agents alone does not decrease uterine activity. Walters and Wood[71] reported that ritodrine, when prescribed by the oral route, was unable to prevent premature labor in patients who required cervical cerclage. Toplis and co-workers[72] reported a prospective randomized controlled study in two groups of patients who required cervical cerclage early in the second trimester of pregnancy. They assessed whether the use of the β-mimetic agent salbutamol had any potential advantages over the use of omnopon, an opioid, during the postoperative period. Plasma levels of 13,14-dihydro-15-keto-PGF were measured by radioimmunoassay, and a significant rise was noted in all patients following cervical cerclage. The 24-hour plasma prostaglandin metabolite levels of both groups of the patients showed no significant difference from the basic levels.

In view of the elevated serum PGF metabolites levels following a cerclage, administration of prostaglandin synthetase inhibitors has been suggested to prevent premature cervical dilatation and effacement. One must, however, be cognizant of the fact that indomethacin and related compounds, which inhibit prostaglandin synthesis, can have adverse effects on the fetal circulation. To date, neither the duration of indomethacin therapy nor the dosage required to achieve a minimal risk-to-benefit ratio has been delineated. It is conceivable that short-term use of indomethacin and other nonsteroidal antiinflammatory agents is efficacious and has no deleterious effect on the fetus; but until more information is available it is advisable not to prescribe them to patients who had a cerclage.

The use of progestational agents to inhibit uterine activity during the perioperative period has been reported; but despite such anecdotal reports, there are no available controlled studies that demonstrate their efficacy. Lind and associates[73] have reported success with the use of salbutamol to preempt labor in the early part of the second trimester. Both tocolytic and progestational agents are not expected to prevent a complication that appears to be primarily cervical. Any benefit from such agents is probably psychological in nature because of extra interest and attention.

The routine use of prophylactic antibiotics is not recommended. Kuhn and Pepperell[11] reported that 154 of their patients received prophylactic antibiotics. They found no apparent difference in the fetal salvage rate between the patients who received antibiotics and those not so treated, irrespective of the technique used (Shirodkar or McDonald procedure). Harger,[74] in a review of the literature, found no evidence to support the concept that such measures as the use of prophylactic antibiotics, uterine relaxants, or progestins during the postoperative management improve the prognosis.

Despite such views, there are certain circumstances where prophylactic antibiotics are prescribed, such as when the cerclage procedure is done with bulging membranes. In such instances, swabs for microbiologic culture are obtained from the exposed membranes and the appropriate antibiotic prescribed. This topic is further discussed under the section dealing with complications.

After discharge from hospital, the patient is encouraged to lead a reasonably normal existence. She is seen at intervals of 2 weeks, at which time a speculum examination is indicated. It allows cervical assessment and reassurance of the patient. Most physicians proscribe coitus, as seminal fluid contains a large amount of prostaglandins. The topic of sexual intercourse must be discussed with the patient, and she must be informed that total abstinence is not necessary. She is told that some reduction in the frequency is suggested and that her partner must use a condom to avoid any possible prostaglandin-mediated stimulation of the uterus.

Fetal maturation is progressively monitored by ultrasonography; and when the risk of preterm labor has passed at 38 weeks' gestation, the suture can be removed and spontaneous labor allowed to take its course. If transabdominal cerclage has been performed, delivery is accomplished by cesarean section when maximum fetal maturity has been attained as demonstrated by ultrasonography and amniotic fluid analysis.

Complications

The optimum time to undertake a cerclage operation is early during the second trimester when the suture can be inserted prophylactically below the sac. After that time cerclage can still be attempted; but if the membranes bulge, the situation is much less favorable, and the membranes may inadvertently rupture during placement of the suture. It is often difficult to avoid this complication when one has to push the fetal membranes into the uterine cavity. Even when amniorrhexis does not occur, a cerclage procedure conducted in the presence of a protruding bag of membranes, when the gestation is more than 18 weeks, is associated with a 30% risk of premature rupture of the membranes or chorioamnionitis.

Olatunbosun and Dyck[75] are in accord with the general opinion that cervical cerclage is best performed when the gestation is between 14 and 16 weeks of age. They indicated that on occasion a patient fails to present for prenatal care until the second trimester and the fetal membranes are herniating through an effaced and dilated cervix. Under such circumstances, unless the cervix is closed by suture, it is unlikely that the pregnancy will result in a viable infant. Of the 12 women on whom they performed an emergency cerclage procedure, 10 had liveborn infants, and in no instance was it necessary to repeat the cerclage procedure.

When the cerclage is performed in the patient with a gestation of more than 20 weeks' duration, there is a 10% chance that the suture will be displaced. A second cerclage is then required, but the success rate is much less and risks of sepsis are increased. Robboy[76] reported a pregnancy failure rate of 50% when the cerclage was performed after the 20th week of gestation compared to 11% in patients who had the operation prior to the 20th week. This report drew attention to the fact that *cervical dystocia* during labor can be a sequela of a cerclage operation. The cervical dystocia is the result of fibrosis but in most instances is not sufficient to warrant cesarean section. On some occasions the excessive fibrous reaction in the cervix may result in extensive lacerations and necessitate cesarean hysterectomy. Wren[77] reported that one-third of his patients who were successfully treated required an abdominal delivery because a rigid fibrous band had developed in the cervix subsequent to the insertion of a mersilene band. Additional complications of labor and delivery related to cervical cerclage are listed in Table 1-2.

Infection is an important consideration in the prognosis of any surgical procedure. Careful surgical technique and bacterial control are prerequisites of any surgical procedure. Lipshitz,[14] in a retrospective study of cervical cerclage in 71 patients, reported that most of the late postoperative complications were the result of infection. Thirty-one (43.6%) of the patients were so afflicted. His report included a maternal death due to gangrene of the uterus and abdominal wall as a sequela to cesarean section. The cerclage operation in this patient was performed because of cervical dilatation in a multiple pregnancy. The patient was treated for a vaginal discharge after the cerclage operation, but it is not possible to ascertain if the operation had any role in the infectious process that was responsible for the patient's demise.

Weekes and associates[84] contended that there is no justification for indiscriminate cerclage operations in the management of

TABLE 1-2. Complications of labor and delivery related to prior cervical cerclage procedures.

Cervical dystocia[76]
Cervical laceration[78]
Uterine rupture[77,79]
Chorioamnionitis[80,81]
Endomyometritis[10,12,80,82]
Disseminated intravascular coagulopathy[79]
Maternal death from sepsis[83]
Endotoxic shock[10]

multiple pregnancies. They compared the outcome of 60 twin pregnancies treated by bed rest, when the duration of the gestation was between 28 and 34 weeks, with 37 women with twin pregnancies who had a cerclage procedure when the pregnancy was a mean of 197 days' (SD ± 12.2 days) duration and another 36 patients with twin pregnancies who had no specific therapy. They found that the incidence of spontaneous labor prior to the 36th week of pregnancy was comparable in all three groups. Furthermore, the mean birth weight of the first and second twins, the mean gestational age at delivery, and the incidence of low-birth-weight infants were similar in all three groups.

Dor and associates,[85] in a randomized trial of cervical cerclage in women with multiple pregnancies, found that the use of a prophylactic suture did not prevent preterm delivery. They randomized 50 patients with multiple pregnancies into two groups when the gestation was at 13 weeks: One group was given routine prenatal care, and the other had a cerclage operation. The data showed no significant differences between the two groups in regard to the incidence of premature rupture of fetal membranes, preterm labor, or early neonatal death. The results of these studies did not support the concept that cervical cerclage is efficacious in the management of multiple pregnancy.

Kuhn and Pepperell[11] stressed in their report the importance of sepsis as a complication of cervical cerclage. Antepartum, intrapartum, or postpartum sepsis complicated 45 (18.6%) of the 242 pregnancies that required cervical cerclage. Of the 45 patients who became infected, 32 (71%) had either premature rupture of the fetal membranes or labor prior to the 28th week of pregnancy. This study indicated that amniorrhexis prior to removal of the cerclage suture was associated with an increased risk of sepsis. There was, however, no significant difference between the patients who did and did not develop sepsis in regard to the time interval from rupture of the fetal membranes to delivery. Removal of the cervical suture immediately after rupture of the fetal membranes was associated with sepsis in approximately one-third of the patients when the duration of pregnancy was 28 weeks or less. Kuhn and Pepperell reported that infection was more likely if the suture was not removed. Furthermore, excessively prolonged membrane rupture could not account for the increased risk of infection observed when the suture was not immediately removed at the time of amniorrhexis. In view of the high incidence of infection when spontaneous labor or premature rupture of the membranes occurred prior to removal of the sutures, they questioned the safety of inhibiting labor as well as the use of corticostertoids to accelerate fetal lung maturation.

Infection at the site of the cerclage may result in chorioamnionitis in the presence or absence of rupture of the fetal membranes. It is of interest that the antibiotic therapy does not circumvent the risk of a serious infection and supports the concept that there is no place for perioperative prophylactic antibiotics in the management of patients with cervical incompetence.

With surgical treatment of cervical incompetence, sepsis is a major concern during the ensuing weeks of the pregnancy, and the possibility of infection is ever present. Heinemann and associates[86] recorded the risks of serious sepsis associated with placental bacteremia as a complication of the Shirodkar technique. The true incidence of infectious complications of cerclage cannot be assessed from the literature, especially because few series report on emergency operative failures. There is unanimity in the reports that the Shirodkar cerclage procedure is associated with a higher postoperative infection rate than the McDonald technique. The higher sepsis rate associated with the Shirodkar operation presumably is due to the greater dissection required for displacement of the bladder.

Aarnoudse and Huisjes[87] stated that preterm premature rupture of the fetal membranes was the most frequent complication encountered in patients who had had a cerclage operation. They suggested that a subclinical intracervical infection, subsequent to the operative procedure, could play a role in amniorrhexis in such patients. They considered that intrauterine infection is difficult to prevent, as the suture is a foreign body in the cervical tissue that can intrude into the vaginal lumen. This condition is ideal for the development of infection, and in most cases the foreign body is associated with a profuse vaginal discharge. These authors indicated that although intrauterine infection is a fairly common complication of cervical cerclage it has received little attention because the pregnancy outcome is almost invariably good.

Charles and Edwards[80] reported that the incidence of chorioamnionitis was increased 2.6-fold and premature rupture of the fetal membranes threefold if the cerclage procedure was performed after the 19th week of the pregnancy. In their study of 115 patients who required a cerclage procedure, they reported that only 10 of the 69 patients (14.9%) in whom the operation was attempted or performed between the 14th and 18th weeks of pregnancy developed chorioamnionitis. Among them, two of the infections were the result of inadvertent rupture of the fetal membranes at the time of operation, and three others were associated with spontaneous rupture of the membranes within 5 days of the operation. Chorioamnionitis occurred in the remaining five patients in association with premature rupture of the fetal membranes before the 32nd week of pregnancy. On the other hand, chorioamnionitis occurred in 18 of the 46 patients (39.1%) in whom the cerclage procedure was instituted between the 19th and 26th weeks of the pregnancy. Such findings suggest that the suture material may be an etiologic factor in chorioamnionitis and premature labor when the cerclage operation is performed during the latter part of the second trimester. The prognosis for the pregnancy attaining 36 weeks is poor when a diagnosis of cervical incompetence is made after the 26th week of pregnancy and treated by a cerclage procedure. Murphy and associates[82] emphasized that the optimal success rate was attained when cerclage for an incompetent cervix was performed between the 14th and 16th weeks of pregnancy. They reported that of 17 pregnancies where the operative procedure was conducted between the 27th and 32nd week of pregnancy seven (41.1%) ended in failure. There is no reason for performing a cerclage procedure after the gestation is more than 30 weeks' duration. In cases where there is no demonstrable abnormality of the cervix or when the history is atypical, most physicians would not perform a cerclage procedure.

Crombleholme and associates[88] held contrary views and claimed that there is a statistical improvement of fetal outcome even in the absence of any cervical aberration. Furthermore, they believed that their data demonstrated a maternal morbidity and infection rate of only 8% associated with the procedure, this being low enough to be outweighed by the benefits. Their study, which comprised 73 patients, was not randomized, and in its evaluation they excluded previous pregnancy terminations. In the subgroup of patients whose obstetric history suggested recurrent premature labor alone as the cause of pregnancy loss but who had short or prematurely effaced cervices, a cerclage procedure did not improve fetal outcome.

Preterm delivery following cervical cerclage may be the result of underlying cervical pathology or due to cerclage suture material. The latter can be the site of colonization by such organisms as group B streptococcus, *Staphylococcus aureus,* or *Escherichia coli.* In the author's experience, the cerclage material can be the nidus of infection, which can lead not only to failure of the cerclage procedure but also chorioamnionitis and intrauterine fetal infection. There is no doubt that once the suture material has been colonized by one of these organisms it can set in motion a series of events that have a deleterious effect on the prognosis of the pregnancy. Because of those observations, we recommend obtaining cervical cultures at least three times during the pregnancy after a cerclage operation. A culture of the cervix is not useful in many instances but could conceivably be critically important in patients with uterine tenderness associated with fever and leukocytosis early during the third trimester of the pregnancy. It is in such instances that we perform amniocentesis; and after Gram stain and culture, specific antimicrobial therapy can be prescribed.

Cerclage procedures must not be performed in the presence of vaginal infections. Operative intervention is contraindicated when there is any evidence of genital herpetic infection. Nahmias and associates[89] reported disseminated fetal herpetic infection following the performance of a Shirodkar procedure when the fetal membranes were intact. When such basic safeguards to maternal and fetal well-being are neglected, infectious complications ensue.

Russell[81] observed a close correlation between chorioamnionitis and cervical cerclage in a tertiary care referral hospital where cerclage was frequently performed as a therapeutic measure in patients with a history of recurrent pregnancy loss. He noted a sixfold increase in chorioamnionitis in patients with cervical cerclage when compared with a control group. The difference in the two groups was most marked during the last weeks of the second trimester, and Russell found no patient free of intrauterine infection if parturition occurred prior to 33 weeks. In a further written communication in 1986, Russell stated that 100% of the patients had acute chorioamnionitis when they had a cervical suture and went into premature labor. Furthermore, about 25% of the patients who had a cervical suture delivered prematurely or aborted. It was impossible to determine to what extent the fetal loss was occasioned by the underlying pathology and how much was influenced by the mechanical presence of a cervical stitch. He stated that these observations caused a change in policy in the hospital because the obstetricians became more circumspect about indicating cervical ligatures for the purpose of preventing the hazards of "functional cervical incompetence." Furthermore, Russell indicated that serious consideration was being given to the use of prophylactic antibiotics. The bacteria encountered in acute chorioamnionitis associated with cervical cerclage were comparable to those associated with chorioamnionitis in the absence of a cervical ligature and, in turn, matched the bacterial flora of the cervicovaginal ecosystem.

In a review of 147 cervical cerclage procedures, Barford and Rosen[78] found that the immediate morbidity following the operation was low, but significant morbidity at the time of delivery related to cervical scarring was observed. They found poor correlation between clinical history, physical findings, adjunct tests for cervical incompetence, and pregnancy outcome subsequent to cerclage. They are in accord with our view that, despite the plethora of uncontrolled studies, prospective randomized studies are essential, not only to evaluate the efficacy of cervical cerclage but to determine what morbidity occurs when the procedure is undertaken to prevent pregnancy loss due to a relatively rare cause of abortion. Harger,[74] in a review of the morbidity and success rates of cervical cerclage, indicated that interpretation of the results are unreliable because the diagnostic criteria for cervical incompetence and the indications for the placement of a cerclage are varied.

Conclusions

It is an accepted fact that a small number of late abortions may result from a basic failure of the cervix to maintain an intrauterine pregnancy. The problem of therapy remains, assuming an ability to accurately diagnose cervical incompetence. Tocolytic agents cannot be expected to prevent a complication that primarily involves the cervix; as a result, cervical cerclage has become the primary modality of treatment. Simple bed rest, however, preferably in hospital, may be the most efficacious form of therapy.

Cervical incompetence is probably one of the most overdiagnosed entities in obstetrics. A small group of patients, however, do have cervical incompetence that is responsible for second trimester abortion and early third trimester pregnancy loss. Such patients can benefit from a cerclage procedure when it is performed between the 14th and 18th weeks of pregnancy. Such a procedure is not, however, a panacea for all gravid women who have a history of second trimester abortion or preterm labor.

The liberal use of cerclage procedures can no longer be justified because randomized controlled studies have demonstrated that, contrary to expectation, cerclage does not improve outcome of pregnancy by decreasing the rate of preterm births. Instances of abortion have been observed following chronic infection in the area of the cerclage. It is a paradox that the less well the patients are selected for the operative procedure (i.e., the more unsuitable are the patients who are included) the better the results appear. It seems that in women with a moderate risk of preterm delivery cerclage may actually be associated with a high incidence of clinical intervention. There is no evidence to support the assumption that cerclage improves the prognosis of multiple pregnancy.

Sufficient doubt surrounds the value of cervical cerclage that well conducted, large-scale, prospective, randomized studies are necessary for appropriate evaluation. Such studies are now in progress, and what is required can best be summed up by the words of John Milton: "Where there is much desire to learn, there of necessity be much arguing, much writing, many opinions—for opinion in good men is but knowledge in the making."

References

1. Warburton D, Stein Z, Kline, J, et al. Chromosome abnormalities in spontaneous abortion: data from the New York City data. In Porter TH, Hook EB (eds): Human Embryonic and Fetal Death. New York, Academic Press, 1980;261–286.
2. Lash AF, Lash SK. Habitual abortion: the incompetent internal os of the cervix. Am J Obstet Gynecol 1950;59:68–76.
3. Gream GT. Dilation or division of the cervix uteri. Lancet 1865;1:381–386.
4. Aschoff L. Zur cervix frage; nachtrag zu der arbeit des Herrn Dr Hohmeier. Monatsschr Geburtsh Gynak 1905;22:611–615.
5. Stieve H. Der halsteil der menschlichen gerbartmutter, seine veranderungen wahrend der geburt und des wochenbettes und ihre bedeutung. Z Mikrosk Anat Forsch 1927;11:290–301.
6. Danforth DN. The fibrous nature of the human cervix and its relation to the isthmic segment in gravid and non-gravid uteri. Am J Obstet Gynecol 1947;54:541–557.
7. Leppert PC, Keller S, Cerreta, J, et al. Conclusive evidence for the presence of elastin in human and monkey cervix. Am J Obstet Gynecol 1982;142:179–182.
8. Leppert PC, Cerreta J, Mandl I. Orientation of elastic fibers in the human cervix. Am J Obstet Gynecol 1986;155:219–224.
9. Palmer R, Lacomme M. La beance de l'orifice interne, cause d'avortements a repetition? Une observation de dechirure cervicoisthmique reparee chirurgicalement avec gestation a terme consecutive. Bull Assoc Gynecol Obstet Fr 1948;47:905–909.
10. Toaff R, Toaff ME, Ballas S, et al. Cervical incompetence: diagnostic and therapeutic aspects. Isr J Med Sci 1977;13:39–49.
11. Kuhn RJP, Pepperell RJ. Cervical ligation: a review of 242 pregnancies. Aust NZ J Obstet Gynaecol 1977;17:79–83.
12. Barter RH, Dusbabek JA, Riva HL, et al. Surgical closure of the incompetent cervix during pregnancy. Am J Obstet Gynecol 1958;75:511–524.
13. Ritter HA. Surgical closure of the incompetent cervix: 15 years experience. Int J Gynaecol Obstet 1978;16:194–196.
14. Lipshitz J. Cerclage in the treatment of incompetent cervix. Suid Afr Tydsk 1975; 49:2013–2015.
15. Stromme WB, Haywa EW. Intrauterine fetal death in the second trimester. Am J Obstet Gynecol 1963;85:223–233.
16. Mann EC, McLaren WD, Hayt DB. The physiology and clinical significance of the uterine isthmus. Am J Obstet Gynecol 1961;81:209–213.
17. Sherman AL. Hormonal therapy for control of the incompetent os of pregnancy. Obstet Gynecol 1966;28:198–205.
18. Palmer R. Le role de la beance de l'isthme uterin dans l'avortement habituel. Rev Gynecol Obstet 1950;45:218–233.

19. Heinonen OP, Slone D, Monson RR. Cardiovascular birth defects and antenatal exposure to female sex hormones. N Engl J Med 1977;296:67–70.
20. Jennings CL. Temporary submucosal cerclage for cervical incompetence: report of forty eight cases. Am J Obstet Gynecol 1972;113:1097–1102.
21. Peters WA, Thiagarajah S, Harbert GM. Cervical cerclage: twenty years experience. South Med J 1979;72:933–937.
22. Singer MS, Hochman M. Incompetent cervix in a hormone-exposed offspring. Obstet Gynecol 1978;51:625–626.
23. Goldstein DP. Incompetent cervix in offspring exposed to diethylstilbestrol in utero. Obstet Gynecol 1978;52(suppl 1):73s–75s.
24. Nunley WC, Kitchin JD III. Successful management of incompetent cervix in a primigravida exposed to diethylstilbestrol in utero. Fertil Steril 1979;31:217–219.
25. Kaufman RH, Binder GL, Gray PM, et al. Upper genital tract changes associated with exposure in utero to diethylstilbestrol. Am J Obstet Gynecol 1977;128:51–59.
26. Haney AF, Hammond CB, Soules MR, et al. Diethylstilbestrol induced upper genital tract abnormalities. Fertil Steril 1979;31:142–146.
27. Cousins L. Cervical incompetence 1980: a time for reappraisal. Clin Obstet Gynecol 1980;23:467–479.
28. Barnes AB, Coulton T, Gundersen MD. Fertility and outcome of pregnancy in women exposed in utero to diethylstilbestrol. N Engl J med 1980;302:609–613.
29. Sandberg EC, Riffle NL, Higdon DV, et al. Pregnancy outcome in women exposed to diethylstilbestrol in utero. Am J Obstet Gynecol 1981;140:194–205.
30. Ben-Baruch G, Menczer J, Mashiach S, et al. Uterine anomalies in diethylstilbestrol-exposed women with fertility disorders. Acta Obstet Gynecol Scand 1981;60:395–397.
31. Weber T, Obel EB. Pregnancy complications following conization of the uterine cervix. Acta Obstet Gynecol Scand 1980;58:259–263.
32. Buller RE, Jones HW III. Pregnancy following cervical conization. Am J Obstet Gynecol 1982;142:506–512.
33. Larsson G, Grundsell H, Gullberg B, et al. Outcome of pregnancy after conization. Acta Obstet Gynecol Scand 1982;61:461–466.
34. Leiman G, Harrison NA, Rubin A. Pregnancy following conization of the cervix: complications related to cone size. Am J Obstet Gynecol 1980;136:14–18.
35. Moinian M, Andersch B. Does cervix conization increase the risk of complications in subsequent pregnancies? Acta Obstet Gynecol Scand 1982;61:101–103.
36. Kantor HI. Cervical incompetence. Obstet Gynecol 1974;44:312–313.
37. Ratten GJ, Beischer NA. The effect of termination of pregnancy on maturity of subsequent pregnancy. Med J Aust 1979;1:479–480.
38. Rath W, Theobald P, Kuhnle H, et al. Changes in collagen content of the first trimester cervix uteri after treatment with prostaglandin $F_{2\alpha}$ gel. Arch Gynecol 1982; 231:107–110.
39. Asplund J. The uterine cervix and isthmus under normal and pathological conditions: a clinical and roentgenological study. Acta Radiol 1952;91(suppl):3–76.
40. Borell V, Fernstrom I. The sphincter mechanism of the isthmus uteri: a radiological study. Acta Obstet Gynecol Scand 1953;32:7–12.
41. Youssef AF. The uterine isthmus and its sphincter mechanism, a radiographic study. II. The uterine isthmus under abnormal conditions. Am J Obstet Gynecol 1958;75:1320–1332.
42. McDonald IA. Cervical cerclage. Clin Obstet Gynaecol 1980;7:461–479.
43. Jeffcoate TNA, Wilson JK. Uterine causes of abortion and premature labor. NY State J Med 1956;56:680–690.
44. Sarti DA, Sample WF, Hobel CJ, et al. Ultrasonic visualization of a dilated cervix during pregnancy. Radiology 1979;130:417–420.
45. Brook I, Feingold M, Schwartz A, et al. Ultrasonography in the diagnosis of cervical incompetence in pregnancy: a new diagnostic approach. Br J Obstet Gynaecol 1981;88:640–643.
46. Bernstine RL, Sung Lee SHM, Crawford WL, et al. Sonographic evaluation of the incompetent cervix. J. Clin Ultrasound 1981;9:417–420.
47. Zemlyn S. The length of the uterine cervix and its significance. J Clin Ultrasound 1981;9:267–269.
48. Costantini S, Valenzano M, Venturini PL, et al. Ultrasonic evaluation of cervical incompetence. Biol Res Pregnancy Perinatol 1986;7:11–16.

49. Jackson G, Pendleton HJ, Nichol B, et al. Diagnostic ultrasound in the assessment of patients with incompetent cervix. Br J Obstet Gynaecol 1984;91:232–236.
50. Vitsky M. Simple treatment of the incompetent cervical os. Am J Obstet Gynecol 1961;81:1194–1197.
51. Bayer H. Einige neve Gesichtspunkte bei der prophylaxe und therapie der drohenden fruhgeburt. Zentralbl Gynakol 1977;99: 547–551.
52. Seyffarth K. Unblutige zerklage mit strutzpessaren zur prophylaxe und therapie der fruhgeburt. Zentralbl Gynakol 1978; 100:1566–1570.
53. Wilson JG, Brent RL. Are sex hormones teratogenic? Am J Obstet Gynecol 1981; 141:567–580.
54. Shirodkar VN. A new method for operative treatment for habitual abortions in the second trimester of pregnancy. Antiseptic 1955; 52:299–300.
55. McDonald IA. Suture of the cervix for inevitable miscarriage. J Obstet Gynaecol Br Emp 1957;64:346–350.
56. Shirodkar VN. Habitual abortion in the second trimester. In: Contributions to Obstetrics and Gynaecology. Edinburgh, Livingstone, 1960;1–7.
57. McDonald IA. Incompetence of the cervix. Aust NZ J Obstet Gynaecol 1978;18:34–37.
58. McDonald IA. Cervical cerclage. Clin Obstet Gynaecol 1980;7:461–479.
59. Hunter RG, Henry GW, Judd CS. Physiology of dysfunctional incompetence of the cervix: diagnosis and surgical treatment in the nonpregnant woman. Am J Obstet Gynecol 1961;83:1183–1190.
60. Benson RC, Durfee R. Transabdominal cervicouterine cerclage during pregnancy for the treatment of cervical incompetence. Obstet Gynecol 1965;25:145–155.
61. Watkins RA. Transabdominal cervicouterine suture. Aust NZ J Obstet Gynaecol 1972; 12:62–64.
62. Mahran M. Transabdominal cervical cerclage during pregnancy: a modified technique. Obstet Gynecol 1978;52:502–506.
63. Novy MJ. Transabdominal cervicoisthmic cerclage for the management of repetitive abortion and premature delivery. Am J Obstet Gynecol 1982;143:44–53.
64. Cole SK. Cervical suture in Scotland: strengths and weaknesses in the use of routine clinical summaries. Br J Obstet Gynaecol 1982;89:528–535.
65. Lazar P. Gveguens S, Dreyfus J, et al. Multicentered controlled trial of cervical cerclage in women at moderate risk of preterm delivery. Br J Obstet Gynaecol 1984;91:731–735.
66. Rush RW, Isaacs S, McPherson K, et al. A randomized controlled trial of cervical cerclage in women at high risk of spontaneous preterm delivery. Br J Obstet Gynaecol 1984;91:724–730.
67. Bibby JG, Brunt J, Mitchell MD, et al. The effect of cervical encerclage on plasma prostaglandin concentrations during early human pregnancy. Br J Obstet Gynaecol 1979;86:19–22.
68. Ellwood DA, Mitchell MD, Anderson ABM, et al. In vitro production of prostanoids by the human cervix during pregnancy: preliminary observations. Br J Obstet Gynaecol 1980; 87:210–214.
69. Green K, Christensen N, Bygdeman M. The chemistry and pharmacology of prostaglandins, with reference to human reproduction. J Reprod Fertil 1981;62:269–281.
70. Levin S, Brook I, Feingold M, et al. The effect of cervical encerclage on uterine dimensions. Acta Obstet Gynecol Scand 1980;59:189–190.
71. Walters WAW, Wood CA. A trial of ritodrine for the prevention of premature labour. Br J Obstet Gynaecol 1977;84:26–30.
72. Toplis PJ, Shepherd HJ, Youssefmejadian E, et al. Plasma prostaglandin concentrations after cerclage in early pregnancy. Br J Obstet Gynaecol 1980;87:669–671.
73. Lind T, Godfrey KA, Gerrard J, et al. Continuous salbutamol infusion over 17 weeks to pre-empt premature labour. Lancet 1980; 2:1165–1166.
74. Harger JH. Cervical cerclage: patient selection, morbidity and success rates. Clin Perinatol 1983;10:321–341.
75. Olatunbosun OA, Dyck F. Cervical cerclage operation for a dilated cervix. Obstet Gynecol 1981;57:166–170.
76. Robboy MS. The Management of cervical incompetence: UCLA experience with cerclage procedures. Obstet Gynecol 1973;41:108–112.
77. Wren BG. Cervical incompetence: aetiology and management. Med J Aust 1973;2:1146–1147.
78. Barford DAG, Rosen MG. Cervical incompetence: diagnosis and outcome. Obstet Gynecol 1984;64:159–163.

79. Lindberg BS. Maternal sepsis, uterine rupture and coagulopathy complicating cervical cerclage. Acta Obstet Gynecol 1979;58:317–319.
80. Charles D, Edwards WR. Infectious complications of cervical cerclage. Am J Obstet Gynecol 1981;141:1065–1071.
81. Russell P. Inflammatory lesions of the human placenta. I. Clinical significance of acute chorioamnionitis. Am J Diagn Gynecol Obstet 1979;1:127–141.
82. Murphy H, Lillie EW, Harrison RF. The incompetent cervix in pregnancy. J Irish Med Assoc 1978;71:152–154.
83. Dunn LJ, Robinson JC, Steer CM. Maternal death following suture of incompetent cervix during pregnancy. Am J Obstet Gynecol 1959;78:335–339.
84. Weekes ARL, Menzies DN, DeBoer CH. The relative efficacy of bedrest, cervical suture and no treatment in the management of twin pregnancy. Br J Obstet Gynaecol 1977;84:161–164.
85. Dor J, Shaley J, Mashiach S, et al. Elective cervical suture of twin pregnancies diagnosed ultrasonically in the first trimester following induced ovulation. Gynecol Obstet Invest 1982;13:55–60.
86. Heinemann M-H, Tang C-K, Kramer EE. Placental bacteremia and maternal sepsis complicating Shirodkar procedure. Am J Obstet Gynecol 1977;128:226–228.
87. Aarnoudse JG, Huisjes HJ. Complications of cerclage. Acta Obstet Gynecol Scand 1979;58:255–257.
88. Crombleholme WR, Minkoff HL, Delke I, et al. Cervical cerclage: an aggressive approach to threatened or recurrent pregnancy wastage. Am J Obstet Gynecol 1983;146:168–174.
89. Nahmias AJ, Alford CA, Korones SB. Infection of the newborn with herpes virus hominis. Adv Pediatr 1970;17:185–226.

2
Induced Abortion

JAMES R. DINGFELDER

Contemporary interest in the techniques and safety of induced abortion appears to have begun with the stimulating 1967 article of Kerslake and Casey.[1] The article reviewed Kerslake's British experience with the vacuum aspiration technique, which she first observed in eastern Europe. The method was already in widespread use in the USSR and China, where it was first popularized during the late 1950s. Prior to this report, standard textbooks described cervical dilation and sharp curettage as a method of treating incomplete abortion, but induced abortion received scant or no attention in most texts. The development of improved technology paralleled liberalized worldwide views about induced abortion. At the beginning of the century, induced abortion was prohibited in almost every country. After World War II, most eastern European countries adopted liberalized abortion laws permitting first trimester abortion on request. Most Western countries, including the United States, adopted similar laws and techniques during the late 1960s and early 1970s.

Three-fourths of the world's population now lives in a country where abortion is legal. Another one billion people, most of them from fundamentalist Islamic nations, Latin America, or developing African countries, still have no legal access to abortion.

Incidence

About 25 million of the approximately 33 million legal abortions performed worldwide each year are done in China and the Soviet Union. The global abortion ratio (the proportion of all viable pregnancies that are terminated by induced abortion) probably ranges from 24 to 32 abortions per 100 diagnosed pregnancies. The abortion rate worldwide may be higher, up to 55 abortions per 1000 women age 15 to 44, if illegal procedures are included. Abortion rates vary substantially for a variety of reasons. Underreporting is a significant factor in some countries. The officially reported abortion rate in Japan is 22 abortions per 1000 women; however, one estimate by a Japanese analyst in 1975 indicated an abortion rate as high as 84 per 1000. A study in Bangledesh indicated that in 1 year there were 69,000 reported legal terminations and ten times as many unreported illegal abortions. There is evidence that women from countries such as Ireland, which has one of the most restrictive abortion laws, travel outside their native country to obtain an abortion. The result is an Irish abortion rate equaling that of more liberalized countries in western Europe.

The United States abortion rate is higher than that of most developed countries (27 abortions per 1000 U.S. women 15 to 44 years of age in 1983) but lower than that of

Japan or eastern European countries. Statistics for the United States collected by the Centers for Disease Control (CDC) showed a rising rate of abortions during the 1970s that plateaued at around 1.5 million abortions annually during the mid-1980s.

Mortality Trends

Since its legalization in the United States in 1972, induced abortion has become one of the most common surgical procedures. In 1982 CDC statistics showed that 1,303,980 abortions had been performed, culminating a decade of continued increases. The 1983 statistics showed a 2.7% reduction to 1,268,987 procedures, the first decline since the CDC began to record incidence figures in 1972. Since 1976 the annual percentage of increase in the number of abortions has been declining. Although the actual numbers increased yearly from 1972 through 1982 (and decreased slightly in 1983) the abortion rate or number of abortions per 1000 women aged 15 to 44 has remained stable at 22/1000 to 24/1000 since 1977. Thus the actual numbers and the rate of abortions appear to have plateaued.

The characteristics of women seeking abortion have changed little in a decade. The patients remain young (66% under age 25), unmarried (77%), nulliparous (57%), and white (69%).[2] They seek a first abortion (71%)[2] at 10 weeks' gestation or earlier (78%) performed by dilatation and evacuation (D&E) (95%) in a nonhospital clinic (71%).

Mortality rates have shown a dramatic reduction since the CDC began to record statistics, particularly since the mid-1980s, and there has been an eightfold decrease in the death/case ratio since 1972. The current rate of 0.5 (one death per 200,000 cases) may be compared to the sevenfold higher childbearing mortality rate (standardized for age and race).[3] Nevertheless, certain factors continue to present a higher abortion mortality risk. Nonwhite races have a fourfold higher risk than whites, a difference that has increased owing to a greater reduction in mortality among white than nonwhite women during the decade.[3] Women aged 30 and greater are at twice the risk of those who are younger.

An as yet unexplained regional variation in mortality rates has existed throughout the years of statistical reporting. The western United States has had the lowest rates since abortion was legalized, and the disparity between its rates and those of the South has increased during the reporting period. The sixfold drop in Western mortality rates has resulted in a threefold greater mortality rate by comparison with the South, with the central and northeastern areas in an intermediate position. These regional differences are not explained by differing ethnic, social, or age characteristics.[3]

The welcome decline in mortality rates appears to be due to a dramatic reduction in deaths attributable to certain specific causes. The most dramatic change is that in mortality associated with the length of gestation, as mortality increases with increased gestational age. The percentage of abortions done at 8 weeks' gestation or less increased from 34% to 51% during the decade, whereas abortions performed at 13 weeks or later declined from 18% in 1972 to 10% in 1981.[3] A gradual trend toward replacement of instillation methods (saline, urea, prostaglandins) with the operative D & E methods during the 13 to 16 weeks' gestation interval has further reduced the mortality risk.

Of growing concern is the relative increase in the anesthesia-related deaths in recent years.[4] Whereas anesthetic deaths represented only 8% of abortion deaths during the 1972—1975 interval, they claimed 29% over the 1980—1985 years. General anesthesia represents an increased risk at any stage of gestation but especially during the second trimester when nonanesthetic complications such as uterine perforation, hemorrhage, and bowel injury and subsequent death related to these injuries are more common. Many anesthesia-related deaths appear preventable. In 23 of 27 deaths reported since 1972, endotracheal tubes were not used and the deaths were attributed to hypoxia or hypoventilation.[4] General anesthesia is chosen rather than local anesthesia for a variety of nonmedical indications, including operator preference, patient preference, and location of the facility (hospital versus outpatient clinic). The threefold greater incidence of anesthesia deaths in the northeastern United States compared to the remaining United States reflects at least in part the higher proportion of general anesthetics administered in the Northeast.[5]

Provision of Services

Although organized opposition to elective abortion has continued since legalization in 1973, in theory, abortion services are available to all women who request them. In some countries almost all abortions are provided at little or no cost to the patient. In the United States access to abortion continues to be somewhat limited by its cost and by the lack of abortion services in some geographic areas.

In 1977 the Supreme Court ruled that states are not required to provide Medicaid funds for abortions that are not "medically necessary," and that public hospitals are not required to provide abortions. The Hyde Amendment suspended Medicaid payments for most abortions, excepting those cases where the mother's life was endangered or in some cases of incest or rape.

During the first year after abortion was legalized (1973) more than one-half of the procedures took place in hospitals. Currently, more than 80% are performed in clinics or physician's offices, and the remaining hospital abortions are generally done on an outpatient basis. Reasons for the predominance of nonhospital services include the for-profit basis of many facilities and the reluctance of traditional health care providers and hospitals to respond to the change in legal status of abortion. In 1982 only 26% of non-Catholic, short-term hospitals and 16% of public hospitals provided this service.[6] By 1980 as few as 379 clinics and 80 hospitals performed 76% of all legal abortions. These facilities, although comprising only 16% of the total providers, on the average served 50 patients weekly, and each performed more than 1000 abortions yearly. Such a concentration of cases usually increases the competence of the operators and the overall quality of care but acts to limit ready access to services for those living some distance from urban centers.

Shortly after abortion was legalized, a survey was conducted among university obstetrics/gynecology departments to evaluate the role of residency programs in training physicians to provide what was assumed would be an increasing patient load. The conclusion was that "many university departments have not fully assimilated the elective abortion as an integral part of the service and educational responsibility of obstetrics and gynecology. The pressing demand for abortion services will require many obstetricians and gynecologists to be prepared."[7] Five years later another survey noted that "between 20 and 40% of all residents are estimated to have had no clinical experience in performing first trimester abortions."[8] A current review has cited little change over the last decade. About 23% of residency programs include abortion training as a routine part of the training program, and 50% offer it only as an option. The 28% that offer no training are predominantly Catholic-affiliated or military training programs.[9] Because only 20% of abortions are now performed in hospital facilities it is clear why only 10% of training programs collectively perform more than ten procedures per week.

An interesting gender difference among providers has emerged as female graduates enter obstetrics and gynecology in greater numbers. In 1986 women comprised 42% of residents in the specialty. Women have been cited as more likely to enter the specialty because of a desire to improve women's health care and are more likely to agree with the concept of abortion on request.[10] They also are more likely to perform abortions; but because of practice-related variables—primarily a reduced tendency to be in solo practice in contrast to a practice with a multispecialty group—women tend to do fewer procedures than men.

Menstrual Extraction

Menstrual regulation, menstrual extraction, and menstrual induction are terms that describe a so-called minisuction procedure utilizing Karman-type soft cannulas of up to 8 mm diameter (Fig. 2-1). The procedure is usually reserved for gestations of 50 days or less from the last period. These euphemistic terms originate from the era when less sensitive pregnancy tests often did not permit a conclusive pregnancy diagnosis until 2 weeks after the missed period. When pregnancy was suspected but not proved, this technique could be utilized to "induce a period" without formally designating the procedure as an abortion. The procedure as originally described by Karman utilized a 50-ml syringe vacuum source, usually did not require cervical dilation to introduce the 5- or 6-mm soft, flexible, plastic cannula, and was occasionally performed without paracervical block anesthesia.

Advantages of the procedure included patient satisfaction with an early resolution of the unwelcome amenorrhea and avoidance of the psychological stigma of an abortion procedure. In some jurisdictions where laws did not permit office abortion procedures unless the facility was certified as an approved abortion facility, it offered a transparent method of performing early abortion procedures. Countering these advantages was the statistical evidence that nearly one-half of patients with 1 week of amenorrhea and 25% of patients with 14 days of amenorrhea could subsequently be shown to have not been pregnant. In addition, up to 2% of patients failed to be aborted, probably because the conceptus was so small it was missed by the procedure.[11] Consistent with the euphemistic subterfuge, the uterine aspirate was rarely examined to confirm pregnancy, thereby missing the opportunity to suspect early ectopic gestations.

With the advent of sensitive pregnancy tests that are invariably positive 10 to 14 days after conception, almost no physician currently performs this procedure without first obtaining a positive pregnancy test as well as an informed patient consent for early abortion. Aspirated tissues are carefully examined to confirm intrauterine pregnancy. The technique and instrumentation are still valuable in that gestations up to 7 to 8 weeks of amenorrhea can be quickly and

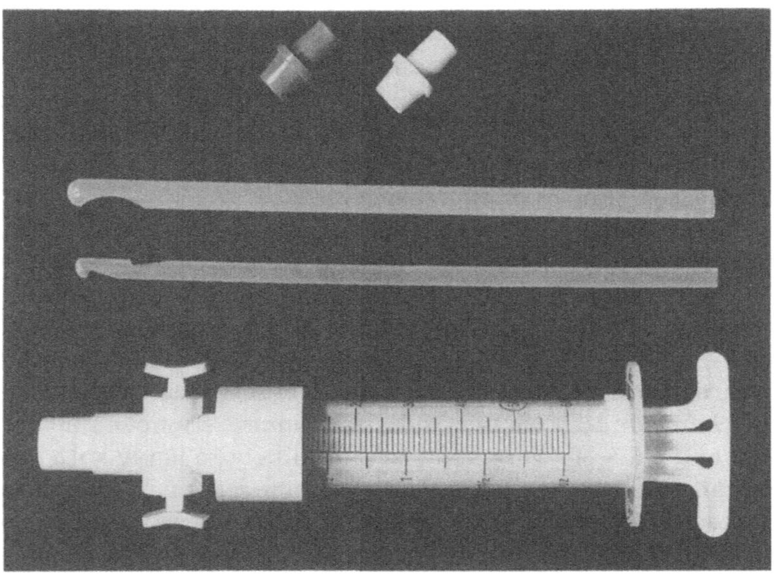

FIGURE 2-1. Karmen-type suction cannulas (6 and 8 mm) with a 50-ml aspiration syringe. (Courtesy of International Projects Assistance Services, 123 West Franklin Street, Chapel Hill, North Carolina 27516.)

atraumatically terminated without need for preliminary cervical dilation. Some physicians refuse to do the procedure before 8 weeks of amenorrhea because of concern about the high procedure failure rate at earlier gestational dates.

This point is of minor concern if the operator has developed experience identifying placental villi. Such identification is easily accomplished in cases of 8 weeks' gestation or less by suspending the aspirated material in normal saline and floating it in a petri dish or clear plastic container. The tissues are then examined under a low power of the microscope or with a hand lens. A characteristic branching pattern is easily recognized.[12] Some examiners suspend the tissues in 5% acetic acid, which causes villous material to become whiter.

Failure to confirm an intrauterine gestation by identifying chorionic villi mandates that the uterine cavity be rechecked immediately by a brief curettage. Failing to retrieve any further tissue, the patient is then given a clear and detailed explanation of the findings and the possibility that an ectopic pregnancy may exist. Time permitting, an ultrasonographic procedure may be ordered to search for possible adnexal masses as well as for the presence of an intrauterine gestational sac. Occasionally one is discovered in a severely retroverted uterus or in one horn of a bicornuate uterus. Blood is then drawn and submitted for a quantitative human chorionic gonadotropin (hCG) determination, to be used on a comparative basis with another determination made a few days later when the pathology report becomes available. Rising hCG titers in serial specimens, in the absence of a demonstrated intrauterine gestational sac, are strong evidence that an extrauterine ectopic pregnancy exists and needs to be ruled out, usually by diagnostic laparoscopy. In most instances the serial hCG titers have declined by this time, and the patient can be assumed to have been in the early stages of a spontaneous abortion. Other cases may be explained by fragmentation of the placental tissues during the vacuum aspiration process such that their architecture becomes unrecognizable to the operator as well as to the pathologist.

An interesting pharmacologic approach has been proposed as an alternative method of early pregnancy termination. Mifepristone (RU486), a progesterone receptor blocker, and Epostane, a progesterone synthesis inhibitor, have been evaluated in preliminary studies for abortifacient effect in gestations of less than 50 days. Both drugs were given orally for 1 week, and each resulted in complete abortions in about two-thirds of patients.[13] Although some patients prefer a medical solution over a surgical one, most physicians hesitate to endorse such agents because one-third of the patients given these drugs required surgical intervention after a week's therapy had proved futile.

First-Trimester Dilatation and Curettage

Preoperative Assessment and Preparation

Prior to the preoperative counseling session, each patient submits a freshly voided urine specimen for pregnancy testing. After the counseling session, blood is obtained to determine the patient's hematocrit, blood type, and Rh factor. A complete history with emphasis on reproductive, cardiovascular, and allergic events is then elicited. If the history is suggestive, an electrocardiogram may be ordered.

The general physical examination, including a pelvic examination with cervical Papanicolaou smear and gonorrhea culture, is done. Depending on operator preference and the routines of the particular facility, the physical examination may be done a day or two before the procedure or concluded only minutes before the abortion is begun. In either case, it is vitally important that corroboration of the patient's estimated length of gestation be made by bimanual physical assessment prior to beginning the abortion procedure. Some facilities routinely perform dating of gestation by ultrasound assessment in all patients, believing that any slight delay or inconvenience caused by this policy is more than offset by a reduction in the complications associated with inaccurate assessment of uterine size. Such a policy also permits early recognition of twin gestation, molar pregnancy, severely retroverted uterus, bicornuate configurations, missed abortion, and an occasional ectopic gestation. Immediate access to ultrasonography is invaluable for the management of incomplete second trimester abortions, permitting precise assessment of the volume and placement of retained tissue and aiding the operator in its removal under direct ultrasound guidance.[14]

The absolute hematocrit or hemoglobin value rarely influences the decision for abortion but may preclude its performance outside a hospital, where there would be more immediate access to cardiac monitoring, anesthesia assistance, and transfusion service in the event of unexpected or excessive blood loss. Individuals who are found to be Rh-negative receive Rh-immune globulin regardless of a potential history of previous Rh sensitization. Determining prior sensitization in the few patients who might be affected adds unnecessary cost and delay and may actually prevent subsequent administration of Rh-immune globulin to patients who need it. First trimester patients are given a "mini" dose (one-sixth the standard pregnancy dosage) immediately after the abortion procedure, and second trimester abortion patients receive a standard pregnancy dose.

The results of cytologic screening and gonorrhea cultures are usually not available prior to the abortion procedure, but they are important laboratory tests that yield higher than usual positive results in this group that is at high risk for sexually transmitted disease. Most clinics now supply prophylactic antibiotics, which are begun immediately after the procedure and continued for several days. The dosage and duration of antibiotic coverage varies widely. Cost considerations favor the current wide use of generic tetracycline, although the synthetic tetracyclines doxycycline and minocycline may be better tolerated and more effective. Both asymptomatic cervical gonorrhea and chlamydial infection are usually covered by tetracycline, although individuals who subsequently prove to have positive gonorrhea cultures must be recultured for both gonorrhea and *Chlamydia* at their postoperative visit.

Most medical conditions, such as well controlled insulin-dependent diabetes, seizures disorders, mitral valve prolapse, and other stable chronic disorders, are amenable to treatment in an outpatient clinic or office setting performed with local paracervical block anesthesia during the first trimester. Conversely, second trimester procedures and patients with acute illness, cyanotic heart disease, or profound anemia are more properly treated in a hospital or adjacent outpatient facility with immediate access to emergency services.

Anesthesia

Paracervical block anesthesia is the preferred method of pain relief in nearly all cases of first and second trimester abortion. Its advantages include satisfactory pain relief, ease of administration and accessibility, and low cost of medication and supplies. In addition to the increased risk associated with general anesthesia cited earlier, second trimester procedure complications have been shown to be reduced when local paracervical anesthesia is employed. This reduction is thought to be due to the gentler, more careful technique employed in patients under local anesthesia. Specifically, uterine perforations and injuries to the lower uterine segment through overly vigorous extraction of fetal parts are greatly reduced, as awake patients are aware of the additional trauma and offer verbal protest when the operator manipulates the myometrium inappropriately.

The technique of paracervical block anesthesia varies among operators but generally involves injection of up to 30 ml of local anesthetic agent divided among four sites (3 o'clock and 9 o'clock positions and uterosacral ligament insertions at the 5 and 7 o'clock positions). The uterosacral insertions may be more easily identified by lifting the cervix upward after placing the tenaculum on the anterior cervix. The local anesthetic agent most often chosen by experienced operators is 1% lidocaine with or without epinephrine 1:200,000, although chloroprocaine is preferred by some. The addition of a vasoconstrictor has several advantages, including prolongation of anesthetic effect, reduction of vasovagal episodes, and virtual elimination of rapid absorption of the anesthetic agent, a phenomenon that has been responsible for several deaths due to cardiac arrhythmia.[15] As much as 30 ml of the agent may be used without fear of toxicity. One investigator showed improved overall anesthetic effect when at least 20 ml was used.[16] Visible blanching of the mucosa underneath the needle top indicates that the anesthetic agent is being injected into the correct tissue plane. Injection into the deeper stroma of the cervix is less effective and may promote unwelcome systemic absorption, which patients may recognize by noting transient numbness of the lips and tongue.

Suction Curettage

After completing preoperative evaluation and counseling and having obtained an informed consent, the patient is brought to the operative area by the nursing assistant, who then briefly summarizes the mechanics of the procedure as they will be experienced by the patient. Alternatively, the physician may choose to provide this information as a means of establishing rapport and easing the inevitable anxiety and fears experienced by most patients at this point. A brief physical examination of the heart, lungs, and abdomen is performed, and the patient is draped in a dorsal lithotomy position.

A thorough pelvic examination is done, emphasizing particularly the size and position of the uterus. Size and date discrepancies must be resolved before the procedure is begun. Ultrasonographic assessment is particularly valuable at this point if any questions remain about the patient's uterine size or position or if obesity, previous pelvic-abdominal surgery, or inability to relax the abdominal wall in any way compromises the gynecologic assessment. Uterine perforations with subsequent hemorrhage or bowel injury are often due to overestimation of the gestation length and uterine size, which promotes excessive cervical dilation consequent to the increased cervical resistance. Conversely, underestimation of uterine size and weeks of gestation may lead to insufficient dilatation and selection of suction curets too small to evacuate the uterine cavity completely. Too often the predictable outcome is an incomplete procedure with retained secundines, which promotes subsequent hemorrhage and infection and requires a second procedure for completion.

To begin the procedure, a tray of suitable sterile instruments is arranged in front of the operator (Fig. 2-2). With the patient in dorsal lithotomy position, a warmed speculum is inserted and the cervix visualized. Cytology and culture specimens are now obtained if not done previously. If the operator had inserted a laminaria for preliminary dilation, it is removed. The cervix is wiped clean of secretions and discharge with a gauze square moistened with povidine-iodine solution or other antiseptic. The effectiveness of antiseptic solutions in the prevention of postoperative infection is unproved and may be entirely due to a bacterial dilution effect (in which case saline or tap water would be equally effective). The anterior cervix is infiltrated with a few milliliters of anesthetic solution and the same area grasped with a single-toothed tenaculum. The infiltration reduces tenaculum pain and serves a hemostatic role as well. Many operators attach the tenaculum in a vertical direction at the twelve o'clock position with the lower jaw of the instrument attached inside the cervical os. This maneuver positions the instrument in the connective tissue substance of the cervix rather

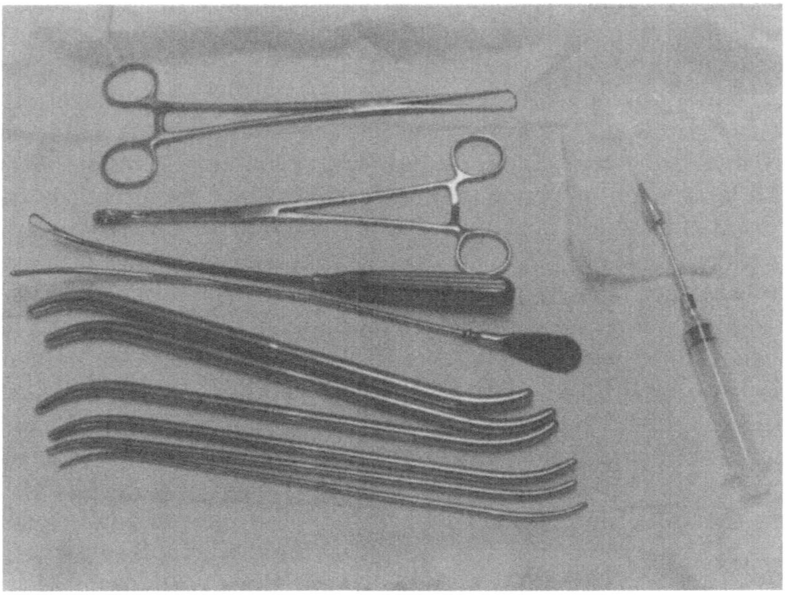

FIGURE 2-2. Instruments for first-trimester D&E procedure include Pratt dilators (13–35 French), Sims sound, sponge forceps, curet, single-toothed tenaculum, and 22-gauge needle attached to anesthesia-extension. (Courtesy of M. Torrington.)

than the superficial anterior cervical mucosa and effectively prevents cervical lacerations. While maintaining firm traction on the tenaculum, the paracervical block anesthetic is administered. Effective anesthesia is usually immediate but can be improved by waiting a few minutes for wider tissue absorption.

While maintaining cervical traction, a Sims sound or the smallest cervical dilator is cautiously advanced through the cervical canal until the tip encounters the uterine fundus. The initial sounding maneuver is a critical one that requires only minimal force in nearly all cases. The correct plane of the canal is sometimes found by gentle rotation of the sound or dilator until the normal avenue is discovered. In cases where cervical stenosis is encountered or an open channel cannot be demonstrated, it may be helpful to carefully advance graduated probes such as lacrimal duct dilators (or dilators as small as 1 mm in diameter) in search of the correct plane. The operator must resist all temptation to advance the sound or dilator blindly in the general direction of the fundus, as this method promotes establishment of a "false passage" and may preclude discovery of the true cervical canal during this operative session. Ultrasound guidance of the dilation process is employed by some operators in difficult cases to facilitate passage of the dilators or sound into the uterine cavity. Failing to sound or dilate the cervical canal, the patient is best served by abandoning the procedure at this point and rescheduling it, preferably in a facility where simultaneous laparoscopy may be employed to guide and assist the surgeon in the subsequent attempt.

Assuming the uterine cavity has been entered with a sound, the cervical os is then progressively dilated with tapered Pratt or Dennison dilators until the cervical canal is large enough to admit a suction curet with a diameter 1 to 2 mm smaller than the estimated weeks of gestation (i.e., at 10 weeks' gestation, a 27 Pratt dilator allows easy introduction of an 8- or 9-mm suction cannula). The suction cannula is advanced until it touches the uterine fundus, then withdrawn slightly before attaching the vacuum source (usually a transparent plastic hose). Touching the fundus with the cannula tip is a final safety check to guard against inadvertent perforation having occurred during the dilation phase. A vacuum source with a negative pressure of 500 to 700 mm Hg is engaged and the suction cannula moved in a simultaneous rotary and to-and-fro motion between the level of the internal os and the midportion of the fundus. The suction evacuation is continued so long as decidual material and uterine contents are seen emerging through the cannula and transparent hose. Appearance of a pinkish-red, bubbly froth in the tubing usually heralds complete emptying of the fundus. The suction is temporarily stopped by disengaging the ring valve situated at the operator's end of the vacuum hose. The cannula is withdrawn just enough to inspect the tip and to free it of obstructing tissues that may be present. A medium-size Sims curet is then gently advanced to explore the fundus for residual tissue. If necessary, the cannula is reintroduced for suction removal of remaining tissues. Skilled operators minimize patient discomfort by restricting unnecessary contact with the sensitive fundal apex by the cannula or curet.

Satisfied that the cavity is empty, the tenaculum is removed, the vagina is cleared of blood clots and tissue fragments, and the speculum is gently closed and withdrawn. Bleeding from tenaculum punctures invariably stops after 1 to 2 minutes of constant pressure with a gauze and sponge stick.

The patient is placed in a supine position, and vital signs are measured. If there is no sign of hypotension or bradycardia, she can be moved to the recovery area for further monitoring. After an hour's observation, the patient may be discharged if she is able to ambulate.

Laminaria

Preoperative cervical softening and dilation with botanic *Laminaria japonicum* or synthetic Lamicel dilators is essential to the safe performance of the second trimester D&E procedure. This technique is occasionally used for first trimester procedures as well. Time constraints may preclude its routine use during the first trimester, although the synthetic laminaria achieve maximal effects within 3 to 4 hours and may be an especially effective adjunct to termination of intermediate length gestations (i.e., 13–15 weeks). In most clinics laminaria are routinely used in gestations of more than 14 weeks and are inserted sometime during the afternoon of the day prior to the procedure.

Before insertion of the laminaria tents, the patient has completed the preabortion counseling and has undergone an ultrasound evaluation to validate the gestational length. With the patient in the dorsal lithotomy position, a speculum is inserted and a specimen is obtained for preoperative cytology and culture. A tenaculum is employed to stabilize the cervix, and the external os is wiped clean with an antiseptic-soaked gauze. Laminaria tents are then inserted with a sponge forceps, sliding each tent along the cervical canal until the stringed-end is just visible at the external os. Depending on the length of gestation, as many as four to six medium to large laminaria can be inserted. Although some operators place one or two folded gauze squares against the external os to hold the laminaria in place, it is probably unnecessary and may even be a potential absorptive wick for infectious organisms.

In cases where only one laminaria can be inserted, the patient may be kept in the facility for a few hours and the dilated laminaria then removed, permitting insertion of two or three new laminaria to provide additional overnight dilation. The next day the laminaria are removed immediately before the paracervical block is placed, prior to introduction of the dilators. in most cases, if enough laminaria have been inserted, the cervix is dilated enough to admit the appropriate suction curet.

Patients are told to expect moderate overnight cramping and minimal bleeding while the laminaria are in the cervix. Ibuprofen is generally sufficient for relief of this discomfort. The first dose of prophylactic tetracycline is started on the evening of the laminaria insertion. A few patients experience spontaneous rupture of membranes during insertion or during the overnight waiting period. Such cases are treated expectantly and are not terminated ahead of schedule unless significant bleeding or persistent labor-like cramping intervenes.

Second Trimester D&E Procedure

Information from surveillance studies performed by the CDC has determined that the D&E method is safer and more commonly used for pregnancy termination between 12 and 16 weeks than the former policy of avoiding the D&E procedure after 12 weeks' gestation and delaying the termination until 16 weeks in favor of an amnioinfusion method employing saline, urea, or prostaglandins.[17] Between 16 and 20 weeks' gestation the morbidity of the two choices (D&E versus amnioinfusion) is comparable. Currently, D&E is chosen twice as often as saline or urea-prostaglandin infusion. The surgical solution offers several advantages, including savings in time and cost, because it does not require an overnight hospitalization; moreover the procedure is concluded within less than 30 minutes with minimal patient discomfort. In contrast, the amnioinfusion procedure generates significant uterine cramping that may persist for hours. Psychologically, D&E spares the patient the emotional distress of delivering a premature fetus although the operators and surgical assistants must handle extracted fetal parts and tissues.

In some respects the technique of second trimester D&E parallels the first trimester procedure, but important differences make it considerably more difficult to perform and statistically more dangerous. The most critical preoperative safety measures are precise gestational dating by ultrasound measurement of the fetal biparietal diameter and adequate preoperative cervical softening and dilation by laminaria insertion. A third factor of importance is the selection of equipment and instruments appropriate for the gestational length.

Special equipment has been designed for termination procedures for gestations of more than 14 weeks. (Prior to this point a standard caliber vacuum hose and curets up to 12 to 13 mm are sufficient.) The most important instruments are the Sopher, Bierer, and Hern extraction forceps, which are used to grasp, crush, and extract fetal tissues. In addition, large Pratt or Dennison cervical dilators sufficient to achieve 16 mm of dilation should be available, although they are rarely needed if sufficient preoperative laminaria have been placed the day before. The final equipment modification is the addition of a 15- or 16-mm suction cannula and a larger transparent suction hose. The additional instruments utilized for second trimester procedures are pictured in Figures 2-3 and 2-4.

There is no substitute for experience in the safe performance of second trimester abortion by D&E, and most residents and practicing gynecologists do not have a sufficient case load to learn the procedure well or to maintain proficiency. Many jurisdictions restrict the performance of second trimester procedures to certified clinic and outpatient facilities. This restriction serves to concentrate such cases in the hands of experienced operators.

On the date of the procedure the second trimester patient with laminaria is generally scheduled as the initial case of the morning, having already received appropriate counseling and instructions. In the dorsal lithotomy position, a speculum is gently manipulated around the enlarged laminaria and opened. The laminaria strings are withdrawn and the laminaria tents discarded. The cervix is then cleansed with antiseptic and a paracervical block instilled. General anesthesia is sometimes employed, especially by more experienced operators who may not need the warnings an awake patient provides when protesting at too forceful or traumatic instrumentation (which may portend an ensuing perforation or cervical tear).

After adequate anesthesia is given, the suction cannula appropriate for the week of gestation is gently inserted to test the cervi-

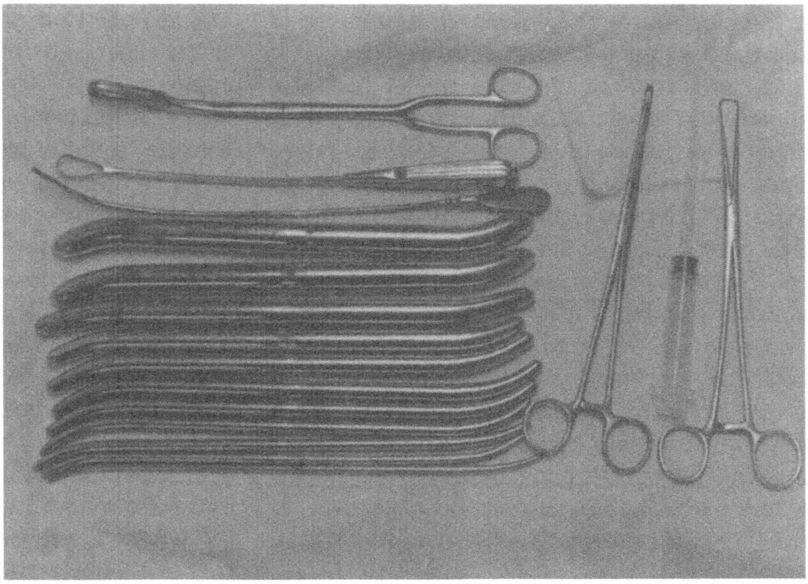

FIGURE 2-3. Additional instruments for second-trimester D&E procedure include Pratt dilators up to 49 French and Sopher forceps for extraction of fetal tissue. (Courtesy of M. Torrington.)

cal dilation. Usually it slides effortlessly into the uterine cavity, and suction is begun immediately. Failing to pass the cannula may indicate that the laminaria did not reach the internal os level, and additional mechanical dilation is required. This technique proceeds as for first trimester procedure, starting with a uterine sound and progressing through the graduated dilators. If the operator cannot be assured that the sound and the dilators have traversed the cervical canal into the fundus, the sound is passed cautiously using continuous ultrasound guidance. Failing at this point, the procedure must be abandoned and rescheduled in an operating room under laparoscopic observation by an assistant. If the procedure is already taking place in this environment, laparoscopy is immediately employed. There is no danger associated with rescheduling the procedure as much as 24 hours later. In fact, there have been instances in which laminaria were removed up to 24 hours after insertion at the patient's request in order to continue with the pregnancy—with no ill effects.

The general principle of the second trimester D&E procedure is piecemeal removal of fetal tissues after they have been brought to the lower uterine segment by the suction curet. Initially the amniotic fluid is aspirated, and occasionally the placenta and some fetal tissues can be aspirated through the cannula. At some point no further tissue can be aspirated. At this point, the cannula is removed rather than persisting with unrewarding manipulations. The special tissue forceps is then employed to grasp, crush, and extract the fetal tissues, placenta, and membranes. Alternating between brief suction to bring tissue within the reach of the forceps

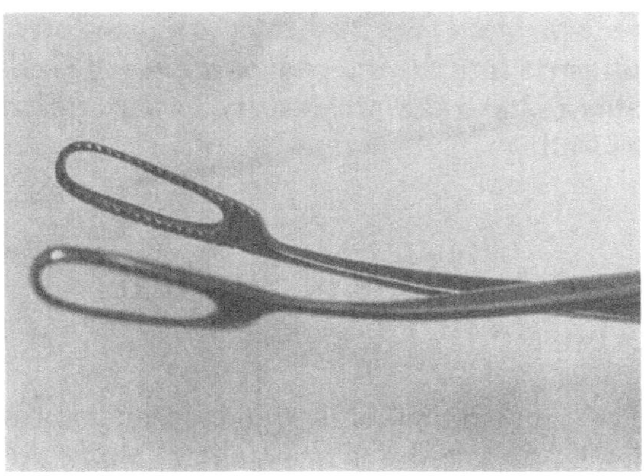

FIGURE 2-4. Close-up of Sopher forceps shows the serrated jaws, which assist in fragmentation and extraction of tissues (Courtesy of M. Torrington.)

and insertion of the forceps for removal of the tissues, the cavity is emptied. Crushing and removing the calvaria is the last and sometimes most difficult maneuver. If it cannot be isolated, a large curet is introduced to explore the cavity and identify its location. Ultrasonography may be useful at this point. If repeated efforts fail to deliver the calvaria, the instruments are removed and the patient is allowed to rest in the supine position for 15 to 30 minutes. Oxytocin or methylergonovine (Methergine) is then given to promote uterine contractility. In nearly all cases, reexamination finds the missing tissue within easy grasp at the internal os, although it may be necessary to wait up to 2 hours for the process to be completed.

After a final exploration with the large, sharp curet and a final suctioning to empty the fundus of any free-floating clots or tissue fragments, the instruments are removed and all the recovered tissues immediately inspected before the patient is transferred from the operative area. The calvaria, spinal column, and extremities must be identified, or resuctioning is performed to recover the missing tissues.

Opinion varies as to the value of preoperative administration of methylergonovine or oxytocin for reducing intraoperative and postoperative blood loss. The author routinely gives methylergonovine 0.2 mg IM just prior to the procedure, believing that it reduces blood loss, facilitates expulsion of fetal parts into the internal os area, and probably reduces the prospensity to perforation as well.

Complications of Surgical Abortion

The complications of first and second trimester D&E procedures are similar. Many are avoidable if sufficient emphasis has been placed on the important safeguards of precise assessment of gestational length and uterine position and avoidance of excessive force. Postabortion syndrome (PAS) is a common entity resulting from an accumulation of blood clots in the uterus secondary to uterine hypotonia.[18] It can occur at any week of gestation with symptoms appearing within minutes to several days after the procedure. The patient complains of unusually severe uterine cramping, and examination reveals an enlarged, soft, tender uterus. Treatment consists in immediate resuctioning, which may yield up to several hundred milliliters of blood and clots. Uterine pain disappears within minutes. When the syndrome develops slowly over several days, the patient may seek attention in another facility where a physician who is not familiar with PAS may misdiagnose the condition as retained tissues, endometritis, or pelvic infection. Tubal reflux of the accumulated blood into the abdomen may simulate an acute abdomen and yield nonclotting blood on culdocentesis. Patients have been subjected to laparoscopy on the mistaken assumption that an ectopic pregnancy or undiagnosed uterine perforation existed. A resuctioning procedure must be the first diagnostic move whenever a postabortion patient presents with uterine enlargement and tenderness.

Retained pregnancy tissues predispose to hemorrhage and infection and must be strongly considered whenever there is unusual postabortion bleeding, fever, or tenderness. Once again, resuctioning is mandatory and usually curative. Endometritis without retained tissue is relatively uncommon with the near-universal use of prophylactic antibiotics but is still the most frequent postoperative complication. If an infection is diagnosed at an early stage, oral antibiotics such as ciprofloxacin, doxycycline, or amoxicillin may be employed, but patients with more serious pelvic infections with fevers of 102°F or higher require hospitalization for intravenous antibiotic therapy.

Cervical lacerations from tenaculum tears are rarely encountered if the tenaculum is placed vertically. These tears and those produced during cervical dilation are easily repaired with 2-0 chromic sutures and heal without incident. It is important to recognize that such injuries usually do not occur with the ordinary forces employed for dilation. Routine employment of excessive force eventually leads to uterine perforation and potential disaster, especially during second trimester D & E procedures.

Uterine perforation is a complication encountered by nearly every gynecologist. In most cases the outcome is uneventful and the treatment expectant. When diagnosed during the initial phase of sounding the fundus or dilating the cervical canal during first trimester procedures, expectant treatment with close monitoring of vital signs during the first few postoperative hours is sufficient. Laparoscopy to confirm a perforation diagnosis is unnecessary unless the patient exhibits unusual bleeding or deteriorating vital signs. An experienced operator may elect to continue and complete the procedure after reexamining the patient to confirm the correct inclination of the cervical canal and cautiously resounding the uterus. Intraoperative ultrasound confirmation of proper intrauterine placement of the dilators and suction curet assists the operator in completing the procedure.

Perforations of the second trimester uterus are potentially more hazardous, as the myometrial wall is thinner and less resistant, and the perforating instruments are of larger caliber. The 16-mm suction cannula easily engulfs the ovaries, uterine tubes, and loops of bowel within seconds; and the appearance of these structures in the suction tubing may be the first warning that a surgical disaster has already occurred. For this reason, most clinicians who encounter perforation during the second trimester complete the operation under laparoscopic assistance. Laparoscopy is mandatory if the perforation is discovered after a suction curet has been introduced into and perhaps beyond the fundus. If perforation is diagnosed prior to aspiration, ultrasonographic monitoring may be employed by experienced surgeons to empty the second trimester uterus under continuous ultrasonic observation.

Hysterotomy and Hysterectomy

The indications for hysterotomy have diminished in recent years with the recognition that surgical D&E is statistically safer than instillation methods up to 16 weeks' gestation. At 16 to 20 weeks the D&E procedure may be selected when amnioinfusion methods are contraindicated, as in patients who have previously undergone a classic cesarean section or when medical complications (heart disease, asthma, ulcer disease, renal compromise) preclude saline or prostaglandin amnioinfusion. Formerly such patients may have been candidates for second trimester hysterotomy, but currently they are best served by the laminaria/D&E method. The occasional hysterotomy procedure still done usually occurs after a uterine perforation or failed D&E procedure. Most surgeons select a vertical incision identical to a classic cesarean incision, although a low-transverse incision has been advocated by Mukerjee on the grounds that it is less susceptible to rupture during a subsequent pregnancy or labor.[19]

There appears to be no place for elective hysterotomy as a primary abortion method, as it is associated with an unacceptably high incidence of postoperative infection, hemorrhage, thrombophlebitis, and coagulopathy.[20] Additional drawbacks include a need for hospitalization and prolonged patient discomfort.

A few experienced surgeons have demonstrated that vaginal hysterectomy as a primary abortion procedure is associated with no higher morbidity than the procedure in the nonpregnant state.[21,22] The decision for hysterectomy in such cases must be based on demonstrable uterine pathology rather than a mere desire for sterilization, as a combined D&E and sterilization procedure offers far lower morbidity and mortality rates.[23] Abdominal hysterectomy has even higher reported morbidity, presumably because it may be used in more advanced stages of gestation.[24]

Effects on Subsequent Fertility

Hogue and colleagues reviewed the world literature and outlined the available data regarding the effect of induced abortion on subsequent fertility and pregnancy outcome.[25] Nearly all studies concluded that induced abortion has a negligible effect on subsequent fertility and does not increase the risk of subsequent ectopic pregnancy. Cervical dilation and sharp curettage (D&C) has been shown in a few studies to be associated with a threefold increased risk of spontaneous midtrimester abortions in subsequent pregnancies, compared to the now-predominant method of D&E vacuum aspiration. The increased risk may be due to the wider dilation needed for the former procedure. The risk of absolute infertility related to previous abortion seems to be insignificant and related only to the rare case of post-abortion endometritis-salpingitis.

There are few available statistics concerning reproductive outcome after second trimester operations, but the data available suggest minimal impact on the measured parameters of premature delivery and low birth weight. Comparative studies suggest that the method of cervical dilation associated with instillation methods and laminaria pretreatment is associated with little if any additional risk, whereas greater degrees of instrumental second trimester dilation showed a higher rate of subsequent low birth weight (although the risk did not reach statistical significance).[26]

Conclusion

Surgical abortion has become one of the most frequent and safest gynecologic operations. The sustained reduction in morbidity and mortality is due to continued diligence in observing the time-honored surgical principles of careful preoperative assessment, gentle manipulation of tissues, adherence to proved surgical methods, and reasoned management of intraoperative complications.

References

1. Kerslake D, Casey D. Abortion induced by means of the uterine aspirator. Obstet Gynecol 1967;30:35–45.
2. Tietz C. Induced Abortion: A World Review. New York, The Population Council, 1981.
3. Binkin NJ. Trends in induced legal abortion: mortality and morbidity. Clin Obstet Gynecol 1986;13(1):84.
4. Atrash HK, Cheek TG, Hogue C Jr. Legal abortion mortality and general anesthesia. Am J Obstet Gynecol 1988;158:420.
5. Chavkin W, Fernandez L, Harris M, et al. Maternal deaths associated with barbiturate anesthesia—New York City. MMWR 1986; 35:579.
6. Henshaw S. Induced abortion: a worldwide perspective. Fam Plann Perspect 1986; 18:250–254.
7. Burkman RT, King TM, Burnett LS, et al. University abortion programs: one year later. Am J Obstet Gynecol 1974;119:131.
8. Lindheim BL, Cotterill MA. Training in induced abortion by obstetrics and gynecology residency programs. Fam Plann Perspect 1978;10:24.
9. Darney PD, Landy U, Macpherson S, Sweet RL. Abortion training in U.S. obstetrics and gynecology residency programs. Fam Plann Perspect 1987;19:158.
10. Margolis AJ, Greenwood S, Heilbron D. Survey of men and women residents entering US obstetrics and gynecology programs in 1981. Am J Obstet Gynecol 1983;146:541.
11. Atienza MF, Burkman RT, King TM, et al. Menstrual extraction. Am J Obstet Gynecol 1975;121:490.
12. Burnhill MS, Armstead JW. Reducing the morbidity of vacuum aspiration abortion. Int J Gynaecol Obstet 1978;16:204.
13. Birgerson L, Odlind V. Early pregnancy termination with antiprogestins: a comparative clinical study of RV 486 given in two dose regimens and Epostane. Fertil Steril 1987; 48:565.

14. Goldenberg RI, Davis RO, Hill D. The use of real-time ultrasound as an aid during difficult therapeutic abortion procedures. Am J Obstet Gynecol 1984;148:826.
15. Grimes DA, Cates W. Deaths from paracervical anesthesia used for first trimester abortion 1972–75. N Engl J Med 1975;295: 1397.
16. Smith GM, Stubblefield PG, Chirchirillo L, McCarthy MJ. Pain of first trimester abortion: its quantification and relation with other variables. Am J Obstet Gynecol 1979; 133:489.
17. Grimes DA, Schultz KF, Cates W Jr, Tyler CW. Midtrimester abortion by dilation and evacuation: a safe and practical alternative. N Engl J Med 1977;296:1141.
18. Sands RS, Burnhill MS. Post-abortal uterine atony. Obstet Gynecol 1974;43:595.
19. Mukerjee AK. Modification of uterine incision for abdominal hysterotomy. Obstet Gynecol 1973;42:149.
20. Cates W, Grimes DA, Tyler CW. Safety of legal abortion. Lancet 1980;1:198.
21. Laufe LE, Kreutner AK. Vaginal hysterectomy: a modality for therapeutic abortion and sterilization. Am J Obstet Gynecol 1971;110:1096.
22. Ballard CA. Therapeutic abortion and sterilization by vaginal hysterectomy. Am J Obstet Gynecol 1974;118:891.
23. Schulman H. Major surgery for abortion and sterilization. Obstet Gynecol 1972;40:738.
24. Stumpf PG, Ballard CA, Lowerjohn R. Abdominal hysterectomy for abortion-sterilization. Am J Obstet Gynecol 1980;136:714.
25. Hogue C Jr, Cates W Jr, Tietze C. The effects of induced abortion on subsequent reproduction. Epidemol Rev 1982;4:66.
26. Hogue C Jr, Peterson W. Outcome of pregnancy following mid-trimester abortion by dilation and curettage. Presented at the Risk Management Seminar, National Abortion Federation, September 1984.

3 Septic Abortion and Septic Thrombophlebitis

RICHARD L. SWEET

Despite the introduction of a multitude of broad-spectrum and potent antimicrobial agents, infection is a major cause of morbidity and an all too often factor in mortality among obstetric and gynecologic patients Whereas genital tract infection remains one of the most common clinical problems in obstetrics and gynecology, bacteremia is fairly uncommon.[1–6] This finding is in contradistinction to medical and surgical services, where gram-negative bacteremia continues to be a common and potentially life-threatening problem. It is estimated that between 71,000 and 330,000 cases of nosocomial gram-negative bacteremias occur in the United States each year, and that more than 30% of these cases are fatal. Of special note is the recognition that the prognosis for bacteremic patients is correlated with the development of shock. Kreger et al. reported that at the Boston City Hospital the mortality in bacteremic patients associated with the development of septic shock was 47%, compared to a mortality of only 7% among patients in whom shock did not develop.[7,8]

In addition to the rate of bacteremia being low in obstetric and gynecologic patients (8–10%),[1–5] the incidence of septic shock complicating bacteremia in obstetric and gynecologic patients is also much lower than that seen in medical or surgical patients and ranges from 0 to 4% (Table 3-1). Fortunately, mortality also is relatively infrequent among obstetric and gynecologic patients in whom septic shock develops (Table 3-1). The incidence of mortality due to sepsis in obstetrics and gynecology is reported to be 0 to 3% versus a 10 to 81% mortality incidence among other patient groups.[1,2,6–10] Gonik suggested several reasons for this lower mortality rate among obstetric and gynecologic patients.[11] They included the younger age, the transient nature of the bacteremia, the type of organisms involved, the fact that the primary site of infection is more amenable to surgical intervention, and the lack of underlying disease that adversely affects the prognosis.

The remainder of this chapter discusses two of the major infections in obstetrics and gynecology that are associated with a significant risk for the development of bacteremia and septic shock: septic abortion and septic pelvic thrombophlebitis.

Septic Abortion

Septic abortion and its sequelae have long been recognized as serious, life-threatening complications in obstetrics and gynecology. In 1956 Studdiford and Douglas first drew attention to the syndrome of endotoxic shock in association with septic abortion.[12] Following their description, physicians became increasingly aware of this entity, and during the late 1950s and 1960s septic abortion and its complications ranked as the leading

TABLE 3-1. Mortality associated with septic shock in obstetrics and gynecology.

Study	Incidence of bacteremia per 1000 pts.	No. of pts. with bacteremia	Patients with septic shock	Mortality (No.)
Blanco et al.[1]	176[a]	0	0	
Ledger et al.[2]	7	139	6 (4%)	4 (3.0%)
Bryan et al.[4]	2	92	—	4 (4.3%)
Chow & Guze[b,9]	—	32[c]	4 (12%)	0
Weinstein[d,6]	5.3 Ob 4.6 Gyn	38	—	1 (2.6%)

[a]Obstetric patients only.
[b]Chow AW, Guze LB. Bacteroidaceae bacteremia: clinical experience with 112 patients. Medicine (Baltimore) 1974; 53:93–126.
[c]*Bacteroides* bacteremia.
[d]Weinstein MP, Murphy JR, Reller MP, Lichtenstein KA. The clinical significance of positive blood cultures: a comprehensive analysis of 500 episodes of bacteremia and fungemia in adults. II. Clinical observations, with special reference to factors influencing prognosis. Rev Infect Dis 1983; 5:56–70.

cause of maternal death in many areas of the United States. At that time it was estimated that septic abortion occurred in 25% of patients with incomplete abortions and that approximately 3 to 17% of septic abortions were complicated by septic shock.[13,14] Prior to the liberalization of abortion laws, Schwartz estimated that the number of deaths per year from septic abortion was between 5,000 and 10,000.[15]

Since the early 1970s there has been a tremendous reduction in the number of women with septic abortion. This dramatic change has been attributed to the implementation of liberal abortion laws and the availability of hospital- or clinic-based pregnancy termination. Concomitant with the decrease in the number of septic abortions there has been a reduction in mortality from sepsis and septic shock secondary to abortion. However, infection as a complication of abortion has persisted.[16–18] In fact, the Centers for Disease Control (CDC) have reported that death from abortion remains a problem and that infection is the leading cause.[16] The CDC noted that pelvic infection was diagnosed in 1% and fever in 2% of women undergoing legal abortion.[16] A variety of risk factors for complications after abortion have been summarized by Sweet and Gibbs.[19] They include greater duration of pregnancy and use of termination procedures other than suction curettage or transcervical evacuation.[17] Whereas the major complication rate was less than 0.5% and the risk of death was 1.3 per 100,000 abortions for suction curettage during the first trimester, with intrauterine injections these rates rose to 2% and 12.5 per 100,000, respectively. For either hysterotomy or hysterectomy the incidence of mortality was 41.3 per 100,000 (32 times greater than for curettage). Because of the increased risk of complications for second trimester abortions performed with intrauterine injection, the use of dilatation and evacuation (D&E) procedures for pregnancy termination during the second trimester have been introduced into clinical practice. In the hands of experienced operators and with the use of laminaria (or dilapan), special instruments, and real-time ultrasonography (D&E) of second trimester pregnancies has been found to have significantly fewer complications, including infection.[20]

With the availability of pregnancy termination, septic abortion, the classic community-acquired infection, has become another nosocomial infection. Although postabortion infection still occur, the scope of the problem is different. The absence of stigma regarding the circumstances of abortion prevents the delays in seeking care that resulted in extensive infection, and the patients present early with symptoms of fever, pain, and bleeding. The patient with a septic abortion can present serious diagnostic and therapeutic problems for the physician. Prompt diagnosis and management are crucial to prevention of the major complications of septic abortions, i.e., septic shock, leading to death, and pelvic abscesses.

Etiology

Pelvic infection, particularly endometritis, is a well known complication of elective abortion. There is an increase in postabortion infection that parallels the increase in gestational age.[17,21] In patients having elective first trimester suction curettage, the reported rate of infection or febrile morbidity has ranged from 0.1 to 5.2%.[17,21–24] For second trimester terminations with intraamniotic hypertonic saline or prostaglandin injection, febrile morbidity has been reported in 1.5 to 18.1% of cases.[17,21,24,25]

The use of D&E for termination of second trimester pregnancies has been reported to markedly decrease the risk for postabortion infection.[20] At the San Francisco General Hospital the postabortion infection rate for second trimester pregnancies terminated by D&E is less than 1% (P Darney and RL Sweet, unpublished data).

The presence of untreated endocervical gonorrhea has been associated with a threefold increase in the risk for developing postabortion endometritis.[26] More recently, a similar increased risk of postabortion endometritis has been reported in patients with untreated *Chlamydia trachomatis* infection of the cervix.[27,28] Both of these sexually transmitted organisms are more frequently recovered from women who are less than 24 years of age, have multiple sexual partners, are noncontraceptors (or use nonbarrier methods of contraception), live in urban inner city areas, and are in the lower socioeconomic group. Thus it is not unexpected that many young women seeking elective pregnancy termination are infected with *Neisseria gonorrhoeae*, *C. trachomatis*, or both. This fact has led several authors to recommend routine antibiotic prophylaxis (usually a tetracycline regimen) for young women undergoing elective abortion procedures.[29] An alternative approach is to screen all per-pregnancy termination patients for *N. gonorrhoeae* and *C. trachomatis* and to treat only those infected with one or both of these organisms.

Spontaneous abortion has been an infrequent source of infected abortion and maternal deaths secondary to infection. The lowest rates of infection are associated with curettage methods of pregnancy termination, and rates are lower with suction curettage than with sharp curettage. In early reports this low rate appears to hold true even for D&E procedures for midtrimester abortion.[20] Infection following abortion is generally an ascending process by which microorganisms from the vagina or cervix gain access to the usually sterile upper genital tract. There are three basic mechanisms of pathogenesis for infection. Most commonly, uterine infection is secondary to retained products of conception. Such necrotic tissue as retained products serves as an ideal nidus for infection with vaginal and cervical microorganisms, which gain access to the uterine cavity at the time of the procedure. Second, a virulent pathogen may be introduced into the operative site at the time of surgery. This situation may occur when the clinician is the source of an organism such as the group A β-hemolytic streptococcus. Third, infection may result from surgically related trauma to the uterus. Uterine perforation results in damage to the myometrium, which provides a favorable environment for bacterial growth and subsequent endomyometritis. If perforation is unrecognized at the time of surgery, patients may later present with extrauterine infections such as infected hematomas or pelvic abscess.

Intraamniotic injection of saline or prostaglandin is associated with a greater risk of infection than curettage termination. The factors that contribute to this increased risk include (1) placental retention, which is the most frequent cause; (2) introduction of bacterial contamination into the amniotic sac (infrequent); and (3) inadvertent injection of solutions into the myometrium, leading to tissue damage and necrosis and producing an environment for bacterial growth, especially anaerobes. Hysterotomy is associated with a high incidence of severe infectious complications due to the presence of necrotic myometrium, foreign body (suture), blood clot, and bacterial contamination. Thus it is an infrequently used method of pregnancy termination.

With the availability of legalized abortion, community-acquired septic abortions are an infrequent occurrence; however, physicians still need to be aware of clinical manifestations of community-acquired infections and septic shock. The pathogenesis of infection is similar to that described for suction elective pregnancy termination with sharp curettage. Many techniques have been utilized in attempts to terminate pregnancy illegally. Instrumentation with unsterile instruments such as catheters leads to the introduction of pathogens from the external environment; most cases are associated with incomplete abortion, resulting in the presence of infected, necrotic tissue. In the past a rare but often lethal method was the use of soap douches as abortifacients. Use of these agents results in extensive pelvic tissue necrosis and severe infections. Potassium permanganate tablets often resulted in severe burns.

During the mid-1970s serious infections causing sepsis and death were reported in pregnant women with an intrauterine device (IUD) in place. Several investigations have incriminated the IUD in septic midtrimester abortions.[30–33] Christian[34] in 1974 reported five deaths, four of which were associated with the Dalkon Shield. In these cases, the initial symptoms were relatively innocuous, but infection progressed rapidly to sepsis and death within hours. Cates and co-workers[35] reported that the risk of death from spontaneous abortion was 50 times greater for those who continued their pregnancies with an IUD in place than for those who did not, and that risk was greater with the Dalkon Shield than with other devices.

The mechanism by which IUDs cause infected abortion, amnionitis, sepsis, and death has not been proved. It has been suggested that the tail string can facilitate ascension of bacteria from the vagina or cervix into the normally sterile uterine cavity.[36,37] Second, the IUD as a foreign body may alter the local host defense mechanisms that normally protect the uterine cavity from the bacteria of the vagina and cervix. Recognition of the increased risk for infection and mortality among pregnant women who continued their pregnancies with an IUD in place had led to the recommendation that if women conceive with an IUD in place, if the string is visible, and if the pregnancy is desired, the IUD must be removed. If the string is not visible, elective pregnancy termination or close monitoring of the woman for early signs of infection is necessary. With the onset of symptoms or signs of infection, the pregnancy is terminated promptly.

Microbiology

Except for *Neisseria gonorrhoeae, Chlamydia trachomatis,* and the group A β-hemolytic streptococcus, the microorganisms that are pathogens in upper genital tract infections, e.g., infected abortions, arise from the normal flora of the cervix and vagina. These organisms become pathogens when appropriate environmental conditions exist in the upper genital tract, i.e., when traumatized tissue, retained necrotic tissue, or foreign bodies are present. All of these factors may be present following induced or spontaneous abortions.

A major problem in identifying the pathogens involved in septic abortion is our inability to obtain cultures from the site of infection that are not contaminated by the bacterial flora of the vagina and cervix. In the past we have relied on positive blood cultures to identify the significant pathogens associated with septic abortion. More recently the use of protected swabs or double-lumen catheters has allowed us to obtain a reasonably uncontaminated transcervical culture specimen from the endometrium.

During the late 1950s, the microbiologic emphasis was on gram-negative facultative bacteria such as *Escherichia coli, Klebsiella,* and *Enterobacter* and their relation to gram-negative sepsis and endotoxic shock. During this era, anaerobic bacteria were not considered significant pathogens in septic abortion, except for the recognition of the lethal potential of pelvic soft tissue infections caused by *Clostridium perfringens*. Clostridial septic abortions were a feared complication of criminal abortions. With the legalization of abortion, the incidence of this lethal form of sepsis has dramatically decreased.

Coincident with the recognition of the significant role of anaerobic bacteria in pelvic infections, anaerobes were also identified as significant pathogens in septic abortion. In a study reported in 1969, Rotheram and Schick[38] recovered anaerobes from the endocervical cultures of nearly all patients with septic abortion and from the bloodstream of most septic abortion patients with bacteremia. Chow et al.[39] confirmed that both aerobic and anaerobic bacteria are associated with septic abortions. As reported by Sweet, the incidence of mixed aerobic-anaerobic bacterial involvement in septic abortion is similar to that described for pelvic infections in general.[40] The most frequent organisms involved are the gram-negative facultative bacteria, especially *E. coli,* facultative streptococci, and anaerobic bacteria, especially *Peptostreptococcus* and *Bacteroides* (including *B. fragilis*). *Bacteroides bivius* and *B. disiens* have been identified as frequent and important pathogens in soft tissue pelvic infections, including postabortion infection.[41] Similar to *B. fragilis* these anaerobes produce β-lactamase enzymes that make them resistant to penicillins, ampicillin, and first-generation cephalosporins.

Of special note is the organism *Clostridium perfringens,* which is a gram-positive, anaerobic, spore-forming rod capable of producing potent exotoxins. Although the incidence of clostridial septic abortions has decreased since the legalization of abortion, it is a potentially lethal infection, and the clinician must be on alert to identify patients at risk for the development of clostridial septicotoxemia (a combination of septicemia and toxemia). The trauma secondary to instrumentation associated with abortion or retained necrotic products of conception provides an ideal nidus for the development of clostridial infections of the endometrium or uterus.

Clinical Manifestations

Patients with infected abortion usually present with tachycardia, fever, chills, lower abdominal pain. pelvic tenderness, and a history of passage of products of conception. However, the clinical presentation varies according to the etiologic conditions. If a virulent pathogen (e.g., group A β-hemolytic streptococcus) is introduced into the operative site at the time of abortion, the patient may experience a sudden rise in temperature to 103°F (39.4°C) or more within 24 hours. There is a lack of localizing signs on abdominal or pelvic examination. A gram-stained smear of uterine exudate reveals many gram-positive cocci in chains, and no fecal odor is present in the exudate. A second pathogen associated with early-onset postabortion infection is *Staphylococcus aureus*. Toxic shock syndrome due to a toxin produced by strain(s) of *Staph. aureus* occurs in a nonmenstrual form. Toxic shock syndrome is a multisystem disease associated with a wide range of symptoms, signs, and laboratory findings.[41–43] It characteristically presents abruptly with the sudden onset of high fever, chills myalgias, vomiting, diarrhea, hypotension, and a generalized "sunburn-like" rash.

Infected abortion secondary to retained products of conception is usually late in onset, becoming evident several days after abortion. If infection follows legal abortion, the patient usually seeks medical care promptly and presents with fever, bleeding, purulent exudate from the cervical os, and uterine tenderness. In contradistinction, patients with infection following illegal attempts at termination of pregnancy seek medical care late in the disease process. They have more extensive infections, often beyond the uterus, and are at greater risk for development of sepsis or septic shock.

Patients whose infection is secondary to unrecognized uterine perforation usually present several days after abortion with fever, uterine tenderness, and a tender mass contiguous with the uterus. In addition, these patients are at risk for traumatic bowel injury. Thus they may present with signs and symptoms of peritonitis, perforated viscus, or bowel obstruction.

With illegal infected abortion, a history of instrumentation may be obtained; however, in most instances patients deny instrumentation. With illegal infected abortion, the major concern is that the patient will become seriously ill with septic shock leading to death. A lack of demonstrable early clinical or laboratory signs in patients at high risk for sepsis can lead to clinical confusion. If septic shock occurs, the classic host responses to infection may be modified; not infrequently, patients are hypotensive, hypothermic, bradycardic, and leukopenic. On examination, the patient is usually febrile; however, the temperature may be normal or subnormal, which in association with severe infection, may indicate incipient septic shock. Suprapubic tenderness and localized or generalized abdominal rebound tenderness are present. Pelvic examination may reveal evidence of instrumentation, e.g., tenaculum marks on the cervix or a foreign body in the cervical canal. Foul-smelling discharge may be coming from the cervical os. Products of conception may be present in the cervical canal or vagina. There is marked uterine tenderness, and the uterus is often enlarged and soft.

Septic shock has been reported in 3 to 17% of septic abortions.[13,14] When present in its full-blown form, endotoxic shock is readily detectable, but the initial manifestations may be subtle. Usually bacteremia gives rise to chills, fever, nausea, vomiting, and prostration. Bacteremia due to gram-negative organisms results in septic shock in approximately 20% of cases compared to 5% for gram-positive organisms.

The pathophysiology of septic shock is complex. With the onset of bacteremia, host defense mechanisms such as natural barriers, phagocytes, and opsonins compete against bacterial virulence factors such as endotoxins or exotoxins. Release of endotoxin, a complex lipopolysaccharide, from the cell walls of gram-negative bacteria as they are disrupted is the initiating step in the development of septic shock.

Endotoxin has a variety of direct effects that involve complement activation, intrinsic clotting, the cardiovascular system, the hematologic system, prostaglandins, and cachectin (tumor necrosis factor) production.[45–52] The pathophysiology is summarized in Figure 3-1.

One of the initial effects of endotoxin is activation of Hageman factor (factor XII), which in turn initiates the intrinsic clotting cascade and directly activates factor VII, thereby stimulating the extrinsic coagulation pathway. Activation of the coagulation system results in concurrent activation of the fibrinolytic system. Plasmin then acts on Hageman factor to produce fragments termed prekallikrein activators, which activate prekallikrein to form kallikrein, causing conversion of plasminogen to plasmin and directly activating Hageman factor. These reactions in turn result in the development of disseminated intravascular coagulation. Kallikrien activation also leads to the production of inflammatory mediators such as hydrogen peroxide, free radicals, and bradykinin. These substances cause intense inflammatory injury to vital organs.[44]

Inflammatory injury also occurs as a result of endotoxin-induced activation of the complement cascade. This sequence may be activated through the classic pathway, the alternate (properdin) pathway, and direct activation of the first component of complement.[45,46] Complement activation produces three major effects. First, a chemotactic effect results in migration of leukocytes into

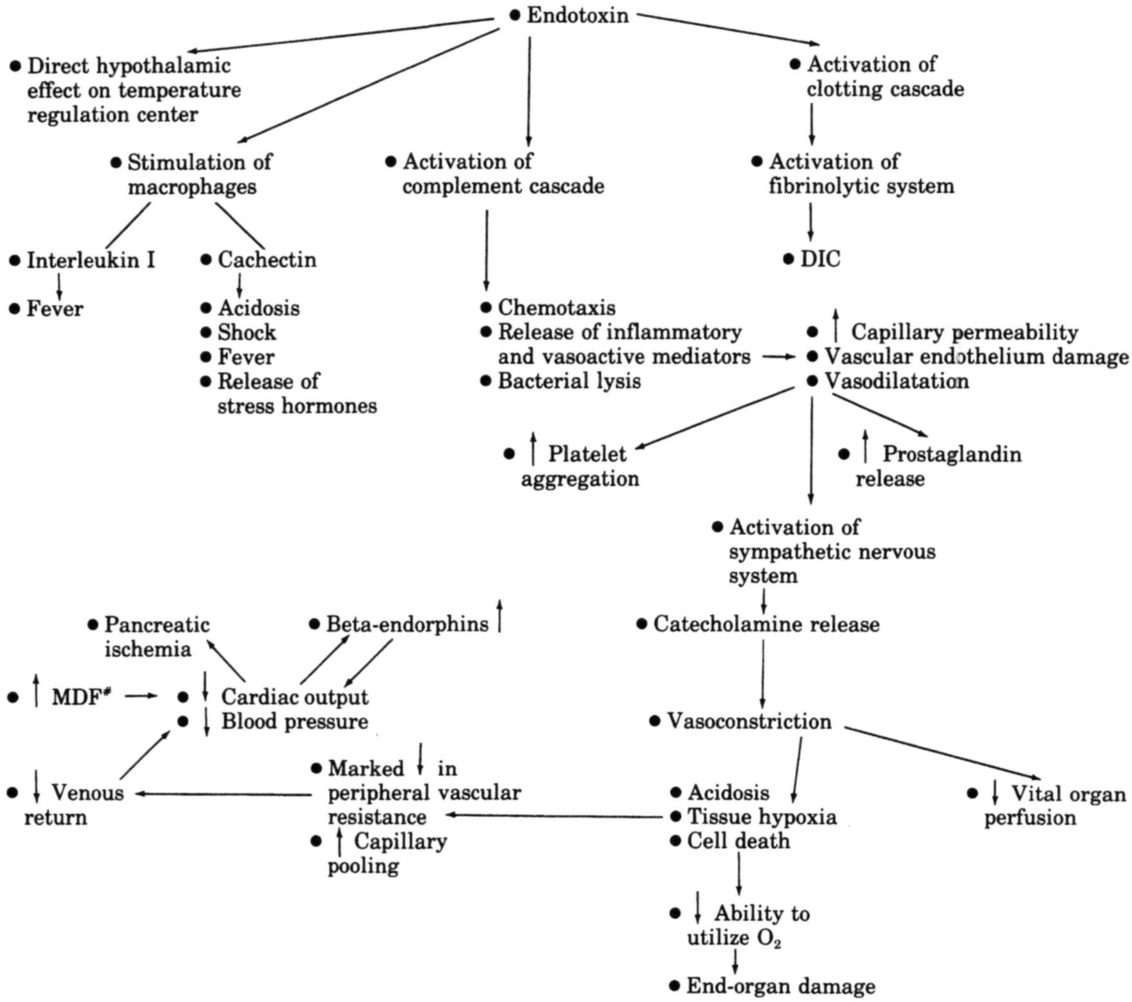

FIGURE 3-1. Pathophysiology of septic shock. #MDF = myocardial depressant factor. Adapted from Gonik B. Septic shock in obstetrics. Clinics in Perinatology 1986;13:751. With permission W. B. Saunders Company.

injured tissue with accelerated inflammation, bacterial lysis, and enhanced release of endotoxin. Second, release of inflammatory mediators by leukocytes causes injury to vascular endothelium. Third, some complement components (C_{3a} and C_{5a}) interact to cause degranulation of mast cells and release of histamine.

Histamine release has several pathophysiologic effects, including disruption of endothelial integrity, increased capillary permeability, decreased plasma volume, vasodilation, and hypotension. In response to this initial state of hypotension, there is an increase in heart rate and cardiac output. In this early stage of sepsis, the patient appears to be warm, flushed, and maximally vasodilated; this stage of septic shock is referred to as "warm shock."[45,46]

The early stage of vasodilation, decreased peripheral resistance, and increased cardiac output is usually short-lived. As sepsis progresses, the sympathetic nervous system responds by increased catecholamine release. This sympathetic discharge leads to generalized vasoconstriction. The release of prostaglandins from injured endothelial tissue contributes to the process of vasoconstriction.[47,48] This phase of vasoconstriction and tissue hypoperfusion is often referred to as "cold shock."

Generalized vasoconstriction produces a profound decrease in perfusion of vital organs, localized tissue hypoxia, and lactic acidosis. This metabolic acidosis results in relaxation of smooth muscle in arterioles and constriction of smooth muscle in venules. As a result, extensive pooling, increased hydrostatic pressure, and transudation of intravascular fluid into the extravascular space occur. Because of generalized capillary pooling, circulating blood volume decreases, with a resultant decrease in venous return to the heart, a progressive decrease in cardiac output, and falling systemic blood pressure.[47]

Two additional factors play a role in producing a decrease in cardiac output. The levels of endogenous β-endorphins are increased significantly. These peptides are released from the pituitary gland along with adrenocorticotropin hormone (ACTH) in response to stress; their principal hemodynamic effect is to lower blood pressure.[49,50] In addition, studies have reported the existence of a myocardial depressant factor, which arises from the ischemic pancreas during evolution of the shock state.[53,54]

When macrophages are stimulated by microorganisms or their toxins, they produce biologically active mediators known as cytokines.[55,56] These cytokines can produce many of the responses characteristic of sepsis and septic shock.[56] Beutler and Cerami first suggested that cachectin (formerly called tumor necrosis factor) is the primary cytokine mediating the responses associated with sepsis and septic shock.[51] Endotoxin is the most potent stimulus for the production of cachectin.[57] Infusion of cachectin into animals produced physiologic changes similar to those noted in animals with gram-negative septicemia.[58,59] Michie et al. reported the detection of circulating cachectin after endotoxin administration in human volunteers.[52] These authors noted that an elevated cachectin concentration was associated with the onset of chills, headache, myalgias, nausea, fever, tachycardia, and an increase in the stress hormones ACTH and epinephrine.

At the endpoint in evolution of the shock state, irreversible circulatory failure develops characterized by intractable hypotension, severe metabolic acidosis, and extensive cellular autolysis. Clinically, this end state is manifested by complete vasomotor collapse, coma, renal failure, and cardiopulmonary failure.

Although the major effects of endotoxin are on the cardiovascular system, this substance also has other pathophysiologic effects. Endotoxin exerts a prominent effect on the temperature-regulating center of the hypothalamus. Because the hypothalamic center may be depressed during the early phase of sepsis, hypothermia may be present. Later endotoxin activates the hypothalamic center and causes elevation of body temperature, an effect mediated by endogenous pyrogen released from host leukocytes.[60]

Endotoxin also frequently causes severe damage to the pulmonary system, directly damaging the endothelium of the pulmonary vasculature. As a result, fragmented collagen is exposed and platelets adhere to the area of injury. Release of platelet factor 3 activates the intrinsic coagulation cascade, resulting in microembolization and stasis of blood flow in the pulmonary microcirculation.[61] Release of prostaglandins from platelets and injured endothelium causes pulmonary vasoconstriction, with a subsequent increase in pulmonary vascular resistance. The net effect is impairment of pulmonary perfusion, ischemia, and progressive injury to the vascular endothelium. Endotoxin-mediated endothelial injury also leads to activation of the complement cascade, which causes further damage to the lung by enhancing leukocyte aggregation into the pulmonary bed. This situation leads to release of additional inflammatory mediators and intrapulmonary leukostasis. Continued inflammation leads to increased capillary permeability, transudation of fluid into the interstitium and alveolus, and destruction of surfactant.[71,72]

The initial or warm-hypotensive phase of septic shock is one of vasodilatation. The patient is hypotensive, alert, and apprehensive. Her face is flushed and her skin warm. Moderate tachycardia is present. Urinary output is adequate. With the liberation of endotoxin, there is catecholamine release that results in constriction of postcapillary sphincters. There is a resultant back-pressure that causes the opening of many shunt channels, the overall effect of which is to maintain tissue perfusion and skin warmth. Endotoxin also causes capillary endothelial damage, which results in extravasation of fluid into the interstitial space. This situation produces increased interstitial pressure and decreased blood flow through tissue. There is a fall in venous return to the heart, which leads to hypotension.

At a later stage in septic shock, the predominant picture is one of extreme vasoconstriction with an increase in systemic vascular resistance, a decrease in cardiac output, and a decrease in stroke volume. This stage is characterized by hypotension, tachycardia, hypothermia, and oliguria. The patient also becomes less alert. The clinical findings in this stage are due to the following: (1) sympathetic activity to compensate for the fall in venous return and hypotension; (2) release of vasoactive amines; and (3) obstruction of microcirculation by diffuse intravascular coagulation (DIC). All these effects are moderated by endotoxin and result in extreme vasoconstriction and reduced tissue perfusion. In turn, this situation leads to further diminished tissue flow, decreased venous return, decreased cardiac output, and lowered blood pressure.

If this cycle is not checked, prolonged cellular hypoxia and anaerobic metabolism are manifested by metabolic acidosis and excessive serum lactate levels. The clinical picture is one of coma, anuria, cardiac distress, and respiratory distress. Unless prompt corrective action is instituted, irreversible shock may occur.

Prior to the legalization of abortion a special concern to obstetrician-gynecologists was the overwhelming sepsis caused by *Clostridium perfringens*. This organism has been recovered from 8 to 27% of patients with postabortion sepsis. The full-blown syndrome of clostridial septicemia has been reported to be fatal in 50 to 85% of cases.[62–64]

Once the uterus has been invaded, the clostridial organisms produce several grades of infection. The mildest and by far the most common, is a low-grade endometritis with or without gas formation and with or without bacteremia. These patients present with vaginal discharge, uterine tenderness, and fever. There are no signs of systemic toxemia. The diagnosis is suggested by the identification of large gram-positive rods and confirmed by culture.

In 0.5 to 1.0% of patients with clostridial infection, severe endometritis and myometritis with necrosis, perforation, peritonitis, and septicemia develop. Most of these severe infections result from criminally induced abortion. In such instances, the organisms rapidly multiply in the necrotic tissue and elaborate potent exotoxins. At 1 to 3 days after abortion, there is fever, uterine and abdominal pain, and foul, often bubbling discharge from the uterus. There is tachycardia out of proportion to the fever, and leukocytosis is marked. A Gram stain of the discharge reveals the characteristic large gram-positive rods of clostridia. This early stage may be followed by striking intravascular hemolysis, jaundice, hemoglobinemia, hemoglobinuria, hypotension, and renal failure. As the liver, kidneys, and blood vessels are damaged by exotoxins, hepatorenal syndrome and DIC develop.

On the other hand, the initial clinical manifestations of septic abortion in patients with an IUD in place are relatively innocuous, i.e., fever and chills and flu-like illness. Following the vague early symptoms, there is rapid progression to sepsis and death. The serious infections associated with IUDs in pregnancy tend to occur during the second trimester.

Diagnosis

A diagnosis of infected abortion is entertained in every febrile pregnant woman who is bleeding during the first or early second trimester and in any woman who recently underwent legal termination of pregnancy and who presents with fever, pain, or bleeding. Additionally, chorioamnionitis and sepsis are suspected in a pregnant woman who has an IUD in place and presents with fever or vague abdominal pain.

The diagnosis of septic abortion may be confirmed by the presence of (1) a tender, soft, enlarged uterus; (2) foul-smelling exudate from the cervix; and (3) an open cervical os with evidence of passage of the products of conception. There may be evidence of attempted instrumentation, such as tenaculum marks on the cervix or the presence of a foreign body. The patient is usually febrile but, with severe infection and septic shock, may be hypothermic. Leukocytosis with a shift to the left is usually present, but leukopenia may occur if there is destruction of leukocytes by endotoxin or aggregation of leukocytes in association with DIC.

After the diagnosis of septic abortion is made, the patient must be critically evaluated for factors that place her at risk for serious infection. At the time of pelvic examination, the extent of infection is assessed; infection extending beyond the uterus has a less favorable prognosis. Gram stains of cervical and endometrial smears are prepared to rule out the presence of clostridial infection and to provide a clue to the predominant pathogen. Radiographs of the abdomen are obtained to determine the presence of intraperitoneal air if a perforation is suspected. With severe clostridial myometritis, gas may be seen in the myometrium.

Clinicians must be alert to the increased risk for the development of septic abortion, chorioamnionitis, and sepsis in pregnant women with an IUD in place. The diagnosis of amnionitis or sepsis in these patients is difficult because of the vagueness of their early symptoms. They present with fever and a flu-like syndrome, but the infection progresses rapidly to sepsis and death. The diagnosis may be confirmed by the presence of (1) fever, (2) leukocytosis with a shift to the left, (3) a tender uterus, and (4) an IUD in utero. In the absence of any specific disorder (e.g., pyelonephritis, pneumonia, urinary tract infection, or viral illness), a pregnant woman with an IUD in place who presents with fever must be assumed to have chorioamnionitis. If the string is not visible, ultrasonography is employed to identify the in utero presence of the IUD.

In cases of septic abortion the diagnosis of clostridial infection is suggested by the identification of large gram-positive rods on Gram stain and is confirmed by the growth of clostridia in culture specimens. Patients with severe endomyometritis due to clostridial infection present with marked abdominal and uterine pain, tachycardia out of proportion to the fever, and marked leukocytosis.

The full-blown syndrome of clostridial septicotoxemia manifests as hemolysis, jaundice, hemoglobinemia, hemoglobinuria, hypotension, renal failure, and DIC. A radiograph of the abdomen that reveals myometrial gas, particularly when layered, is synonymous with extensive necrosis of uterine tissue and infection with *Clostridium perfringens*.

A major concern in patients with septic abortion is the development of septic shock. Although full-blown endotoxic shock is readily detectable, the initial manifestations are subtle, and early diagnosis depends on maintaining a high index of suspicion. A fall in blood pressure, decreased urine output, an inappropriately low white blood cell count, and low or normal temperature in association with obvious infection are signs that appear late in the progression of septic shock.

Initially, the picture in septic shock is one of vasodilatation. The patient is hypotensive, alert, and apprehensive. The face is flushed and the skin warm. Temperature is usually higher than 102°F (39°C), and chills are present. The urine output is adequate. Respiratory alkalosis is present and represents an attempt to expire carbon dioxide to compensate for the tendency to develop metabolic acidosis. Tachypnea is therefore noted.

At a later stage the clinical picture is one of extreme vasoconstriction and the characteristic tetrad of hypotension, tachycardia, hypothermia, and oliguria. The patient becomes less alert. Metabolic acidosis is present, and blood lactate levels are elevated. If this stage is prolonged, coma, anuria, and cardiac and respiratory distress occur.

In many patients a deficiency of several clotting factors occurs secondary to the development of DIC. It is usually manifested by thrombocytopenia, decreased levels of fibrinogen, decreased levels of factors V and VIII, and evidence of fibrinolysis with elevated levels of fibrin degradation products.

Treatment

The goal of treatment of patients with infected abortion is to prevent sepsis, septic shock, and pelvic abscess formation. Because the mortality rate associated with septic shock is high, treatment must be vigorous and prompt.

Use of antibiotics is indicated in every patient with septic abortion; complications such as septic pelvic thrombophlebitis, septic pulmonary emboli, bacteremia, and pelvic abscess are more frequent when antibiotics are delayed. Broad-spectrum systemic antibiotics are used to provide early coverage for the patient at risk for septic shock.

Generally, septic abortions are mixed aerobic-anaerobic infections and necessitate antibiotic coverage against both gram-negative facultative bacteria (especially *E. coli*), facultative streptococci such as group B streptococci, and anaerobic bacteria, predominantly *Peptostreptococcus* and *Bacteroides*. Although many of these infections are mixed, incomplete antibiotic coverage has yielded good results. Equivalent therapeutic results have been obtained with various regimens using either a single broad-spectrum antibiotic or multiple antibiotics.[39,65,66]

Our initial antibiotic regimen is determined by the clinical presentation. With mild to moderately severe infection either a single agent or a combination of agents that cover the major aerobic and anaerobic bacteria associated with pelvic infections can be employed. Table 3-2 lists suggested regimens and dosage schedules. In patients with suspected sepsis or septic shock the antibiotic regimen employed must be a combination that covers all the potential pathogens. Most commonly we recommend a combination of clindamycin 900 mg IV every 8 hours, gentamicin 1.5 mg/kg IV every 8 hours, and vancomycin 0.5 g IV every 12 hours. Such a combination covers not only the anaerobes, gram-negative facultatives such as *E. coli, Klebsiella* sp. and *Proteus* sp., but also enterococci, group B streptococcus, and *Staph. aureus,* including methicillin-resistant strains.

With septic abortions due to retained necrotic tissue, good clinical results have been obtained with the administration of antibiotics ineffective against *Bacteroides fragilis,* despite the frequent recovery of this organism from the site of infection. This fact suggests that in most incomplete infected abortions it is not the choice of antibiotic(s) but the operative removal of necrotic tissue that is the crucial factor of recovery. With early antibiotics and dilatation and curettage (D&C), recovery is rapid in most cases.

In patients in whom the onset of infection is within 24 hours after abortion and in whom infection is due to introduction of a virulent pathogen into the operative site, treatment with antibiotics alone is usually sufficient. If a secondary temperature elevation occurs, a D&C is performed to rule out the presence of retained products of conception.

With endometritis following elective pregnancy termination, antibiotic therapy alone

TABLE 3-2. Antibiotic regimens for the treatment of mild to moderately severe postabortion infection.

Single agent
Cefoxitin 2g IV every 6 hours
Cefotetan 2 g IV every 12 hours
Cefotaxime 2 g IV every 8 hours
Ceftizoxime 2 g IV every 8 hours
Imipenem 0.5 g IV every 6–8 hours
Piperacillin 4.5 g IV every 6 hours
Mezlocillin 4.0 g IV every 6 hours
Sulbactam/ampicillin 3 g IV every 6 hours
Combination agents
Clindamycin 900 mg IV every 8 hours + gentamicin 1.5 mg/kg IV every 8 hours
Metronidazole 500 mg IV every 6 hours + gentamicin 1.5 mg/kg IV every 8 hours

is sufficient where there is no suggestion of retained products of conception. However, if the cervical os is open, if bleeding occurs per cervical os, or if tissue if passed, D&C is indicated in addition to antibiotics. If after 24 to 48 hours of antibiotic therapy no response has occurred, a D&C is performed to exclude retained products.

During the preantibiotic era, a conservative approach to operative intervention was the rule. Host defense mechanisms were relied on to wall off the infection, and curettage was not employed until the patient was afebrile for 24 hours. Recognition of (1) endotoxic shock secondary to gram-negative sepsis and (2) fatal septicotoxemia due to clostridial exotoxins led to a change in this strategy and a more rapid intervention to remove the nidus of infection.

Most often, septic abortion is secondary to retained products of conception, and removal of infected necrotic tissue is the cornerstone of treatment. Therapy with antibiotics is of no help if this tissue is not removed. Antibiotics are instituted immediately to protect patients from curettage-associated septicemia, and the uterus is evacuated once adequate serum levels are obtained. We prefer suction curettage because it is less vigorous than sharp curettage.

A major problem in the surgical management of septic abortion patients is early determination of the extent of necrotic tissue in the pelvis so that appropriate operative intervention can be utilized. In most cases infection is limited to the endometrial cavity, and curettage suffices to remove the nidus of infection. Patients not responding to appropriate antibiotic therapy and D&C are evaluated by pelvic examination plus ultrasonography or computed tomography (CT) scans (or both) to rule out the presence of an abscess that requires surgical drainage. When no suggestion of an abscess is present, the diagnosis of septic pelvic thrombophlebitis is entertained and a trial of heparin therapy instituted (see Septic Pelvic Thrombophlebitis, below). However, if the infectious process has progressed to a stage of microabscess formation in the myometrium or of extrauterine infection, more extensive surgical intervention may be required. If radiographic studies of the pelvis and abdomen reveal intrauterine gas, an intraperitoneal foreign body, or free air in the peritoneal cavity suggestive of uterine perforation, immediate exploratory surgery is indicated to determine the extent of damage. Laparotomy is indicated if there is (1) a history suggestive of widespread pelvic infection (i.e., soap douche), (2) laboratory evidence of massive infection (i.e., intravascular hemolysis), and (3) clinical evidence of sepsis and deterioration with fever, hypotension, or renal failure. When infection extends beyond the uterus, hysterectomy and salpingo-oophorectomy are often necessary.

Although septic abortions have decreased in number, patients with septic shock secondary to septic abortion still present a serious therapeutic problem to the physician. Most patients with septic abortion respond rapidly to aggressive, early antibiotic treatment and curettage. However, in a small percentage, septic shock complicates the situation.

Management of Septic Shock

Early, aggressive management of infected patients with suspected septic shock is crucial to achieving a successful result. The initial management of patients with septic shock is based on five general goals: (1) correction of hemodynamic abnormalities and improvement in circulating intravascular volume; (2) accurate monitoring of patient status; (3) ensurance of adequate oxygenation; (4) identification of the septic focus; and (5) institution of empiric broad-spectrum antibiotic therapy against the most likely pathogens.

Correction of Hemodynamic Abnormalities and Restoration of Vascular Volume

The first priority and mainstay of treatment for suspected septic shock is correction of the hemodynamic dysfunction. It requires that the patient's effective circulating blood volume be restored. Adequate and aggressive volume expansion to correct hypovolemia is a critical determinant for the prognosis of patients with septic shock.[19,67] A large-bore intravenous catheter is inserted immediately. If acute blood loss coexists with sepsis, the circulating volume can be restored by infusing either packed red blood cells (preferably) or whole blood. Otherwise, the plasma volume is restored by infusion of either isotonic crystalloid solution (normal saline or lactated Ringer's) or colloid solution (plasmanate). As a result of pathophysiologic alterations associated with septic shock, e.g., profound vasodilatation, increased capillary permeability, and extravasation of fluid extravascularly ("third spacing"), it is not unusual for a large volume of fluid to be required.

Fluid resuscitation in the septic patient requires close monitoring. The ideal means to evaluate cardiac function in the septic patient is with the use of a flow-directed pulmonary artery catheter (i.e., Swan-Ganz catheter). When this method is not possible, a central venous pressure catheter is used. In addition, and arterial line is inserted to assess blood pressure and arterial oxygenation on a continuous basis. A urethral catheter is inserted to monitor urine output.

Correction of intravascular volume and hemodynamic function is guided by measurements of pulmonary artery wedge pressure (PAWP). A level in the range of 10 to 12 mm Hg must be maintained.[61] Shubin and associates suggested a method for monitoring the fluid replacement in the septic patient.[47] They recommended the use of "fluid challenges" by administering a bolus of 50 to 200 ml in 10 minutes with continuous monitoring of PAWP. If PAWP increases more than 7 mm Hg above the baseline level, the infusion is temporarily halted. On the other hand, if after the "fluid challenge" the PAWP does not rise more than 3 mm Hg above baseline a second fluid challenge of 50 to 200 ml in 10 minutes is administered. This method is known as the "7–3" rule for fluid replacement in patients with septic shock. The baseline PAWP determines the volume to be used for the "fluid challenge." With PAWP readings of less than 12 mm Hg, 12 to 20 mm Hg, or more than 20 mm Hg, the volume of the bolus is 200, 100, or 50 ml, respectively. If central venous pressure monitoring is used, the decision levels are 5 cm H_2O and 2 cm H_2O.

Several authors have suggested that while fluid replacement is being initiated the use of shock trousers (MAST suit) improves cardiac function.[68,69] The pneumatic trousers mobilize blood that has pooled in the lower extremities and the lower abdomen and pelvis. This situation can decrease the amount of fluid required and reduce the risk of iatrogenic pulmonary edema. However, use of pneumatic trousers is a temporary measure, and they must be deflated slowly as volume replacement reestablishes normal hemodynamic function.

When fluid resuscitation fails to restore normal cardiovascular function and adequate tissue perfusion, administration of vasoactive agents is necessary. Dopamine, a drug with dose-dependent α-adrenergic and β-adrenergic effects is the agent most commonly used and has proved to be a valuable adjunct for correcting the hemodynamic abnormalities associated with septic shock.[70] In low doses (< 15/μg/kg/min) dopamine exerts a weak β-mimetic effect that results in increased myocardial contractility, heart rate, and cardiac output without increasing myocardial oxygen consumption. In addition, dopamine (at low doses) produces selective vasodilatation and thus increases blood flow in renal, mesenteric, coronary, and cerebral vessels. Simultaneously, low dose dopamine causes vasoconstriction in skeletal muscle. Thus the net effect of low dose dopamine is increased blood flow to the kidneys, myocardium, brain, and bowel. However, with doses above 15 μg/kg/min, the principal action of dopamine is α-adrenergic and results in marked vasoconstriction and even further reduction in tissue perfusion. This effect is similar to that associated with the use of older vasopressor agents such as ephedrine, metaraminol, and levophed and is undesirable in patients with septic shock.

Dopamine is administered by continuous

intravenous infusion. The recommended starting dose is 2 to 5 μg/kg/min. The solution for infusion is prepared by mixing 800 mg of dopamine (200 mg per 5-ml ampule × 4) in 500 ml of fluid, which results in a dopamine concentration of 1600 μg/ml. The dopamine infusion is titrated to obtain adequate blood pressure, cardiac output, and tissue perfusion.

In patients with congestive heart failure complicating septic shock, digitalization is indicated. Generally, a loading dose of digoxin 0.75 mg in three divided doses is given 4 to 6 hours apart, followed by a daily maintenance dose that provides a serum digoxin level in the 0.5 to 2.5 ng/ml range.

Respiratory System Support

Prevention of respiratory failure is a major goal in the management of patients with septic shock.[71,72] Patients with septic shock tend to be hypoxic and acidotic, and so oxygen by nasal cannula or face mask must be promptly instituted. It is crucial that arterial blood gases are frequently monitored to detect early onset of respiratory failure.[71,72] At the earliest suggestion of pulmonary compromise, intubation and mechanical ventilation are rapidly accomplished to prevent irreversible hypoxic damage to the pulmonary vasculature.

The diagnosis of acute respiratory distress syndrome (ARDS) is made on the basis of progressive hypoxemia, a normal PCWP, diffuse infiltrates seen by chest radiography, and decreased pulmonary compliance. Treatment of ARDS requires intubation and ventilatory support to maintain adequate gas exchange. Positive end-expiratory pressure is a useful adjunct to accomplish this effect. Once ARDS develops, serial monitoring of arterial blood gas is required.

Treatment of Infection

Simultaneous with the above discussed efforts to improve cardiovascular function and provide adequate oxygenation, the source of sepsis must be determined and empiric antibiotic therapy commenced rapidly. For patients with septic shock associated with an infected abortion, the source of infection is the uterine cavity; hence appropriate cultures for aerobic and anaerobic bacterial must be obtained from the endometrial cavity and blood. In patients with septic shock secondary to septic abortion, empiric antibiotic therapy includes coverage against a wide variety of aerobic and anaerobic bacteria. The most common organisms in septic abortion are *E. coli,* facultative streptococci such as group B streptococcus, *Peptostreptococcus* sp., and *Bacteroides* sp., especially *B. bivius, B. disiens,* and *B. fragilis.* Less often the enterococcus or *Staph. aureus* is recovered. In patients with severe life-threatening infections such as septic shock, we prefer combination antibiotic therapy with clindamycin (900 mg IV every 8 hours), gentamicin (1.5 mg/kg IV every 8 hours), and ampicillin (2 g IV every 4–6 hours). This combination provides effective coverage against most pelvic pathogens. In communities where methicillin-resistant *Staph. aureus* is a problem, vancomycin (1 g IV every 12 hours) is substituted for ampicillin.

As is true for infected abortions not complicated by septic shock, any retained products of conception must be eliminated. Hence these patients require a D&C once they are stable and antibiotic levels have been determined. If infection has extended beyond the endometrium, more extensive surgery may be necessary if the patient does not respond within a reasonable time (48–72 hours) or begins to deteriorate. The ability to extirpate the source of infection in septic shock is a crucial determinant for a good prognosis. This factor contributes to the excellent prognosis for septic shock in obstetric and gynecologic patients.

General Treatment Measures

In addition to the above discussed principles, a variety of other measures are required for the management of patients with septic shock. They include management of electrolyte imbalances, correction of metabolic acidosis, management of coagulation abnormalities, and monitoring renal function.

Coagulation abnormalities usually are indicative of DIC. These abnormalities must be corrected by administration of cryoprecipitate, fresh frozen plasma, platelets, or fresh whole blood. It is rare that heparin is required.

Controversial Treatment Modalities

Until recently the use of high-dose corticosteroids for the treatment of septic shock was controversial. Theoretically, the use of corticosteroids in the management of septic shock seem appropriate. These agents stabilize cellular membranes, inhibit complement-induced inflammatory changes, improve myocardial performance, dilate peripheral vasculature, promote gluconeogenesis, and inhibit the release of endorphins from the pituitary and cachectin from macrophages.

Despite previous reports,[73–75] particularly that of Shumer,[73] demonstrating the efficacy of glucocorticoids in the management of septic shock, investigations have failed to confirm any such benefit.[76–78] Sprung and coworkers noted that although there was a high incidence of early shock reversal with the use of corticosteroids within 4 hours of the onset of shock, no benefit accrued with their use for overall survival.[76] Most importantly, two large multicenter studies demonstrated that the use of high-dose corticosteroids provided no benefit in the treatment of severe sepsis and septic shock.[77,78] Bone et al. noted no significant differences in the prevention of shock, reversal of shock, or overall mortality.[77] In fact, these investigators found that among the subgroup with elevated serum creatinine levels at enrollment the mortality rate was significantly higher in the patients who received glucocorticoids. Similarly, the Veteran's Administration Systemic Sepsis Study reported that the mortality rate was similar in the placebo (22%) and glucocorticoid (21%) groups.[78] Thus the current consensus is that high dose glucocorticoid therapy provides no benefit for and should not be used in the management of severe sepsis and septic shock.

Additional adjuvant therapeutic modalities have been proposed for use in the management of septic shock. The finding by McCabe and co-workers that shock and mortality were one-third as frequent in patients with high titers of antibody against an antigen in the cell wall of Enterobacteriaceae[79] has led to the suggestion that human antisera be used to provide passive immunization against *E. coli* for the treatment of gram-negative bacteremia and shock.[80] Ziegler and colleagues[80] noted that, in recipients of antiserum, mortality occurred in 23 of 103 (22%) compared to 42 of 109 (39%) in the control group ($p = 0.011$) and concluded that human antiserum to the lipopolysaccharide core can substantially reduce deaths from gram-negative bacteremia.

Several investigators have advocated the use of prostaglandin synthetase inhibitors to block the role of prostaglandins in the pathophysiology of septic shock.[48,81,82] Fletcher and colleagues reported that pretreatment with indomethacin prevented the hemodynamic derangements seen in the dog infused with endotoxin.[48] Cefalo et al., using a sheep model, noted that indomethacin or ibuprofen blocked the increase in prostaglandins E and F_2, which are responsible for vasoconstriction of pulmonary vasculature.[81] Rao and coworkers demonstrated that aspirin pretreatment in a primate animal model reduced platelet aggregation and thus minimized stasis, microembolization, and release of inflammatory mediators such as prostaglandins and complement.[82]

Also intriguing are the studies suggesting that β-endorphin plays a role in the pathophysiology of septic shock and may have an adverse effect on the outcome.[83–85] Moreover, these reports have described the beneficial use of endorphin antagonists (naloxone) in patients with septic shock.[84,85]

These adjunctive therapies are still experimental, however, and additional clinical research is necessary to establish their role in the management of severe sepsis and septic shock.

Management of Clostridial Infections

The treatment of abortions infected with *Clostridium perfringens* is controversial. Some physicians support aggressive operative therapy with removal of the uterus and adnexae when gram-positive, club-shaped rods are identified.[86] This approach is claimed to result in survival rates of 80 to 90%. In contradistinction, other investigations have shown that curettage and antibiotics suffice and effect a cure in most cases.[87,88] Results of the Gram stain applied to material from the site of infection may provide an early warning of the presence of clostridia but alone are not sufficient justification for immediate laparotomy and extirpation of the uterus and ovaries. It is important to recognize that (1) most clostridial infections are mild and respond to antibiotics and curettage, (2) clostridia are part of the normal microflora of the vagina and cervix, and (3) patients with signs of a full-blown syndrome of clostridial septicotoxemia survive if proper treatment is instituted.[64,87] The patient must be observed closely for the following signs of extensive infection: (1) intravascular hemolysis; (2) hemoglobinemia or hemoglobinuria; (3) renal failure; (4) presence of gas in pelvic soft tissues; and (5) deterioration with sepsis. When any of these signs is present, prompt operative exploration is indicated for adequate débridement of necrotic tissue and removal of the nidus of infection, which is the source of exotoxins. In patients with septic abortion, it requires total abdominal hysterectomy and in many cases bilateral salpingo-oophorectomy.

With clostridial infections, the antibiotic of choice is penicillin 30 to 40 million units IV per 24 hours. In patients allergic to penicillin, cephalosporins, clindamycin, or chloramphenicol may be used.

In women with clostridial sepsis and overt hemolysis who do not die of the initial clostridial toxemia, renal failure secondary to acute tubular necrosis invariably follows. Early hemodialysis has been a crucial adjunct to the management of clostridial sepsis, and the proper management of renal failure may be the major factor contributing to the improved survival rates seen in some series. Hemodialysis is an exchange transfusion that removes the clostridial exotoxins and free hemoglobin as well as potassium and metabolic end-products.

Other modes of therapy such as hyperbaric oxygen and polyvalent antiserum have been recommended. None has been shown convincingly to enhance survival in women with clostridial pelvic infections.

Septic Pelvic Thrombophlebitis

Pelvic thrombophlebitis is a disease process that has been described for more than a century. Since Michaeles' description of the association of fever and obscure thrombosis in women who developed pulmonary emboli, pelvic thrombophlebitis in association with pelvic infection has received intermittent attention. However, only since the 1960s has septic pelvic thrombophlebitis been widely recognized as a serious complication of puerperal, postabortal, and postoperative pelvic infections.

Septic pelvic thrombophlebitis (SPT) is an uncommon complication of infection in obstetrics and gynecology. The reported incidence among puerperal women ranges from 1 in 569 to 1 in 2207 deliveries[89–91]; and for second trimester pregnancy terminations, 1 in 1000 were complicated by SPT.[92] Among patients with postpartum endomyometritis, the risk of SPT has been reported to be in the 1 to 2% range,[93,94] with Sweet and Ledger noting that more than 8% of their postpartum endometritis patients were treated with heparin because of a suspected diagnosis of SPT.[95] As noted by Duff and Gibbs,[96] this disease is primarily an obstetric complication due to several physiologic alterations associated with pregnancy, including relative hypercoagulable state, pelvic venous stasis, and estrogen-induced pelvic vein changes predisposing them to injury. As a result of infection in the pelvis in combination with these physiologic alterations, SPT develops in the pelvic veins.

The major pathologic sequelae of SPT are (1) septic pulmonary emboli and (2) metastatic abscess formation. Autopsy studies have shown that pelvic thrombophlebitis was present in 35% of patients who died of puerperal sepsis, and that 50% of deaths attributed to puerperal sepsis were due to pulmonary emboli. Septic pulmonary emboli have been reported in 31 to 39% of patients with SPT.[91,97] These investigations confirmed the seriousness of SPT as a complication of such pelvic infections as puerperal sepsis or septic abortion.

Etiology

It is not surprising that pelvic vein thrombophlebitis occurs during the postpartum period. The disposition to thrombosis in the puerperium is well known. In 1856 Virchow described a triad of factors that predispose to thrombosis: (1) stasis of blood; (2) biochemical changes in blood that lead to hypercoagulability; and (3) physical changes or injury in the wall of vessels. Each of these factors contributes to the pathogenesis of SPT.

Significant alterations occur in the coagulation system during pregnancy and the puerperium. Ygge[98] described the changes during the puerperium as a gradual increase of factors I, V, VII, VIII, and IX and plasminogen. These alterations result in postdelivery hypercoagulability.

The most significant factor in the pathogenesis of SPT appears to be damage to the vessel wall secondary to trauma or infection. Collins et al.[97] described the pathologic sequence: (1) The intima of a vein is damaged as a result of bacterial invasion by direct extension along the intima or via the perivenous lymphatics. (2) After intimal damage, the clotting processes are initiated. (3) Bacterial invasion of the clot produces suppurative thrombophlebitis with liquefication, fragmentation, and septic emboli.

Clinically, the significance of infection and trauma as antecedent factors in the pathogenesis of SPT is apparent. Cesarean section and septic abortion patients have a tenfold higher incidence of SPT than do patients delivered vaginally. The increased trauma to which pelvic tissues are subjected during cesarean section and septic abortion results in a greater risk of thrombosis and thrombophlebitis secondary to vessel wall injury. Cesarean section leads to a significantly increased incidence of puerperal endometritis, which may progress to SPT as a result of lymphatic spread of bacterial pathogens from the uterus to the pelvic veins. With septic abortion, thrombophlebitis also begins as an endometritis, which leads to lymphatic spread and involvement of pelvic veins.

Because intrauterine infection (postpartum and postabortion) is a major antecedent to the occurrence of SPT, the microorganisms that are pathogenic in this disease process are similar to those generally associated with female genital tract infections. Therefore it is not unexpected that the bacterial organisms associated with SPT are also of the mixed aerobic-anaerobic type seen in pelvic infections in general. Extension of an infectious process such as puerperal sepsis or septic abortion from the uterus initiates the involvement of major pelvic veins in an extensive inflammatory process

It has become apparent that, except for *N. gonorrhoeae*, *C. trachomatis*, and group A β-hemolytic streptococcus, it is the normal flora of the vagina and cervix that become pathogenic when appropriate environmental conditions are present.[99] Thus the traumatization and devitalization of tissue that occurs with cesarean section, septic abortion, or traumatic vaginal delivery produces an environment appropriate for the growth of these bacteria; and this tissue provides access to the upper genital tract for the bacterial flora present in the vagina and cervix. Both aerobic and anaerobic bacteria comprise the normal flora of the vagina and cervix. The common microorganisms involved are gram-negative facultative aerobes, especially *E. coli;* (2) gram-positive aerobes such as streptococci and staphylococci; and (3) the anaerobic bacteria *Peptostreptococcus* and *Bacteroides* species.

The presence of SPT is one of the cardinal signs of anaerobic infection.[99] In the past, anaerobic cocci *(Peptostreptococcus)* have been regarded as the principal offenders in puerperal sepsis and SPT; however, more recently attention has focused on *Bacteroides* species, especially *B. fragilis,* as a cause of serious pelvic sepsis and thrombophlebitis. Felner and Dowell[100] reported that thrombophlebitis was present in 20% of patients with *Bacteroides* bacteremia. In addition, *B. fragilis* has been shown to be capable of heparinase production, which may account for its frequent association with SPT.[100]

Aerobic bacteria have also been associated with SPT. Streptococcal puerperal infection extends rapidly through blood vessels and lymphatics, usually retroperitoneal, to produce SPT. During the preantibiotic era, the group A β-hemolytic streptococcus was the major cause of death in patients with puerperal sepsis. Occasionally, the streptococcus still produces puerperal sepsis an septic thrombophlebitis. Other aerobic bacteria such as *Staph. aureus* and gram-negative enterobacteria (e.g., *E. coli, Proteus,* and *Klebsiella)* have been reported in association with SPT. The isolation of bacteria from resected venous specimens confirms the role of bacterial infection in the etiology of SPT.

The important role of the *Bacteroides* group of organisms in the pathogenesis of SPT is further documented by clinical response patterns in the treatment of postcesarean section endomyometritis. Prior to recognition of the critical need to provide early aggressive antibiotic therapy against resistant gram-negative anaerobes of the *Bacteroides* group, the traditional approach to the treatment of such infections was the use of ampicillin alone or a penicillin plus an aminoglycoside.[94] The use of antibiotics such as clindamycin, cefoxitin, or moxalactam, which are effective against *Bacteroides,* significantly reduced the incidence of SPT.[101]

Clinical Manifestations

The usual clinical presentation of SPT was described by Collins and colleagues in their classic work published in 1951.[97] They reported 375 patients who presented with (1) chills and fever, with a picket-fence spiking pattern; (2) a rapid pulse rate, sometimes out of proportion to the temperature; (3) tachypnea; and (4) hemoptysis and chest pain. The pelvic examination was usually normal, except for tenderness. In approximately one-third of cases, worm-like masses of thrombotic veins were palpated in the parametrial areas. This description is still valid today.

More recently, Gibbs divided cases of SPT into two distinct syndromes with different onset, symptoms, and findings on examination.[102] The first is a syndrome of ovarian vein thrombosis, characterized by early onset (usually within 2 to 3 days after delivery), severe pain, a normal to high temperature, and findings consistent with an acute abdomen. In approximately 50% of cases there is a palpable abdominal mass. Flank pain and tenderness of the costovertebral angle may also be present. In most cases the right side is involved. The abdominal findings are usually so impressive that exploratory laparotomy is performed for a preoperative diagnosis of acute appendicitis, twisted ovarian cyst, or pelvic abscess.

The second syndrome is much less dramatic and usually manifests as an "obscure" fever that persists despite treatment with multiple antibiotics. These patients are ambulatory. They do not appear ill, and they have either no pain or minimal poorly localized discomfort, despite a hectic spiking temperature course. Occasionally, cord-like induration is palpable on pelvic examination. It is important to recognize that an unexplained temperature elevation in the puerperal, postabortion, or postoperative patient may be the only manifestation of SPT. Rarely, the initial manifestation is the occurrence of septic pulmonary embolism.

Diagnosis

The diagnosis of SPT may be difficult, and it is often considered a diagnosis of exclusion. The typical patient has failed to respond to appropriate antibiotic treatment of a pelvic infection (e.g., endometritis, parametritis, or pelvic cellulitis) and now presents with hectic, spiking temperatures and chills. Generally, there are no exact physical findings or laboratory test to aid in making a diagnosis, and early diagnosis depends on maintaining a high index of suspicion. With ovarian vein thrombosis, the clinical signs are more dramatic; abdominal pain is localized, rebound tenderness is present, and a mass may be palpable.

Additional clinical findings that suggest a diagnosis of SPT include failure to identify a source of infection, such as an abscess or septic pulmonary emboli. The occurrence of septic embolism in a patient with apparent pelvic infection is thought to be the most reliable evidence short of inspection of the veins at laparotomy.

The use of heparin as a diagnostic test in suspected cases of SPT has been suggested.[89,91,103,104] Prompt clinical improvement following heparinization is viewed as a confirmatory sign (within 24 to 36 hours from the onset of heparin therapy).

Ovarian vein thrombosis may be suspected on the basis of the clinical findings of abdominal pain, rebound tenderness, fever, and a palpable mass in the postpartum patient. However, because this condition is difficult to distinguish from acute appendicitis (which carries a serious prognosis if rupture occurs during the immediate postpartum period), many cases of ovarian vein thrombosis will continue to be confirmed at laparotomy.

Venography and surgical exploration have been suggested as diagnostic methods for SPT or ovarian vein thrombosis. However, these procedures are associated with significant risks and morbidity and have not been commonly used. The use of CT scanning has been shown to be helpful in the diagnosis of SPT and ovarian vein thrombosis.[105–107] Using CT scans, venous occlusions can be identified by visualization of (1) a low-density mass within the vessel lumen, (2) enlargement of the vessel, and (3) enhancement of the vessel wall following intravenous infusion of a contrast medium.

Treatment

Until the 1960s, a purely surgical approach, as advocated by Collins et al.,[97,104] was the usual treatment for SPT. In 1917 Miller reviewed the literature dealing with ligation of involved pelvic veins: There were 182 operations with 95 deaths, a mortality rate of 52%. In 1951 Collins modified the surgical approach to provide complete ligation of the normal venous return from the pelvis. He recommended ligation of the inferior vena cava and both ovarian veins. After the ligation, all antibiotics were halted; no anticoagulants were used. The immediate results of therapy were good, and long-term follow-up indicated a low incidence of serious sequelae.[97] Of 202 patients treated using this surgical approach, 22 deaths occurred within 6 weeks after surgery; the survival rate was 89%.

Since 1962 the purely surgical treatment of SPT has given way to a medical approach based on the use of anticoagulants and antibiotics. The consensus is that early presumptive diagnosis of SPT and early anticoagulation obviate the need for surgical intervention. Since the initial report by Schulman and Zatuchni describing the successful use of heparin in the treatment of suspected SPT, numerous studies have confirmed the efficacy and safety of anticoagulants as alternative management.[91,103,104] Even patients with ovarian vein thrombosis, diagnosed at exploratory laparotomy, have been shown to respond to anticoagulation and antibiotic therapy.[90,91,103]

Heparin is the anticoagulant of choice for the treatment of SPT. The aim is to maintain either the clotting time or the activated partial thromboplastin time (aPTT) at two to three times normal. Intravenous heparin infusion as a continuous drip via an infusion pump is the optimal method of heparinization. Treatment is begun with 750 units of heparin per hour, and additional increments are added as determined by the clotting time or aPTT levels. Generally, the dosage range is between 750 and 2000 units per hour. The duration of therapy with heparin is 7 to 10 days. However, if pulmonary emboli occur, anticoagulation treatment is continued with an oral agent such as warfarin (Coumadin) for 3 to 6 months to prevent recurrent pulmonary emboli.

Heparin is not only a therapeutic modality but also a useful diagnostic tool for confirming a presumptive diagnosis of SPT. The effect of heparin appears to be specific for thrombophlebitis and does not mask the presence of other sources of infection, such as wound infection, pelvic cellulitis, pelvic abscess, or acute appendicitis.[91,103,104] In addition, this effect of heparin is rapidly manifested, and patients with SPT without associated pelvic infection requiring surgical drainage become afebrile in 24 to 48 hours.

Treatment of SPT must also be based on the recognition that the disease is an infectious process in which the usual initiating factor is a pelvic infection in a postpartum, postabortion, or postoperative patient. Thus antibiotic therapy is the second essential component in the medical management of SPT. The proper choice of antimicrobial agents depends on accurate knowledge of the bacterial pathogens associated with this disease. As a general rule, SPT, like most pelvic infections, is associated with a mixed aerobic-anaerobic bacterial type infection. The most common organisms are (1) gram-negative aerobic enteric bacilli such as *E. coli;* (2) aerobic gram-positive bacteria such as streptococci and staphylococci; and (3) the anaerobic bacteria *Peptostreptococcus* and *Bacteroides* species, especially *B. fragilis.* Therefore antibiotic treatment must provide adequate broad-spectrum coverage for a mixed aerobic-anaerobic infection as described previously.

Surgical intervention is reserved for instances in which medical treatment with antibiotics and heparin has failed. If the patient is either not afebrile within 24 to 48 hours or continues to develop septic pulmonary emboli, surgical exploration is indicated for ligation of the inferior vena cava and both ovarian veins or appropriate surgical extirpation of the infection site.

References

1. Blanco JD, Gibbs RS, Castaneda YS. Bacteremia in obstetrics: clinical course. Obstet Gynecol 1981;58:621–625.
2. Ledger WJ, Norman M, Gee C, et al. Bacteremia on an obstetric and gynecologic service. Am J Obstet Gynecol 1975;121:205–212.
3. Monif GRG, Baer H. Polymicrobial bacteremia in obstetric patients. Obstet Gynecol 1976;48:167–169.
4. Bryan CD, Reynolds KL, Moore EE. Bacteremia in obstetrics and gynecology. Obstet Gynecol 1984;64:155–158.
5. Weinstein MP, Reller LB, Murphy JR, Lichtenstein KA. The clinical significance of positive blood cultures: a comprehensive analysis of 500 episodes of bacteremia and fungemia in adults. I. Laboratory and epidemiologic observations. Rev Infect Dis 1983;5:35–59.
6. Weinstein MP, Murphy JR, Reller MP, Lichtenstein KA. The clinical significance of positive blood cultures: a comprehensive analysis of 500 episodes of bacteremia and fungemia in adults. II. Clinical observations, with special reference to factors influencing prognosis. Rev Infect Dis 1983;5:56–70.
7. Kreger BE, Craven DE, Carling PC, McCabe WR. Gram-negative bacteremia. III. Reassessment of etiology, epidemiology and ecology in 612 patients. Am J Med 1980;68:332–344.
8. Kreger BE, Craven DE, McCabe WR. Gram-negative bacteremia. IV. Reevaluation of clinical features and treatment in 612 patients. Am J Med 1980;68:344–356.
9. Chow AW, Guze LB. Bacteroidaceae bacter-

emia: clinical experience with 112 patients. Medicine (Baltimore) 1974;53:93–126.
10. Young LS, Martin WJ, Meyer RD, et al. Gram-negative rod bacteremia: microbiologic, immunologic, and therapeutic considerations. Ann Intern Med 1977;86:456–471.
11. Gonik B. Considerations in managing septic shock in the obstetric patient. Infect Surg 1988;February:119–122.
12. Studdiford WE, Douglas GW. Placental bacteremia: a significant finding in septic abortion accompanied by vascular collapse. Am J Obstet Gynecol 1956;71:842–851.
13. Weil MH, Shubin H. Diagnosis of Shock. Baltimore, Williams & Wilkins, 1967.
14. Santamarina B, Smith SA. Septic abortion and septic shock. Clin Obstet Gynecol 1970; 13:291–304.
15. Schwartz RH: Septic Abortion. Philadelphia, Lippincott, 1968.
16. Centers for Disease Control, Abortion Surveillance. Annual Summary 1978. US Department of Health and Human Services, November 1980.
17. Tietze C, Lewit S. Legal abortions: early medical complications: an intern report of the Joint Program for the Study of Abortion. J Reprod Med 1972;8:193–197.
18. Grimes DA, Cates W, Selik RM. Fatal septic abortion in the United States, 1975–1977. Obstet Gynecol 1981;57:739–744.
19. Sweet RL, Gibbs RS. Postabortal infection and septic shock. In Infectious Diseases of the Female Genital Tract. Baltimore, Williams & Wilkins, 1985.
20. Grimes DA, Schulz KF, Cates W, et al. Mid trimester abortion by dilatation and evacuation: a safe and practical alternative. N Engl J Med 1977;296:1141–1145.
21. Stewart GK, Goldstein PJ. Medical and surgical complications of therapeutic abortions. Obstet Gynecol 1972;40:539–550.
22. Hodgson JE, Portmann KC. Complications of 10,453 consecutive first-trimester abortions: a prospective study. Am J Obstet Gynecol 1974;120:802–807.
23. Matjamspm BN. Ambulatory abortion: experience with 26,000 cases. N Engl J Med 1972;286:403–407.
24. Evans DA, Gudson JP. Medical and surgical complications of therapeutic termination of pregnancy. NC Med J 1974;35:87–91.
25. Burnett LS, Wentz AC, King TM. Techniques of pregnancy termination. Part II. Obstet Gynecol Surv 1974;29:6–42.
26. Burkman RT, Tonascia JA, Atienza MG, King TM. Untreated endocervical gonorrhea and endometritis following elective abortion. Am J Obstet Gynecol 1976;126:648–651.
27. Moller BR, Abrons S, Laurin J, Mardh P-A. Pelvic infection after selective abortion associated with Chlamydia trachomatis. Obstet Gynecol 1982;59:210–213.
28. Barbacci MB, Spence MR, Kappus EW, et al. Postabortal endometritis and isolation of Chlamydia trachomatis. Obstet Gynecol 1986;68:686–690.
29. Grimes DA, Schulz KF, Cates W Jr. Prophylactic antibiotics for curettage abortion. Am J Obstet Gynecol 1984;150:689–694.
30. Hurt WG. Septic pregnancy associated with Dalkon Shield IUD. Obstet Gynecol 1974; 44:491–495.
31. Shine RM, Thompson JG. The in situ IUD and pregnancy outcome. Am J Obstet Gynecol 1974;119:124–130.
32. Viechnicki MB. Septicemia and abortion with the Copper-7. Am J Obstet Gynecol 1975;121:203–212.
33. Eisenger SH. Second-trimester spontaneous abortion, the IUD, and infection. Am J Obstet Gynecol 1976;124:393–397.
34. Christian CD. Maternal deaths associated with an intra-uterine device. Am J Obstet Gynecol 1974;119:441–444.
35. Cates W, Ory HW, Rochat RW, Tyler CW. The intrauterine device and deaths from spontaneous abortion. N Engl J Med 1976;295:1155–1159.
36. Tatum HJ, Schmidt FH, Phillips D, et al. The Dalkon Shield controversy: structural and bacteriological studies of IUD tails. JAMA 1975;231:711–717.
37. Sparks PA, Purrier BG, Watt PJ, Elstein M. Bacteriological colonization of uterine cavity: role of tailed intrauterine contraceptive devices. Br Med J 1981;282:1189–1191.
38. Rotheram EB, Schick SG. Non-clostridial anaerobic bacteria in septic abortion. Am J Med 1969;46:80–89.
39. Chow AW, Marshall JR, Guze LB. A double-blind comparison of clindamycin with penicillin plus chloramphenicol in treatment of septic abortion. J Infect Dis 1977;135 (suppl):35–39.

40. Sweet RL. Anaerobic infections of the female genital tract. Am J Obstet Gynecol 1975; 122:891–901.
41. Landers DV, Sweet RL. Upper genital tract infections. In Galask R (ed): Clinical Perspectives in Obstetrics and Gynecology. New York, Springer-Verlag, 1985.
42. Wager GP. Toxic-shock syndrome: a review. Am J Obstet Gynecol 1983;146:93–102.
43. Reingold AL, Shands KN, Dans BB, Broome CV. Toxic-shock syndrome not associated with menstruation. Lancet 1982;1:1–4.
44. Mason JW, Kleeberg U, Dolan P, Colman RW. Plasma kallikrein and Hageman factor in gram-negative bacteremia due to gram-negative organisms. Ann Intern Med 1970;73:545–551.
45. McCabe WR. Serum complement levels in bacteremia due to gram-negative organisms. N Engl J Med 1973;288:21–23.
46. Fearon DT, Ruddy S, Schur PH, McCabe WR. Activation of the properdin pathway of complement in patients with gram-negative bacteremia. N Engl J Med 1975;292:937–940.
47. Shubin H, Weil MH, Carlson RW. Bacterial shock. Am Heart J 1977;94:112–114.
48. Fletcher JR, Ramwell PW, Herman CM. Prostaglandings and the hemodynamic course of endotoxin shock. J Surg Res 1976;20:589–594.
49. Guillemin R, Vargo T, Rossier J, et al. Beta-endorphin and adrenocorticotropin are secreted concomitantly by the pituitary gland. Science 1977;197:1367–1370.
50. Morberg GP. Site of action of endotoxins on hypothalamic-pituitary-adrenal axis. Am J Physiol 1971;220:397–400.
51. Beutler B, Cerami A. Cachectin: more than a tumor necrosis factor. N Engl J Med 1987;316:379–385.
52. Michie HR, Manogue KR, Spriggs DR, et al. Detection of circulating tumor necrosis factor after endotoxin administration. N Engl J Med 1988;318:1481–1486.
53. Weisel RD, Vito L, Dennic RC, et al. Myocardial depression during sepsis. Am J Surg 1977;133:512–521.
54. Raffa J, Trunkey DD. Myocardial depression in sepsis. J Trauma 1978;18:617–622.
55. Dinarello CA, Mier JW. Lymphokines. N Engl J Med 1987;317:940–945.
56. Michalek SM, Moore RN, McGhee JR, et al. The primary role of lymphoreticular cells in the mediation of host responses to bacterial endotoxin. J Infect Dis 1980;141:55–63.
57. Ziegler EJ. Tumor necrosis factor in humans. N Engl J Med 1988;318:1533–1535.
58. Tracey KJ, Beutler B, Lowry SF, et al. Shock and tissue injury induced by recombinant human cachectin. Science 1986;234:470–474.
59. Tracey KJ, Lowry SF, Fahey TJ III, et al. Cachectin/tumor necrosis factor induces necrosis factor induces lethal shock and stress hormone response in the dog. Surg Gynecol Obstet 1987;164:415–422.
60. Bernheim HA, Block LH, Atkins E. Fever: pathogenesis, pathophysiology, and purpose. Ann Intern Med 1979;91:261–270.
61. Eskridge RA. Septic shock. Crit Care Q 1980;2:55.
62. Smith LP, McLean APH, Maughan GB. Clostridium welchii septicotoxemia: a review and report of 3 cases. Am J Obstet Gynecol 1971;110:135–149.
63. Ledger WJ, Hakett KA. Significance of clostridia in femal reproductive trace. Obstet Gynecol 1973;41:525–530.
64. Kadner JL, Anderson GV. Septic abortion with hemoglobinuria and renal insufficiency with special reference to Clostridium welchii infection. Obstet Gynecol 1963;21:86–95.
65. Ostergard DR. Comparison of two antibiotic regiments in the treatment of septic abortion. Obstet Gynecol 1970;36:473–474.
66. Josey WE, Farrar WE Jr. Cephaloridine in septic abortion. Am J Obstet Gynecol 1970;106:237–242.
67. Gonik B. Septic shock in obstetrics. Clin Perinatol 1986;13:741–754.
68. Paris PM. Antishock garments. N Engl J Med 1981;305:960.
69. McSwain NE. Pneumatic tousers and the management of shock. J Trauma 1977; 17:719–724.
70. Goldberg LI. Dopamine—clinical uses of an endogenous catecholamine. N Engl J Med 1974;291:707–710.
71. Kaplan RL, Sahn SA, Petty TL. Incidence and outcome of the respiratory distress syndrome in gram negative sepsis. Arch Intern Med 1979;139:867–869.
72. Clowes GHA, Hirsch E, Williams L, et al. Septic shock and shock lung in man. Ann Surg 1975;181:681–692.

73. Shumer W. Steroids in the treatment of clinical shock. Ann Surg 1976;184:333–341.
74. Sladen A. Methylprednisolone—pharmacologic doses in shock lung syndrome. J Thorac Cardiovasc Surg 1976;71:800–806.
75. Kusajima K, Wax S. Webb WR. Effects of methylprednisone on pulmonary microcirculation. Surg Gynecol Obstet 1974;139:1–5.
76. Sprung CL, Caralis PV, Marcial EH, et al. The effects of high-dose corticosteroids in patients with septic shock. N Engl J Med 1984;311:1137–1143.
77. Bone RC, Fisher CJ, Clemmer TP, et al. A controlled clinical trial of high-dose methylprednisolone in the treatment of severe sepsis and septic shock. N Engl J Med 1987;317:653–658.
78. Hinshaw L, Peduzzi P, Young E, et al. Effect of high-dose glucocorticoid therapy on mortality in patients with clinical signs of systemic sepsis. N Engl J Med 1987;317:659–665.
79. McCabe WR, Kreger BE, Johns M. Type-specific and cross-reactive antibodies in gram-negative bacteremia. N Engl J Med 1972;238:261–267.
80. Ziegler EJ, McCutchan JA, Fierer J, et al. Treatment of gram-negative bacteremia and shock with human antiserum to a mutant Escherichia coli. N Engl J Med 1982; 307:1225–1230.
81. Cefalo RC, Lewis PE, O'Brien WF, et al. The role of prostaglandins in endotoxemia. Am J Obstet Gynecol 1980;137:53–57.
82. Rao PS, Cavanagh D, Lamont WG. Endotoxic shock in the primate: effects of aspirin and dipyridamole administration. Am J Obstet Gynecol 1981;140:914–921.
83. Holaday JW, Faden AI. Naloxone reversal of endotoxin hypotension suggests role of endorphins in shock. Nature 1978;275:450–451.
84. Tiengo M. Naloxone in irreversible shock. Lancet 1980;2:690.
85. Peters WP, Friedman PA, Johnson MW, Mitch WE. Pressor effect of naloxone in septic shock. Lancet 1981;1:529–532.
86. Decker WH, Hall W. Treatment of abortions infected with Clostridium welchii. Am J Obstet Gynecol 1966;95:394–399.
87. Pritchard JA, Whalley PJ. Abortion complicated by Clostridium perfringens infection. Am J Obstet Gynecol 1971;111:484–492.
88. O'Neill RT, Schwarz RG. Clostridial organisms in septic abortions. Obstet Gynecol 1970;35:458–461.
89. Schulman H, Zatuchni G. Pelvic thrombophlebitis in the puerperal and postoperative gynecologic patient. Am J Obstet Gynecol 1964;90:1293–1296.
90. Brown TK, Munsick RA. Puerperal ovarian vein thrombophlebitis: a study of 46 cases. Am J Obstet Gynecol 1971;109:263–273.
91. Josey WE, Staggers SR. Heparin therapy in septic pelvic thrombophlebitis: a study of 46 cases. Am J Obstet Gynecol 1974;118:210.
92. Tietze C, Lewit S. Early medical complications of abortion by saline: joint program for the study of abortion. Stud Fam Plann 1973;4:133–137.
93. Gibbs RS, Jones PM, Wilder CJ. Antibiotic therapy of endometritis following cesarean section. Obstet Gynecol 1978;52:31–37.
94. DiZerega G, Yonekura L, Roy S, et al. A comparison of clindamycin-gentamicin and penicillin-gentamicin in the treatment of post cesarean section endomyometritis. Am J Obstet Gynecol 1979;134:238–242.
95. Sweet RL, Ledger WJ. Puerperal infectious morbidity. Am J Obstet Gynecol 1973; 117:1093–1100.
96. Duff P, Gibbs RS. Pelvic vein thrombophlebitis: diagnostic dilemma and therapeutic challenge. Obstet Gynecol Surv 1983;38: 365–373.
97. Collins CG, MacCallum EA, Nelson EW, et al. Suppurative pelvic thrombophlebitis: a study of 70 patients treated by ligation of the inferior vena cava and ovarian vessels. Surgery 1951;30:298.
98. Ygge J. Changes in blood coagulation and fibrinolysis during the puerperium.Am J Obstet Gynecol 1969;104:2–12.
99. Sweet RL. Anaerobic infections of the female genital tract. Am J Obstet Gynecol 1975;122:891–901.
100. Felner JM, Dowell VR Jr. "Bacteroides" bacteremia. Am J Med 1971;50:787–796.
101. Sweet RL. Treatment of mixed aerobic-anaerobic infections of the female genital tract. J Antimicrob Chemother 1981;8:105–114.
102. Gibbs R, Weinstein AJ. Puerperal infection in the antibiotic era. Am J Obstet Gynecol 1976;124:769–787.
103. Dunn LJ, Van Voorhis LW. Enigmatic fever

and pelvic thrombophlebitis: response to anticoagulants. N Engl J Med 1976;276:265–268.

104. Collins CG. Suppurative pelvic thrombophlebitis: a study of 202 cases in which the disease was treated by ligation of the vena cava and ovarian vein. Am J Obstet Gynecol 1970;108:681–687.

105. Angel J, Knuppel R. Computed tomography in diagnosis of puerperal ovarian vein thrombosis. Obstet Gynecol 1984;63:61–64.

106. Cohen M, Pernoll M, Gevirtz C, et al. Septic pelvic thrombophlebitis: an update. Obstet Gynecol 1983;62:83–89.

107. Isada NB, Landy HJ, Larsen JW Jr. Postabortal septic pelvic thrombophlebitis diagnosed with computed tomography: a case report. J Reprod Dis 1987;32:866–868.

4
Ectopic Pregnancy

LUIS A. CIBILS, MAHMOUD A. ISMAIL, and EMILY BAKER

Ectopic gestation is the implantation of a fertilized ovum outside the endometrial cavity. It encompasses implantation in the tubes, ovaries, anywhere in the abdomen, the cervix, and the interstitial-cornual area of the tubouterine anatomy.

There is no question that from the standpoint of prevalence ectopic pregnancy is one of the most important surgical complications of pregnancy in the United States as well as in other countries of the world about which there is information available.[1,2] It occurs at an alarmingly increasing rate that has more than trebled from 1970 to 1983: from 4.5 to 14 per 1000 of reported pregnancies. This trend has continued during the subsequent years according to partial reporting, including the experience summarized here. Moreover, Ectopic pregnancy is now one of the leading causes of maternal mortality in the United States[3] despite the fact that the risk of death from it has been falling steadily through the past several years.

The material of the Chicago Lying-in Hospital accumulated from 1967 to 1987 (21 years) have been reviewed. There were 932 cases confirmed by pathology.

Epidemiology

The number of ectopic gestations has been increasing annually in the United States by about 10%, far outpacing the growth of the female population of reproductive age. Calculated for the total population of reproductive age the rate has increased from 4.2 per 10,000 women in 1970 to 12.6 per 10,000 in 1983 (Table 4-1). However, the actual risk for a pregnant patient is better expressed when calculated for reported pregnancies; and for that population it rose from 4.5 per 1000 in 1970 to 14 per 1000 in 1983. This same trend has been reported for other countries,[4–6] suggesting that it may be a worldwide problem.

Several tentative explanations have been offered to explain this impressive change. Better diagnostic capabilities due to more precise and sensitive beta human chorionic gonadotropin (βhCG) determinations and more refined and accurate ultrasonography techniques and apparatus are frequently mentioned. This line of thought implies that a large number of ectopic gestations were in the past spontaneously arrested and reabsorbed, and now earlier diagnosis finds them before the spontaneous "cure." There is no convincing epidemiologic information to support this point of view.

Other potential factors, such as a chronologically simultaneous increase in the incidence of pelvic infections,[7] may have contributed in some way, but clearly these factors do not represent the sole cause. The widespread use of intrauterine contraceptive devices (IUDs) has also been chronologically coincidental with the increased numbers of ectopic gestations, but there is no solid epidemiologic evidence to substantiate

TABLE 4-1. Numbers and rates of ectopic pregnancies in the United States: 1970–1983.

		Rates		
Year	No. (rounded)	Women (per 10,000) ages 15–44	Live births (‰)	Reported pregnancies (‰)
1970	17,800	4.2	4.8	4.5
1971	19,300	4.4	5.4	4.8
1972	24,500	5.5	7.5	6.3
1973	25,600	5.6	8.2	6.8
1974	26,400	5.7	8.4	6.7
1975	30,500	6.5	9.8	7.6
1976	34,600	7.2	11.0	8.3
1977	40,700	8.3	12.3	9.2
1978	42,400	8.5	12.8	9.4
1979	49,900	9.9	14.3	10.4
1980	52,200	9.9	14.5	10.5
1981	68,000	12.7	18.7	13.6
1982	61,800	11.5	17.0	12.3
1983	69,600	12.6	19.2	14.0
Total	563,300	8.3	11.8	9.2

Adapted from Centers for Disease Control. Ectopic pregnancy in the U.S. 1970–1983. MMWR 1986;35(No. 2SS):29SS.

this point[8,9] in the United States. However, there is a suggestion that for other countries the rise may be explained solely on basis of the large number of pelvic infections.[9] The emergence of voluntary abortions and tubal sterilizations have also been considered important factors, on the same level as corrective tubal surgery and conservative management (preservation of tubes) of ectopic gestations. It may well be that all of these factors contribute to some extent to the rise in rate, but it is difficult to accept that they are the total explanation for such a massive increase.

The breakdown of nationwide figures into subpopulation groups indicates that women over age 35 who become pregnant are at much higher risk of having an ectopic pregnancy than younger women; and if they are black, that risk is doubled.[1] Clearly there is a steady increase in rate with advancingage. However, the total numbers are highest in the under-30 age group because of their much higher pregnancy rate.

More limited reviews of material may give trends for small populations or institutions that have important practical implications. In the present material the rate for the Chicago Lying-In Hospital increased from 1 per 200 deliveries in 1967 to 1 per 40 in 1987, a fivefold rise over 20 years. The implication is that there is a true increase in the rate of ectopic gestations in the population of South Chicago; more important, physicians in this hospital must suspect ectopic pregnancies with much greater frequency to avoid overlooking the diagnosis and the possibly disastrous consequences.

With the predictable decrease in maternal mortality following legalization of abortion, mortality due to ectopic gestation and its complications has come to the fore as one of the leading causes of death. This phenomenon has occurred despite a steady fall in mortality rates due to ectopic gestation to less than 1 per 1000 (Fig. 4-1), the estimate being that about 50 women die in the United States following ectopic gestations from more than 4 million pregnancies. This same trend has been observed in the Chicago population: Since 1975 there has not been a death due to this condition in more than 700

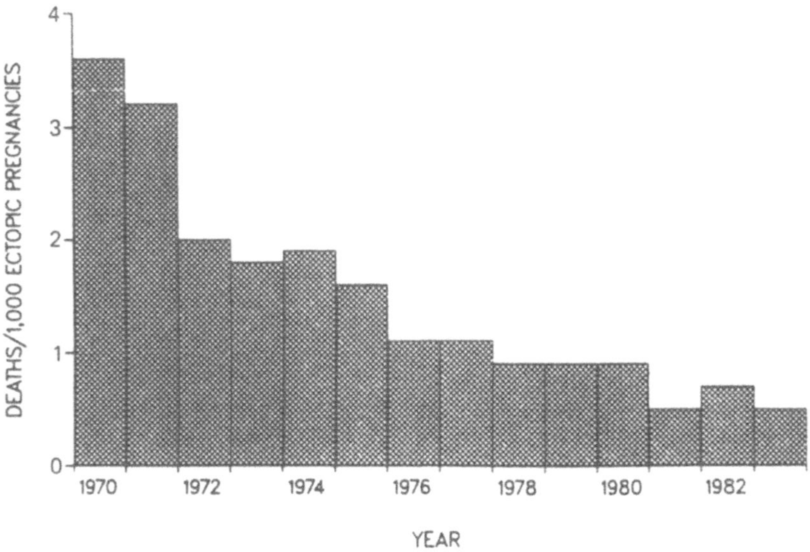

FIGURE 4-1. Mortality rates per thousand ectopic gestations in the United States. From reference 11, with permission.

cases, whereas during the preceding 8 years there was a 1% mortality rate.[10] The mortality rate does not cluster around particular subpopulations. Mortality occurs equally among all ages; there are more deaths among the 20- to 30-year age group because it is the age subpopulation with more gestations. Likewise, the death case rate for black women is significantly higher than among whites, and this factor combined with the higher incidence of ectopic pregnancies among these women predicts a mortality risk of five times above that for whites.[11]

The risk of death for a patient with an ectopic gestation is about ten times higher than for a woman carrying an intrauterine gestation to term and around 50 times above that of a woman having an induced abortion (therapeutic or voluntary).[12] Attempts to find predicting situations have indicated only that blood loss was the most frequent common symptom, and that a high number of women died at home.[12] In the series here evaluated all three women died in the hospital postsurgery after having been admitted in shock and transfused. The immediate cause of death was cardiac arrest in one and pulmonary edema in two.[10] In one report, the age of gestation in the women who died ranged from 4 to 30 weeks, with nearly 25% occurring after 12 weeks, and most involved known dangerous implantations: cornual or abdominal.[13] Failure to diagnose the condition occurred in nearly 50% of the cases.

Etiology

There are likely several factors predisposing to the implantation of an oozyte in an ectopic area. These factors may be divided in two major groups: anatomic factors and functional factors. Other minor or rare possible situations are considered likely causes in isolated cases.

Infections

Alterations of the endosalpinx due to infections are considered to be the most prevalent cause of ectopic gestation. There is indirect evidence that this may be true when one observes the parallel increase in pelvic infections and ectopic gestations since the 1970's. Small population observations have given support to this postulate. Westrom observed a high rate of ectopic gestations (6.9%) among women who had objective evidence of prior acute salpingitis.[14] Furthermore, he observed that the chances of a pregnancy being ectopic increased markedly in proportion to the preceding number of episodes of salpingo-oophoritis. The immediate mechanism is assumed to be adhesions, scarring, and edema of the endosalpingeal folds (also called plicae or fronds) and the interference it creates to the progression of the fertilized egg toward the uterus. Careful histologic observations have documented a high incidence of inflammatory (most likely infectious) reactions in the tubes lodging ectopic implanted pregnancies.[15–17] The more common etiologic pathogens are *Neisseria gonorrhoeae* and *Chlamydia trachomatis*. The importance of the latter has come to the fore only recently but with such strength that in many countries it is the prevailing cause of salpingo-oophoritis, or pelvic inflammatory disease (PID).[18] In many cases the infections due to this bacterium are subclinical and can be detected only by searching for antibody titers. In the experience here reported, 24% of the women had histologic evidence of chronic salpingitis, and 28.5% had a history of clinical acute pelvic infection. Postabortion infections have always been considered important causes of secondary infertility and therefore of potential ectopic gestations.[19] The legalization of abortion in the United States reduced drastically the prevalence of illegal abortion and their infectious sequelae, but it appears that abortion still plays an important role as a predisposing factor to extrauterine gestations.[20] In the present series 38% of the women had had one or more previous abortions.

Intrauterine Contraceptive Devices

There exists the perception that users of IUDs have higher chances of ectopic gestation because of their risk of uterotubal infection. This idea was reenforced when most of the devices were removed from the U.S. market following widespread litigation blaming their use for severe infections and subsequent infertility. Early population surveys done in Europe gave support to that hypothesis.[21–23] More limited case-control studies comparing IUD users with pregnant noncontraceptive users indicated a higher relative risk for the IUD wearers,[24] but there was no difference of relative risk when comparing IUD users to users of other contraceptive methods. It appears that an apparent increased risk of ectopic gestations may be due to the protection provided against intrauterine pregnancies, which does not affect ectopic implantation. Furthermore, ovarian pregnancies seem to occur in IUD users in excess of what it would be expected.[24,25] but without a reasonable explanation. On the other hand, an excellent case-control study conducted in the United States[26] concluded that there is no increased risk of ectopic gestation for IUD users when compared to never users, and furthermore that they in fact have a lower risk. Prolonged use (more than 25 months) appeared to increase the chances of ectopic gestation, compared to that of short-term users; these observations were irrespective of the type of device used. The mild inflammatory reaction produced by the foreign body seems to have an effect on tubal transport only after prolonged use; and among those cases, many may last up to 1 year after removal.[26]

Tubal Surgery

Attempts at preserving the tubes during operations for tubal ectopic pregnancies seem to carry with them an increased risk of recurrence of the same condition, perhaps owing to preexisting pathologic conditions[27,28] in the same or the contralateral tube (when present). A similar observation has been made following the widespread application of microsurgery to repair damaged tubes or to reverse surgical sterilization procedures.[29,30] In other words, the mere reestablishment of patency does not imply that the full function of unimpeded ovum transport has been achieved. The recurrence rate for these two types of conservative surgery is in the range of 8 to 25%. Procedures on the tubes done with the purpose of causing permanent sterilization have a known failure rate, the site of the pregnancy depending on the method applied. The popular Pomeroy technique, usually done postpartum or during a cesarean section, has a failure rate of up to 2%,[31] many of those pregnancies being extrauterine. The also popular tubal cauterization by laparoscopy has a lesser total failure rate of about 0.5%,[32] but almost all of these pregnancies are ectopic.[33] Some explanations offered postulate an active action by the endosalpingeal mucosa[34] promoting recanalization, but the most common reason may be fistulization due to insufficient tissue destruction. Other methods of laparoscopic sterilization such as Falope ring application and the Hulka clip also have failure rates, some of the failures being ectopic pregnancies. In the present material 4% of the patients had a history of tubal sterilization, most by cauterization.

Functional Causes

Abnormal function of the tubal musculature has been postulated as the cause of ectopic gestation in the absence of anatomic alterations of the tubal lumen. The inappropriate development seen in some women with *congenital abnormalities* of unknown etiology strongly suggest that the physiology must also be altered. The induced alterations triggered by in utero exposure to high and sustained doses of *diethylstilbestrol* has been studied in this clearly defined population,[35] and a significant number of ectopic gestations have been observed among them.[36]

The rapid development and clinical application of nonsurgical infertility treatment soon revealed that a number of such treatments were associated with an increased risk of ectopic gestation. *Induction of ovulation* seems to predispose to such a condition,[37] an abnormal hormonal balance being blamed for the problem. *In vitro fertilization* with ovum transfer has a relatively low rate of ectopic implantations, the first reported case having been ectopic[38]; others have reviewed the worldwide experience and have reported rates from 1.8% to 6.0%.[39,40] It seems that minor technical details may play an important role in causing the abnormal implantation.[40]

Other, more esoteric irregularities have been postulated as causing tubal dyfunction and thus favoring ectopic implantation of the ovum. *Late ovulation* and hormonal imbalance have been proposed as major etiologic factors of tubal implantation of a conceptus.[41] Speculations based on unproved dogmas referring to "ovum pick-up" mechanisms *(ovum transmigration)* have also been postulated as probable causes based on observations that a high number of tubal gestations are located in the tube contralateral to the corpus luteum.[17,42,43] The main criticism one may address to this postulate is that there is no way to prove which tube transported the ovum in the intrauterine pregnancy. A hypothesis still more difficult to accept is that of "internal transmigration," by which method of ovum reaches the uterine cavity through one tube and is displaced into the opposite one where it eventually implants.

The increased risk factors for several potential causes have been assessed and found to range from 2.5 for history of infertility, to pelvic infection, prior tubal surgery, and up to 13 for current IUD users.[44] A history of *previous ectopic gestation* constitutes a high risk because there is a high incidence of "repeats,"[45–47] ranging from 12 to 25% of cases. In our own material 13% of patients had a prior similar episode.

Symptomatology

One would expect that *amenorrhea,* being the predominant symptom of pregnancy would be also the most frequent symptom of ectopic gestations. In fact, it was present in less than 90% of our series, the duration of amenorrhea ranging between 6 and 9 weeks in 65% of them. *Irregular bleeding* occurs with great frequency and was present in 74% of our population; these episodes are not to be confused with period-like bleeding that occurs at the expected time. Rather, the episodes follow periods or appear after days or weeks of amenorrhea; the bleeding is generally light, often accompanied by lower abdominal pain.

The most common symptom found in our

population was *lower abdominal pain,* seen in 95% of patients. It was diffuse and bilateral in 60% of patients, the remainder experiencing either right or left lower quadrant pain. In general, the pain starts shortly after the missing period or a bleeding episode interpreted as such by the patient: It may be vague or colicky, or sharp or dull, localized or radiating. There is not a "typical" pain of ectopic gestation perhaps because many anatomic conditions trigger pain. Rectal pressure, often mentioned in textbooks, was present in only 5% of our patients.

Some patients report fainting episodes, nausea and vomiting, and other symptoms of early gestation. However, *there is no pathognomonic symptom of ectopic pregnancy,* and therefore the clinician must suspect the condition in order to orient the examination.

Diagnosis

Ectopic (extrauterine) implantation of a conceptus constitutes an emergency by virtue of the potentially life-threatening complications that may develop at unpredictable times. A rapid diagnosis is therefore of utmost importance once the clinician suspects this condition.

The presenting appearance of the patient depends on the symptomatology, which is related to the type of implantation and complication that prompted the visit to the physician. *Tachycardia* and orthostatic hypotension are seen only in cases of massive intraabdominal bleeding. With more cases diagnosed before rupture these signs are less frequently observed, but when present they suggest the need for prompt action.

On physical examination the common presenting symptom of pain can be evoked without difficulty. *Adnexal tenderness* was present in 71% of our patients and *cervical motion tenderness* in 65%. Of course these same signs are seen with pelvic infections, and so they must be considered only part of the picture. In only 31% was an *enlarged uterus* found, and in only 26% was an *adnexal mass* palpable. In other words, the prevalence of negative findings is high, more so if one considers that *cul-de-sac fullness* was appreciated in only 22% of cases. These findings are in agreement with the observations of others who found that the clinical diagnosis is accurately made in less than 50% of cases.[48,49] It is because of this fact that one must almost always resort to laboratory tests and other diagnostic aids to refine the diagnosis. Fortunately, advances in immunology and ultrasonography have significantly contributed to accuracy in diagnosis.

Pregnancy Test

The almost essential information is provided by a pregnancy test, which is now reliable thanks to the improved methods of measuring quantitatively the β fraction of hCG and the preservation of the reagents, thus increasing the sensitivity and specificity of the test to nearly 100%. The ability to measure low levels of βhCG is critical because most ectopic trophoblastic tissues produce lower levels of βhCG than eutopic gestations for equivalent periods of amenorrhea.[50] The more routine use of serum quantitative tests have demonstrated these differences as well as the abnormal rise (doubling time) in cases of ectopic implantation,[51] a useful ancillary application of this information. Only rarely does one find truly false-negative results with this method.[52] However, qualitative urine pregnancy tests are now able to detect low levels of βhCG and thus are useful for the rapid diagnosis of pregnancy.[53,54] Such a test was carried out in 830 of our patients and was positive in 88%. This experience covers 21 years, and therefore the sensitivity of the tests early in the period was known to be limited—hence the relatively high rate of false negatives.

Sonography

The possibility of imaging just-developing gestational sacs with ultrasound waves has signified an extraordinary advance in the early diagnosis of pregnancy. Conceptuses of 5 weeks' menstrual age (3 weeks post conception) have been clearly identified,[55] thus enabling the clinician to make an early diagnosis. Refinements in equipment and utilization of the vaginal probe have further increased the sensitivity of the method to accurately diagnose the site of pregnancy[56–58] by ruling out intrauterine gestations and more easily *visualizing the adnexae* (Fig. 4–2). This examination has been carried out in 280 of our last 350 cases, with the sensitivity during the last year being above 95%.

Culdocentesis

Culdocentesis has been for many years the standard gesture in the attempt to make a diagnosis in patients who do not warrant an immediate laparotomy. It is reported to be highly informative in the hands of experienced operators, although others believe that it could be misleading and therefore do not recommend it. It is *positive* when nonclotting blood is obtained (with a hematocrit above 15%), *negative* when clear (or barely tinged) fluid is aspirated, and *inconclusive* when it is "dry," i.e., no fluid is aspirated. The clear understanding of the last finding (or no finding) is most important because too many clinicians consider it negative, a misinterpretation that probably fuels the conviction of the skeptics. Positive rates in range of 80% or higher have been reported[10,49,59] for early gestations[60] but apparently with less sensitivity for more advanced cases. Of course it is going to be positive only in cases of ruptured, aborting, or "leaky" ectopic pregnancies, the uncomplicated ones constituting the false negatives. The specificity is also less than 100% because it indicates only the presence of free blood in the pouch of Douglas, not its source. Ruptured corpus luteum, retrograde blood flow during periods, ruptured ovarian cysts, or early miscarriages are among the conditions found to produce this finding.[60] Culdocentesis was positive in 64% of 432 procedures done in our series, negative in 12%, and inconclusive in the remaining 24%. Others have reported inconclusive rates of up to 16%.[61] As the sensitivity of the other ancillary methods of diagnosis increases, the frequency of culdocentesis performed drops in proportion.

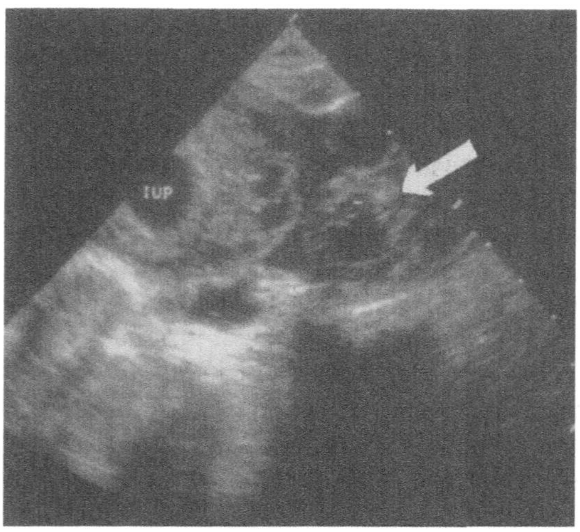

FIGURE 4-2. Heterotopic pregnancy (intrauterine and ectopic) diagnosed by intravaginal sonography. The extrauterine gestation is indicated by the white arrow; a cystic corpus luteum is above it. The intrauterine pregnancy (IUP) is indicated. The diagnosis was confirmed at laparotomy.

Laparoscopy

The widespread use of laparoscopy preceded by a few years the described improvements in more sensitive pregnancy tests and ultrasonography. As a consequence it was widely applied as the definitive means of diagnosis in cases of suspected ectopic gestation. It is still used for this purpose (rightfully so) in centers where those refined techniques are not available or in centers where they are available but not with the expediency sometimes required by some clinical manifestations. For example, highly skilled sonographers handling equipment with high resolution are not available in most emergency rooms in the United States. However, a laparoscopy set and a reasonably trained gynecologist usually are, and therefore laparoscopy is still a much used method utilized in cases of suspicion. In fact, the number of laparoscopies done to rule out ectopic pregnancy are more than double the number of ectopic pregnancies diagnosed. In other words, more than 50% of the laparoscopies do not reveal an ectopic pregnancy but, instead, find acute salpingitis, appendicitis, ruptured or unruptured corpus luteum, ovarian cysts, or even a normal pelvis. During the last year of our series this figure was 35%. In the total experience reported here of nearly 500 cases of diagnosis confirmed by laparoscopy there has not been one false positive, with only one false negative in an early-stage pregnancy, confirmed and operated 1 week later. Higher incidences of false positive and false negatives have previously been reported.[62]

Laparoscopy has been instrumental in facilitating the diagnosis of unruptured ectopic pregnancies. Before 1970 when we introduced its systematic use 35% of our material were unruptured, whereas during the subsequent period this proportion increased to reach 55% in the last year. A similar trend ranging around 50% is observed when reviewing relatively recent publications.[6,63,64]

Despite the progress made and the awareness among gynecologists of the "epidemic" of ectopic gestations, there is still an appalling number of patients who are seen repeatedly by physicians before a diagnosis is made. In fact, among the extreme cases, nearly one-third of those who ended up as fatalities[11,13] had sought care long before their condition became lethal, and some even died without intervention.[11] Few publications mention this critical point—one that is remediable with intensive education of medical personnel who attend the emergency rooms of hospitals. In our series 14% of the women had visited the emergency room or had seen a physician 1 to 5 days before diagnosis, and 28% were seen 6 days or more before that, a great number of them than 10 days prior to admission and surgical intervention.

Pathology

The abnormal site of implantation of a pregnancy is of enormous importance in determining the symptomatology, diagnostic difficulties, severity of complications, operative procedures, and ultimately the possible mortality of a given case.

Standard textbooks state that approximately 95% of the ectopic pregnancies are implanted in various parts of the *Fallopian tube,* the remainder being distributed as ovarian, abdominal, and other miscellaneous, rarer sites. The preceding discussions regarding symtomatology and diagnosis apply almost exclusively to tubal pregnancy. However, because of some overlapping characteristics they may also apply to the more rare sites of implantation.

The unruptured *tubal pregnancy* constitutes, in the more recent series, 45 to 75% of cases treated in large centers.[6,63–65] The pregnancy may or may not be partially separated from its implantation site; when it is, there generally is leakage of blood into the pelvis, and a culdocentesis may be positive. Because they are generally in the early stages and thus small, they offer the best possibilities for a successful conservative surgical procedure. Rarely, a tubal pregnancy remains intact beyond the third month of amenorrhea, but there have been some isolated reports of such observations, and one pregnancy in this series reached term and produced a live infant. The classic type presents with acute abdominal symptomatology of pain and signs of internal bleeding. The amount of blood and clots present in the abdomen may range from a few hundred milliliters to more than 2 liters. In our series 45% of the women had a blood loss of more than 800 ml and 20% of more than 1500 ml with the site of implantation torn and actively bleeding. The most frequent spot in the tube for these implants is the isthmic-cornual region, observed in 16% of this series.

The *abdominal pregnancy* is a rare occurrence but a very dangerous one because it accounts for a disproportionate number of maternal deaths. The average mortality of 0.7 per 1000 for nonabdominal cases rises to over 5 per 1000 in abdominal cases in nationwide surveys.[66] In small series the mortality may be as high as 18%.[67,68] Abdominal pregnancies may be *primary,* that is with original implantation in the abdomen, or *secondary* where the abdominal implantation follows a tubal abortion by the fimbriae or through a rupture of the tube. The generally accepted criteria for defining a primary implantation were stated by Studdiford,[69] who essentially required that the uterus, tubes, and ovaries be intact. It is believed that most abdominal pregnancies are secondary, comprising about 1.5% of ectopic pregnancies.[68] They tend to be diagnosed in a much more advanced stage than in other implantation sites[70] with a peculiar symptomatology that is generally misleading. Of course the amniotic sac is free in the abdomen, and the placenta is implanted over many organs, as tubes, uterus, broad ligament, mesentery, bowels. There is bleeding only in rare instances when the placenta separates. There were six abdominal gestations in our series, none of which produced a viable fetus. The incidence is thus about 0.7% of ectopic gestations, somewhat less than that found in other reports.[68]

Ovarian gestations are contained within the ovary. They are believed to occur when the ovum is not expelled from the follicle and is fertilized within the forming corpus luteum. To consider a pregnancy as primary ovarian, it must fulfill the requirements proposed by Spiegelberg,[71] which essentially state that the pregnancy must be *within* identifiable ovarian tissue and that the tube and fimbria must be intact. Its incidence among all pregnancies is between 1 per 3000 and 1 per 8000, whereas in the incidence among ectopic pregnancies ranges from 2 to 4%.[72] Its relative number has increased following the extensive use of intrauterine contraceptive devices; this fact seems to be due to the protective effect of the devices against intrauterine pregnancies and even tubal gestations but their lack of effect toward ovarian implantations.[73] The preoperative diagnosis has usually been an ovarian mass or ectopic pregnancy, presumed tubal.[72] The ovary, when intact, appears generally cystic; and when it is ruptured it presents with severe intraperitoneal hemorrhage making difficult even an intraoperative diagnosis.[74] In the present series there were seven ovarian gestations for an incidence of 0.8% of ectopics. Each presented as a cystic ovary, and in none was a certain preoperative diagnosis made.

Various *other sites* of ectopic implantation have been described, such as cervical pregnancy[75] or intraligamentous attachment.[76] There were no cases of this type in the series reviewed here. Extremely rare implantations have been reported in the *vagina,*[77] the *spleen*[78] or the *liver.*[79]

Other rare but less infrequent types are *multiple ectopic gestations,* such as bilateral ectopic pregnancies,[80] of which there was one case in this series. The so-called heterotopic pregnancies are also multiple gestations, but with one of them implanted in the uterine cavity and the other usually in one tube. A relatively important number of these pregnancies have been reported[81,82]; two were observed in the present series (Fig. 4–2).

Management

After establishing the diagnosis of ectopic gestation, one must decide on the most appropriate means of treatment. This varies depending on the clinical, anatomic, social, and technologic circumstances peculiar to each case. The limited options once available have expanded dramatically during the 1980s with the array of techniques and choices available. The clinician may offer to the patient surgical procedures (conservative, radical, ultraradical), the use of chemotherapeutic agents, or an expectant attitude with periodic reevaluation of the circumstances.

Surgical Procedures

Surgery has been the only method to treat ectopic pregnancies since the time when R. Lawson Tait performed the first successful operation late in the nineteenth century. However, a number of techniques have been available among the surgical approaches proposed to correct this dramatic problem.

Classically, an urgent laparotomy and *salpingo-oophorectomy* were the correct treatment. This approach was justified because most cases presented with a ruptured tube and extensive pelvic hematocele. This procedure is still the one of choice when the tube is severely damaged owing to previous infections or operations (to correct infertility or previous ectopic pregnancies) and usually immediately follows a diagnostic laparoscopy. A conservative approach under those circumstances would only multiply the chances of a repeat extrauterine gestation.[83,84] Likewise, when the patient desires permanent sterilization or already had a tubal ligation (failed), it is the procedure of choice. In our material 69% had salpingectomy and 18% salpingo-oophorectomy.

The advances in diagnosis previously reviewed as well as the increased awareness by the clinicians have been instrumental in the impressive drop in the number of ruptured cases at diagnosis and, reciprocally, the increase in early unruptured ectopic pregnancies. The latter, ranging between 45 and 75% of cases, produce much less anatomic destruction of the tubes and therefore are amenable to conservative operations or other options to be discussed below. The choice of procedure can usually be made during the laparoscopy carried out to confirm the diagnosis. Less radical procedures (than salpingo-oophorectomy) may be employed depending on the location of the pregnancy and the size of the distended tube. These conservative techniques are used when the patient desires to maintain the possibility of future pregnancies at a maximum.

In those cases *linear salpingostomy* is utilized when the tube is dilated less than 2.5 cm and the pregnancy is near the end of the ampulla or in the fimbriated end. An incision is made, along the antimesosalpinx area of the tube overlying the ectopic gestation, of sufficient length to allow easy removal of the pregnancy sac. The instrument of choice, because it creates hemostasis, is the needle cautery utilizing cutting current. More recently various types of laser have been utilized successfully with the same purpose. The products of conception are then removed, but care must be taken to be not too thorough when cleaning the implantation bed as oozing blood is the inevitable consequence. When the pregnancy is in the ampulla itself the procedure applied is similar, with the possibility of either closing the window that had been made or leaving it open. The few direct observations made as well as follow-up hysterosalpingograms have demonstrated that this window closes spontaneously and that the tube is generally patent. However, this point of the technique has not yet been clearly settled. In our material 9% of the cases underwent salpingostomy, most of them during the second half of the study period.

If the pregnancy is located in the isthmic or cornual portion, *partial salpingectomy* is the procedure of choice. This operation allows a subsequent anastomosis or implantation procedure to restore the patency of the tube. A more difficult problem presents when the ectopic pregnancy is located in the only tube a patient may have: Salpingostomy was followed by repeat ectopic pregnancies in 15 to 17% of cases,[85,86] and salpingotomy (that is closure of the incision by suture) was associated with a 35% repeat rate,[87] indicating the exceedingly high risk of repeated ectopic gestation with either type of procedure. In a case of cornual gestation care must be taken to remove as little myometrium as possible so as to not weaken its wall and risk rupture in case of subsequent intrauterine gestation. End-to-end anastomosis after partial isthmic resection is not advised at the time but, instead, should be done at a later date.

A still more conservative approach has been proposed: *salpingostomy during laparoscopy*. It is an option in cases of small ectopic pregnancies when the operator has adequate training, the patient is stable, and appropriate instrumentation is available. The cutting instruments are similar to those described for laparotomy: needle cautery and laser. From small series it appears that it offers the same chances of intrauterine pregnancy and repeat ectopic as the procedure done by laparotomy.[88] Others have reported shorter operating and recovery time as additional advantages.[89] Persistence of βhCG and retained trophoblastic tissue have been reported as disadvantages.[90] Follow-up of quantitative βhCG blood levels could help in differentiating between normal postoperative findings and persistence of trophoblastic tissue.[91] A small group of laparoscopists have proposed the practice of *partial salpingectomy* during laparoscopy as treatment in selected cases.[88,89,92] It carries the advantage of a much shorter postoperative recovery and hospitalization time. Only five patients in our series were treated in this manner, all without complications.

The ultraradical approach of *total hysterectomy and salpingo-oophorectomy* is another option. Under certain conditions, particularly abdominal pregnancies with extensive placental implantation on the uterus, adnexa, or upper aspect of broad ligaments, it is the only option. It requires a patient with a stable (or stabilized) cardiovascular system. It was used in 4.5% of cases in our series.

Blood Loss and Replacement

Various degrees of blood loss are a natural consequence of ectopic gestations. Even among the unruptured cases it is common to find a significant amount of blood in the pelvis. Severe blood loss is almost always part of the symptomatology of *cornual and isthmic* gestations, often accompanied by cardiovascular collapse and shock. This aspect of the problem is not always discussed in reports on this subject. However, it is of the utmost importance because it is almost exclusively the immediate cause when the patient dies. There is not always a direct relation between the amount of blood in the abdomen and cardiovascular collapse; the collapse depends more directly on the speed with which the treatment of choice is applied. The technique used depends on several factors, particularly the patient's condition, the type and location of the ectopic implant, the skill of the surgeon, and the instrumentation available at the time of surgery. In our material 51% lost more than 500 ml of blood (ranged few ml to 7000). Blood transfusions were given to 47% of patients, but to less than 30% in the last two years.

Medical Procedures

The unexpected observation by many of small, old, ectopic gestations that apparently had regressed spontaneously has stimulated the thought of possibly using *antitrophoblastic substances* for the management of ectopic gestations. Isolated case reports on the successful use of methotrexate for this purpose have been followed by a small series in which five of six patients responded satisfactorily,[93] although two had late complications. This treatment is obviously a path that must be explored.

Abortificient substances have also received a limited trial, but no systematic experience has yet been reported. These substances too must be given a full trial.

The *expectant noninterventionist management* of ectopic pregnancy was studied in the past[94] and was found to have no advantages over the surgical approach. Attempts were later made to assess that conduct, with better selection of patients by means of diagnostic laparoscopy and more precise follow-up using serial quantitative βhCG determination.[95] Despite theoretically possible conveniences, at the present time this method offers no advantages over conservative surgery.

For the moment, conservative surgery is the treatment of choice when the patient wishes to maintain her fertility potential. In cases where the tube is too damaged or the patient elects sterilization, salpingectomy is the procedure of choice.

References

1. Centers for Disease Control. Ectopic pregnancy in the U.S. 1970–1983. MMWR 1986;35(No. 2SS):29SS.
2. Loffer FD. The increasing problem of ectopic pregnancies and its impact on patients and physicians. J Reprod Med 1986;31:74.
3. Atrash HK, Friede A, Hogue CJ. Ectopic pregnancy mortality in the U.S.A. 1970–1983. Obstet Gynecol 1987;70:817.
4. Hemminki E, Heinonen PK. Time trends of ectopic pregnancy. Br J Obstet Gynaecol 1987;94:322.
5. Makinen JI. Ectopic pregnancy in Finland 1967–83: a massive increase. Br Med J 1987;294:740.
6. Gorodeski IG, Bahary CM. Tubal pregnancy—reappraisal of incidence. Eur J Obstet Gynecol Reprod Biol 1987;24:57.
7. Curran JW. Economic consequences of pelvic inflammatory disease in the U.S. Am J Obstet Gynecol 1980;138:848.
8. Rubin GL, Peterson HB, Dorfman SF, et al. Ectopic pregnancy in the U.S. 1970–78. JAMA 1983;249:1725.
9. Chow WH, Daling JR, Cates W, et al. Epidemiology of ectopic pregnancy. Epidemiol Rev 1987;9:70.
10. Helvagioglu A, Long EM, Yang SL. Ectopic pregnancy: an eight year review. J Reprod Med 1979;22:87.
11. Dorfman SF. Epidemiology of ectopic pregnancy. Clin Obstet Gynecol 1987;30:173.
12. Dorfman SF. Deaths from ectopic pregnancy, U.S. 1979–80. Obstet Gynecol 1983;62:334.
13. Dorfman SF, Grimes DA, Cates W, et al. Ectopic pregnancy mortality U.S. 1979–80: clinical aspects. Obstet Gynecol 1984;64:386.
14. Westrom L. Effect of acute inflammatory disease on fertility. Am J Obstet Gynecol 1975;121:707.
15. Kleiner GJ, Roberts TW. Current factors in the causation of tubal pregnancy. Am J Obstet Gynecol 1967;99:21.
16. Bone NL, Greene RR. Histological study of uterine tubes with tubal pregnancy: a search for evidence of previous infection. Am J Obstet Gynecol 1961;82:1166.
17. Pauerstein CJ, Croxatto HB, Eddy CA, et al. Anatomy and pathology of tubal pregnancy. Obstet Gynecol 1986;67:301.
18. Svensson L, Mardh PA, Ahlgren M, et al. Ectopic pregnancy and antibodies to Chlamydia trachomatis. Fertil Steril 1985;44:313.
19. Panayotou PP, Kaskarelis DB, Miettinen LS, et al. Induced abortion and ectopic pregnancy. Am J Obstet Gynecol 1972;114:507.
20. Levin AA, Schoenbaum SC, Stubblefield PG, et al. Ectopic pregnancy and prior induced abortion. Am J Public Health 1982; 72:253.
21. Beral V. An epidemiological study of recent trends in ectopic pregnancy. Br J Obstet Gynaecol 1975;82:775.
22. Erkkola R, Liukko P. Intrauterine device and ectopic pregnancy. Contraception 1977;16: 569.
23. Savolainen E, Saksela E. Ectopic pregnancy: relationship to the preceding contraception. Ann Chir Gynaecol 1978;67:198.
24. McKenna P, Pomeroy L, Gaughan B. Tubal pregnancy in the Rotunda Hospital: a review of 13 years. Ir J Med Sci 1985;154:274.
25. Evans MI, Angerman NS, Moravec WD, et al. The intrauterine device and ovarian pregnancy. Fertil Steril 1979;32:31.
26. Ory HW, The Women Health Study. Ectopic pregnancy and intrauterine contraceptive devices: new perspectives. Obstet Gynecol 1981;57:137.
27. DeCherney AH, Kase N. Conservative surgical management of unruptured ectopic pregnancy. Obstet Gynecol 1979;54:451.
28. Lauger R, Bukowsky I, Herman A, et al. Conservative surgery for tubal pregnancy. Fertil Steril 1982;38:427.
29. Russell JB, DeCherney AH, Laufer N, et al. Neosalpingostomy: comparison of 24 and 72 months follow-up shows increased pregnancy rate. Fertil Steril 1986;45:296.
30. Henderson SR. The reversibility of female sterilization with the use of microsurgery: a report on 102 patients with more than one year follow-up. Am J Obstet Gynecol 1984; 149:57.
31. Prystowsky H, Eastman NJ. Puerperal tubal sterilization: report of 1830 cases. JAMA 1955;158:463.
32. Phillips J, Keith D, Hulka J, et al. Gynecologic laparoscopy in 1975. J Reprod Med 1976;16:105.
33. McCausland A. High incidence of ectopic pregnancy following laparoscopic tubal coag-

ulation failures. Am J Obstet Gynecol 1980;136:97.
34. McCausland A. Endosalpingosis ("endosalpingoblastosis") following laparoscopic tubal coagulation as an etiologic factor of ectopic pregnancy. Am J Obstet Gynecol 1982; 143:17.
35. DeCherney AH, Cholst I, Naftolin F. Structure and function of the fallopian tubes following exposure to diethylstilbestrol during gestation. Fertil Steril 1981;36:741.
36. Herbst AL, Hubby MM, Azizi F, et al. Reproductive and gynecologic surgical experience in diethylstilbestrol-exposed daughters. Am J Obstet Gynecol 1981;141:1019.
37. McBain JC, Evans JH, Pepperell RJ, et al. An unexpectedly high rate of ectopic pregnancy following the induction of ovulation with human pituitary and chorionic gonadotropin. Br J Obstet Gynaecol 1980;87:5.
38. Steptoe PC, Edwards RG. Birth after the reimplantation of a human embryo. Lancet 1978;2:366.
39. Seppala M. The world collaborative report on in vitro fertilization and embryo replacement: current state of the art in January 1984. NY Acad Sci 1985;442:558.
40. Yovich JL, Turner SR, Murphy AJ. Embryo transfer technique as a cause of ectopic pregnancies in in-vitro fertilization. Fertil Steril 1985;44:318.
41. McElin TW, Iffy L. Ectopic gestation: a consideration of new and controversial issues relating to pathogenesis and management. Obstet Gynecol Ann 1976;5:241.
42. Berlind M. The contralateral corpus luteum—an important factor in ectopic pregnancies. Obstet Gynecol 1960;16:51.
43. Breen JL. A 21 years survey of 654 ectopic pregnancies. Am J Obstet Gynecol 1970; 106:1004.
44. Marchbanks PA, Annegers JF, Coulam CB, et al. Risk factors for ectopic pregnancy: a population-based study. JAMA 1988; 259:1823.
45. Corson SL, Batzer FR. Ectopic pregnancy: a review of etiologic factors. J Reprod Med 1986;31:78.
46. Sandvei R, Bergsj P, Ulstein M, et al. Repeat ectopic pregnancy: a twenty year hospital survey. Acta Obstet Gynecol Scand 1987;66:35.
47. Stock RJ. The changing spectrum of ectopic pregnancy. Obstet Gynecol 1988;71:885.
48. Weinstein L, Morris MB, Dotters D, et al. Ectopic pregnancy—a new surgical epidemic. Obstet Gynecol 1983;61:698.
49. Weckstein LN, Boucher AR, Tucker H, et al. Accurate diagnosis of early ectopic pregnancy. Obstet Gynecol 1985;65:393.
50. Romero R, Kadar N, Copel JA, et al. The effect of different human chorionic gonadotropin assay sensitivity on screening for ectopic pregnancy. Am J Obstet Gynecol 1985;153: 72.
51. Kadar N, Romero R. Observations on the log human chorionic gonadotropin–time relationship in early pregnancy and its practical implications. Am J Obstet Gynecol 1987;157:73.
52. Taylor RN, Padula C, Goldsmith PC. Pitfall in the diagnosis of ectopic pregnancy: immunocytochemical evaluation of a patient with false negative serum βhCG levels. Obstet Gynecol 1988;71:1035.
53. Barnes RB, Roy S, Yee B, et al. Reliability of urinary pregnancy test in the diagnosis of ectopic pregnancy. J Reprod Med 1985; 30:827.
54. Nordensjold F, Ahlgren M, Erreth L, et al. A sensitive urine pregnancy test as an aid in the diagnosis of ectopic pregnancy. Fertil Steril 1985;43:748.
55. Nyberg DA, Filly RA, Mahony BS, et al. Early gestation: correlation of hCG levels and sonographic identification. AJR 1985; 144:951.
56. Mendelson EB, Bohm-Velez M, Joseph N, et al. Gynecologic imaging: comparison of transabdominal and transvaginal sonography. Radiology 1988;166:321.
57. Dashefsky SM, Lyons EA, Levi CS, et al. Suspected ectopic pregnancy: endovaginal and transvesical US. Radiology 1988;169:181.
58. Nyberg DA, Mack LA, Jeffrey RB, et al. Endovaginal sonographic evaluation of ectopic pregnancy: a prospective study. AJR 1987; 149:1181.
59. Weckstein LN. Current perspective on ectopic pregnancy. Obstet Gynecol Surv 1985;40:259.
60. Schwab RA. Ultrasound versus culdocentesis in the evaluation of early and late ectopic pregnancy. Am Emerg Med 1988;17:801.
61. Romero R, Copel JA, Kadar N, et al. Value of culdocentesis in the diagnosis of ectopic pregnancy. Obstet Gynecol 1985;65:519.
62. Samuelson S, Sjovall A. Laparoscopy in suspected ectopic pregnancy. Acta Obstet Gynecol Scand 1972;51:31.
63. Sherman D, Langer R, Sadovsky G, et al. Im-

proved fertility following ectopic pregnancy. Fertil Steril 1982;37:497.
64. Makinen J, Nikkanen V, Kivikoski A. Problems and benefits in early diagnosis of ectopic pregnancy. Eur J Obstet Gynecol Reprod Biol 1984;16:381.
65. Seppala M, Purhonen M. The use of hCG and other pregnancy proteins in the diagnosis of ectopic pregnancy. Clin Obstet Gynecol 1987;30:148.
66. Atrash HK, Friede A, Hogue CJ. Abdominal pregnancy in the United States: frequency and maternal mortality. Obstet Gynecol 1987;69:333.
67. Strafford JC, Ragan WD. Abdominal pregnancy: review of current management. Obstet Gynecol 1977;50:548.
68. Delke I, Perez-Veridiano N, Tancer ML. Abdominal pregnancy: review of current management and addition of 10 cases. Obstet Gynecol 1982;60:200.
69. Studdiford WE. Primary peritoneal pregnancy. Am J Obstet Gynecol 1942;44:487.
70. Clark JF, Jones SA. Advanced ectopic pregnancy. J Reprod Med 1975;14:30.
71. Spiegelberg O. Zur Casuistik der Ovarialshwangerschaft. Arkh Gynaecol 1878;13:73.
72. Grimes HG, Nosal RA, Gallagher JC. Ovarian pregnancy: a series of 24 cases. Obstet Gynecol 1983;61:174.
73. Lehfeldt H, Tietze C, Gorstein F. Ovarian pregnancy and the intrauterine device. Am J Obstet Gynecol 1970;108:1005.
74. Hallat JG. Primary ovarian pregnancy: a report of 25 cases. Am J Obstet Gynecol 1982;143:55.
75. Parente JT, Chau-Su O, Levy J, et al. Cervical pregnancy analysis: a review and report of 5 cases. Obstet Gynecol 1983;62:79.
76. Vierhout ME, Wallenberg HCS. Intraligamentary pregnancy resulting in a live infant. Am J Obstet Gynecol 1985;152:878.
77. Duckman S, Suarez J, Spitaleri J. Vaginal pregnancy presenting as a subureteral cyst. Am J Obstet Gynecol 1984;149:572.
78. Yackel DB, Panton ON, Martin DJ, et al. Splenic pregnancy—case report. Obstet Gynecol 1988;71:471.
79. Harris GJ, Al-Jurf AS, Yuh WTC, et al. Intrahepatic pregnancy. JAMA 1989;261:902.
80. Levy JS, Rosenzweig BA, Blumenthal L. Bilateral tubal pregnancies after tubal sterilization. Obstet Gynecol 1988;72:494.
81. Reece EA, Petrie RH, Sirmans MF, et al. Combined intrauterine and extrauterine gestations: a review. Am J Obstet Gynecol 1983;146:323.
82. Laband SJ, Cherney WB, Finger HJ. Heterotopic pregnancy: report of 4 cases. Am J Obstet Gynecol 1988;158:437.
83. Shenker JG, Evron S. New concepts in the surgical management of tubal pregnancy and the consequent postoperative results. Fertil Steril 1983;40:709.
84. DeCherney AH, Silidker JS, Mezer HC, et al. Reproductive outcomes following two ectopic pregnancies. Fertil Steril 1985;43:82.
85. Henry-Suchet J, Tesquier L, Loffredo V, et al. La chirurgie conservatrice de la grossesse extrauterine. In: Oviducte et Fertilite. Paris, Masson, 1979;393.
86. DeCherney AH, Maheux R, Naftolin F. Salpingostomy for ectopic pregnancy in the sole patent oviduct: reproductive outcome. Fertil Steril 1982;37:619.
87. Oelsner G, Rabinovitch O, Morad J, et al. Reproductive outcome after microsurgical treatment of tubal pregnancy in women with a single fallopian tube. J Reprod Med 1986;31:483.
88. DeCherney AH, Diamond MP. Laparoscopic salpingostomy for ectopic pregnancy. Obstet Gynecol 1987;70:948.
89. Brumsted J, Kessler C, Gibson C, et al. A comparison of laparoscopy and laparotomy for the treatment of ectopic pregnancy. Obstet Gynecol 1988;71:889.
90. Bell OR, Awadalla SG, Mattox JH. Persistent ectopic syndrome: a case report and literature review. Obstet Gynecol 1987;69:521.
91. Steier JA, Bergsjo P, Myking OL. Human chorionic gonadotropin in maternal plasma after induced abortion, spontaneous abortion and removed ectopic pregnancy. Obstet Gynecol 1984;64:391.
92. Loffer FD. Outpatient management of ectopic pregnancies. Am J Obstet Gynecol 1987;156:1467.
93. Ory SJ, Villanueva AL, Sand PK, et al. Conservative treatment of ectopic pregnancy with methotrexate. Am J Obstet Gynecol 1986;154:1299.
94. Lund JJ. Early ectopic pregnancy. J Obstet Gynaecol Br Emp 1955;62:70.
95. Mashich S, Carp HJA, Serr DM. Nonoperative management of ectopic pregnancy: a preliminary report. J Reprod Med 1982;27:127.

5 Surgical Diseases of the Endocrine Glands During Pregnancy

MARK H. KADOWAKI and EDWIN L. KAPLAN

Surgical diseases of the endocrine glands are uncommon during pregnancy. However, they are important, as the hormonal abnormalities that do occur often cause profound changes, morbidity, and potential mortality in the mother, fetus, and neonate. This chapter explores the diagnosis and treatment of surgical diseases of the thyroid, parathyroid, and adrenal glands during pregnancy.

Thyroid Diseases

Thyroid diseases are among the most common experienced by women during pregnancy. As shown in Table 5-1, many tests are available to assess the function and anatomy of the thyroid gland. During pregnancy thyroid homeostasis does not change in the mother in a major way, even though changes appear in the thyroid function tests. Presumably due to increased estrogen, there is an increased production of thyroxine-binding globulin (TBG) by the liver, which results in increased levels of circulating bound thyroxine (T_4) and bound triiodothyronine (T_3). However, the concentrations of circulating free T_4 and free T_3, which are the physiologically important thyroid hormones, do not appear to change much during pregnancy. Serum thyroid-stimulating hormone (TSH) levels are also similar to nonpregnant levels after 10 to 12 weeks gestation. Thus measurement of both the free thyroxine index (FTI or T_7) as well as the serum TSH level gives the best assessment of thyroid function during pregnancy.[1]

The placenta produces two chorionic hormones that have minor thyroid-stimulating activity in normal pregnancies: human chorionic gonadotropin (hCG) and human chorionic thyrotropin (hCT). Both may have thyroid-stimulating effects of importance when a hydatidiform mole or a choriocarcinoma is present.

It has commonly been stated that enlargement of the maternal thyroid gland is common during pregnancy. This effect is variable and is possibly due to increased iodide excretion by the kidneys. Thus development of a large goiter or a nodule during pregnancy is abnormal and is worthy of investigation. Finally, although isotope scans of the thyroid are commonly used for the diagnosis of thyroid nodules, they must not be used during pregnancy because of the danger of harming the fetal thyroid gland.

Studies have demonstrated that little thyroid hormone traverses the placenta from maternal to fetal circulations.[1] A marked gradient of T_4 and free T_4 exists from mother to fetus until the 35th to 40th week, after which the fetal level rises dramatically. A large gradient of T_3 also exists throughout the entire pregnancy. Thus the placenta is relatively impervious to transfer of thyroid hormones from mother to fetus.

TABLE 5-1. Thyroid function tests.

Test	Abbreviation
Direct and indirect measures of T_4 and T_3 supply	
Serum T_4	T_4
Free thyroxine	FT_4
Resin T_3 uptake	RT_3U
Free thyroxine index	$FTI(T_7)$
Serum T_3	T_3
Radioactive iodine uptake	RAIU
Serum TSH	TSH
Thyroxine-binding globulin capacity	TBG
Measures of autoimmunity	
Antithyroglobulin antibodies	TGHA
Antimicrosomal antibodies	MCHA
Long-acting thyroid stimulator	LATS
Thyroid-stimulating immunoglobulins	TSI
Measures of thyroid and pituitary responsiveness	
T_3 suppression test	
TRH stimulation test	
TSH stimulation test	
Assessment of thyroid anatomy	
Thyroid isotope scan	Scan
Thyroid fluorescent or computed tomographic (CT) scan	Scan
Ultrasonographic scan	Scan
Fine needle aspiration with cytologic examination	

Maternal Hypothyroidism

Untreated hypothyroidism is associated with infertility and an increased prevalence of abortions, stillbirths, and congenital malformations; treatment corrects these abnormalities. Maternal hypothyroidism also may compromise placental blood flow and fetal nutrition. Thus the detection and optimum treatment of hypothyroidism in women planning a pregnancy or already pregnant are important. The diagnosis is based on the same criteria in pregnant as in nonpregnant women, except that the FTI and serum TSH levels are most important. Optimal thyroxine replacement therapy usually is similar in pregnant and nonpregnant women. However, some pregnant women require a modest increase in their thyroxine dose to normalize the serum TSH concentration.[2]

Maternal Thyrotoxicosis

Thyrotoxicosis refers to a syndrome that is due to oversecretion of thyroid hormones. It may occur as a result of a toxic nodular goiter or a hyperfunctioning adenoma, or rarely it is due to struma ovarii, which means hyperfunctioning thyroid tissue within an ovarian teratoma. However, the most common cause of thyrotoxicosis today in the United States is Graves' disease. It is estimated that this disease occurs in 1 of every 1500 pregnancies.

Graves' Disease

Graves' disease is the eponym used in the English-speaking world to describe the syndrome of diffuse toxic goiter. Graves was an Irish physician who described this syndrome in the early English literature.

The major findings in Graves' disease are illustrated in Table 5-2. In addition to the symptoms and signs of thyrotoxicosis, these patients also may have exophthalmos, a diffuse goiter, and pretibial myxedema, all of which are thought to be manifestations of a generalized autoimmune disease. This triad is not found with the other forms of thyrotoxicosis. Individuals with Graves' disease also have a relative lymphocytosis, lymphoid infiltration of the thyroid parenchyma, and round cell infiltration of the retroocular muscles and fat. The diagnosis, however, may be more difficult during pregnancy because there is a mild state of hypermetabolism during this period even in euthyroid individuals.

The diagnosis is not difficult when the serum FTI value is elevated. Furthermore, the new serum TSH test can demonstrate a suppressed state that is lower than normal. Finally, antithyroid antibodies—against microsomal fragments (MCHA) and against thyroglobulin (TGHA)—are commonly detectable in the serum.

The maternal thyrotoxic state usually does not harm the fetus, as thyroid hormones do not readily cross the placenta. However, maternal weight loss must be prevented, as it may lead to fetal malnutrition. Maternal hypothyroidism also is undesirable, as it compromises placental blood supply and fetal nutrition. In addition, circulating maternal TSH receptor-binding immunoglobulins (thyrotropin binding inhibitory immunoglobulin and thyroid-stim-

TABLE 5-2. Signs and symptoms of Graves' disease.

Symptoms and signs	Thyrotoxic patients (%)	Controls (%)
Symptoms		
Dyspnea on exertion	81	40
Palpitations	75	26
Tiredness	80	31
Preference for cold	73	41
Excessive sweating	68	31
Nervousness	59	21
Increased appetite	32	2
Decreased appetite	13	3
Weight loss	52	2
Weight gain	4	16
Diarrhea	8	0
Constipation	15	21
Excessive menses	3	6
Scant menses	18	3
Signs		
Goiter	87	11
Diffuse	49	11 (slight)
Nodular	32	0
Single adenoma	4	0
Exophthalmos	34	2
Lid lag	62	16
Hyperkinesis	39	9
Finger tremor	66	26
Sweating hands	72	22
Hot hands	76	44
Auricular fibrillation	19	0
Regular pulse over 90	68	19
Average pulse (beats/minute)	100	78

ulating immunoglobulins) cross the placental barrier and may produce fetal or neonatal hypothyroidism or thyrotoxicosis.

Three treatment regimens are available for the nonpregnant woman with Graves' disease: antithyroid drugs, radioactive iodine, and surgery. During pregnancy the options are fewer, and more care must be utilized during treatment. Radioiodine therapy is contraindicated, as it crosses the placenta and has been known to ablate the fetal thyroid gland.[3]

Drug therapy with the thionamide drugs is the first line of treatment. Most physicians prefer to use propylthiouracil (PTU) for treatment with or without added T_4 and to follow the thyrotoxic, pregnant woman carefully. Iodide, which increases the risk of fetal goiter, usually is not necessary. If possible, the PTU dose is kept below 300 mg/day through 7 months and 150 mg/day during the last 6 to 8 weeks. Methimazole has been reported to cause fetal skin defects but is an acceptable alternative drug in women with PTU side effects. Both PTU and methimazole are known to cross the placenta and to affect the fetal thyroid gland as well. The danger, of course, is overtreatment, which might result in fetal hypothyroidism and the development of a goiter, which in turn might lead to tracheal compression.[4]

Subtotal thyroidectomy is also an acceptable treatment for maternal thyrotoxicosis during pregnancy.[5] Usually it is postponed until the second trimester when operation is safest. Antithyroid drug therapy is used to prepare the patient to a euthyroid state. T_4 can be used to prevent hypothyroidism. Propranolol may be useful in helping to prepare the pregnant thyrotoxic woman for surgery, but this drug does not control the hypermetabolic state, and if used near term it may affect uterine contractions and produced neonatal depression.

Subtotal thyroidectomy is followed by rapid resolution of the thyrotoxic state. PTU and propranolol can be stopped at the time of operation in most cases.

Thyroxine therapy is continued during the remaining pregnancy to prevent maternal hypothyroidism. Complications of operation include hypoparathyroidism and recurrent nerve injuries, the incidence of both of which is low in the hands of an experienced surgeon.

Neonatal Graves' Disease

Neonatal Graves' disease may occur in the newborn of a mother with relatively high concentrations of thyroid-stimulating immunoglobulins (TSI). In most instances the disease is transient, lasting 1 to 3 months; the half-life of TSI in the neonate is approximately 2 weeks. The sex ratio for infants with neonatal Graves' disease is 1 : 1, in contrast to the 6 : 1 to 8 : 1 predominance of involved women with adult-onset disease. In some instances neonatal Graves' disease has a delayed onset, a prolonged course, or both. Infants with high transplacental TSI levels can have relatively prolonged disease. Infants born of women with euthyroid Graves' disease or hypothyroid Hashimoto's thyroiditis also have been reported with neonatal Graves' disease due to maternal transfer of TSI during pregnancy. Transient hypothyroidism in the neonate is due to transfer of thyrotropin binding inhibitor immunoglobulins (TBII), which block the TSH receptor. Thus the status of the newborn infant of a woman with autoimmune thyroid disease is the net effect of transplacentally acquired TSI, TBII, and antithyroid drugs.

Thyroxine secreted by the fetus is converted largely to the inactive analog, reverse T_3. After birth, thyroid gland secretion and tissue T_4 to T_3 conversion are stimulated, and postnatal hypertriiodothyroninemia can lead to progressive clinical manifestations of hyperthyroidism over 5 to 10 days. If the intrauterine TSI effect is marked, increased fetal thyroidal T_3 secretion may occur with intrauterine or early neonatal hyperthyroidism. Fetal tachycardia in a pregnant woman with Graves' disease has been considered an indication for maternal antithyroid drug treatment directed at the fetal hyperthyroid state. Thus careful monitoring of the fetal heart rate in women with Graves' disease is desirable.

Neonatal Graves' disease carries a mortality approximating 20% and thus requires vigorous therapy with iodides and PTU. The status of all infants of women with autoimmune thyroid disease must be carefully evaluated at birth and carefully followed during the neonatal period to prevent prolonged hypothyroidism or to provide early treatment of neonatal Graves' disease.[6]

Hyperthyroidism Caused by Trophoblastic Tumors

Trophoblastic tumors, either benign hydatidiform mole or malignant choriocarcinoma, are uncommon complications of normal pregnancy. An excellent review of these disorders has been written by Fisher.[7] Hydatidiform mole occurs in about 1 in 2000 pregnancies in the United States. Choriocarcinoma is less common, 1 in 60,000 pregnancies, and follows hydatidiform mole in about 40% of cases. Molar pregnancy is said to be tenfold more prevalent in Asian countries. Thyroid hyperfunction is known to occur in women with trophoblastic tumors, particularly hydatidiform mole, and it is now clear that the hyperthyroid state is due to elaboration of a humoral thyroid stimulator by the tumor. The identity of the trophoblastic thyroid stimulator is not entirely clear. Trophoblastic tumors, like normal placental tissue, produce hCG, which has inherent thyroid-stimulating activity. Most, but not all authors have reported a correlation between the blood or urinary levels of hCG and the degree of stimulation in patients with trophoblastic disease. Nonetheless, there is the possibility that trophoblastic tumors also secrete a chorionic thyrotropin (hCT) distinct from hCG.

The degree of hyperthyroidism in women with trophoblastic tumors is variable. Severe thyrotoxicosis is uncommon. This tendency to covert rather than overt hyperthyroidism in trophoblastic disease remains unexplained, although some investigators believe that, with time, clinical thyrotoxicosis does occur.

Trophoblastic thyrotoxicosis subsides rapidly with removal of the tumor. The hCG values fall within 48 to 72 hours, and elevated serum T_4 and T_3 concentrations return to normal within 3 to 5 days. Clinical thyrotoxicosis also responds rapidly after chemotherapy of responsive metastatic choriocarcinoma. Thus thyroid ablation with radioiodine for trophoblastic thyrotoxicosis is not indicated. The use of propranolol with or without iodide, an antithyroid drug, or both usually suffices.

Postpartum Thyroid Dysfunction

Up to 5% of women in the United States have either transient hypothyroidism or transient thyrotoxicosis postpartum. Both diseases appear to be due to activation of an underlying autoimmune thyroiditis. A worsening of Graves' thyrotoxicosis postpartum has also been reported.

Nodules of the Thyroid

It is estimated that 4% of the American population have a nodular thyroid gland. Thyroid nodules are more prevalent in women than in men. Hence it is not surprising that many thyroid nodules are diagnosed during pregnancy. It is important that each solitary nodule be properly evaluated even during pregnancy because of the possibility that it may represent a cancer of the thyroid.

Diagnostic Tests for Thyroid Cancer

Except for the fact that isotope scans of the thyroid are not performed during pregnancy, the work-up of a nodule proceeds in the usual way. The lack of isotope scanning is not important, as in the author's opinion the use of needle aspiration with cytologic evaluation offers the most important facet of diagnosis and it can readily be done during pregnancy.

It is important to obtain a careful history and to conduct a physical examination. A history of low dose irradiation to the neck is perhaps the most important fact that can be ascertained. On physical examination most cancers are palpated as hard, irregular lumps. Enlarged, abnormal lymph nodes in the lateral neck or adjacent to the thyroid strongly suggest malignancy.

Thyroid function tests are not useful as diagnostic tests, as most patients with thyroid cancer are euthyroid. However, in the presence of hyperthyroidism, a nodule is less likely to be a cancer.

B-Mode ultrasonography can be used to differentiate solid from cystic lesions. Because most cancers are solid, this technique can be helpful. Core needle biopsy of the thyroid utilizing a Vim-Silverman or a Tru-cut needle yields a core of tissue that can be examined microscopically. However, cytologic evaluation of the aspirate of a nodule obtained by 22-gauge needle (fine needle aspiration) has gained greater usage. Although these techniques are helpful, it must be pointed out that false-negative results have been obtained using either core needle biopsies or needle aspiration cytology. False-positive results are less common.

In summary, of greatest concern are thyroid nodules that are solitary, are recent in origin, and appear solid on ultrasonography. A history of low dose irradiation to the head and neck alerts one to the fact that a cancer of the thyroid is more likely. Finally, a definitive diagnosis of cancer can be made preoperatively by needle aspiration of the nodule with cytologic examination.

Four types of thyroid cancer are known: papillary, follicular, medullary, anaplastic. In most women of childbearing age, papillary cancer is most likely and carries an excellent prognosis. Follicular carcinoma is less likely in young women, and anaplastic thyroid cancer is rare in this age group. Medullary carcinoma must be considered, especially if a familial history is present. Remember that pheochromocytomas and primary hyperparathyroidism accompany medullary carcinomas as part of the multiple endocrine neoplasia, type IIa (MEN-IIa) syndrome (Table 5-3). In the MEN-IIB syndrome, pheochromocytomas, a marfanoid habitus, and neuromas of the tongue, buccal mucosa, and conjunctiva are present (Fig. 5-1).

TABLE 5-3. Characteristics of the MEN-II syndromes.

Type IIa (MEN-IIa)
Bilateral medullary carcinoma of the thyroid gland
Pheochromocytoma(s)
Parathyroid hyperplasia
No specific phenotype
Familial inheritance as a mendelian autosomal dominant trait
Variable rate of MCT[a] progression, more commonly indolent
Type IIb (MEN-IIb)
Bilateral medullary carcinoma of the thyroid gland
Pheochromocytoma(s)
Parathyroid disease (rare)
Specific phenotype
Can occur in mendelian autosomal dominant pattern but more commonly is not familial
Generally MCT progresses rapidly

[a]MCT = Medullary Carcinoma of the thyroid gland.

Details of treatment for thyroid cancer are beyond the scope of this discussion; but it must be noted that if thyroid cancer is diagnosed during pregnancy surgical treatment is performed without great delay, especially during the second trimester. Thyroxine therapy is given postoperatively to suppress maternal TSH levels. Radioiodine therapy, if necessary, is withheld until the postpartum period.

Primary Hyperparathyroidism

Primary hyperparathyroidism is an uncommon disease of pregnancy. One review cited 102 pregnancies in 73 women with this disease since 1938.[8] Undoubtedly many others are missed clinically or were not reported. In the general population, as many as 1 in 500 patients who enter a hospital have been diagnosed as having this condition. Women outnumber men at least 3 : 1. Although most

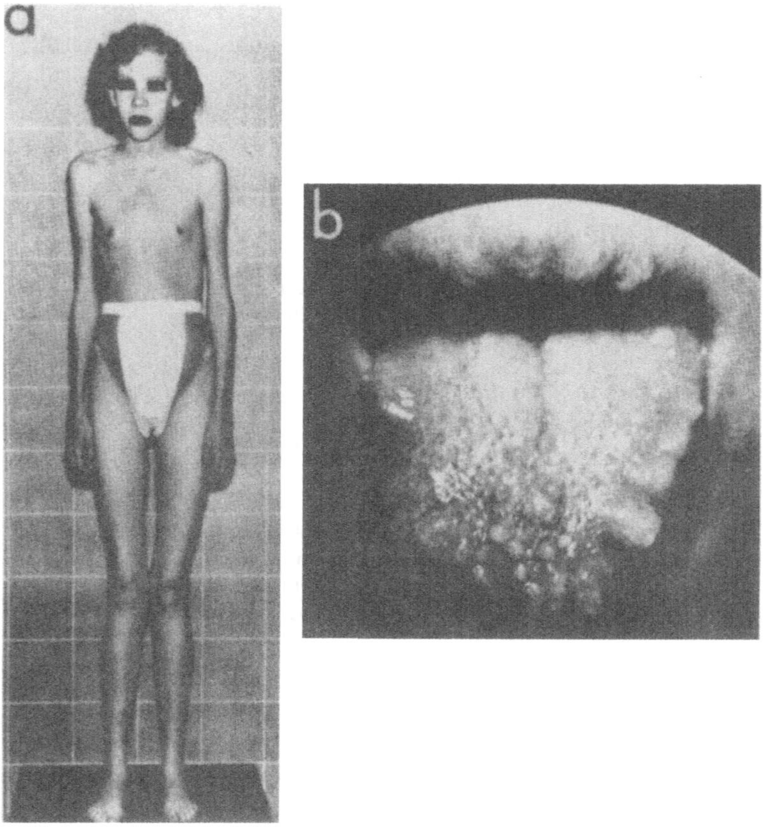

FIGURE 5-1. Patient with MEN-IIB syndrome displaying the characteristic body habitus (a) and multiple neuromas of the tongue (b). (Courtesy of Dr. Glen Sizemore.)

of the women are postmenopausal, many others are within the childbearing years. Diagnosis and treatment during pregnancy are critical, as severe maternal, fetal, and neonatal complications frequently occur. More patients with this condition have been reported in recent years, which suggests increased physician awareness.

General Features

Hyperparathyroidism is a condition in which there is oversecretion of parathyroid hormone. In primary hyperparathyroidism the serum calcium level is elevated, usually due to an adenoma (80%) or hyperplasia (10–20%); rarely, a carcinoma of the parathyroid glands is present. These same pathologic entities, including several carcinomas, have been found during pregnancy as well.[9]

Patients with primary hyperparathyroidism present with hypercalcemia, hypophosphatemia, and increased calcium and phosphate in their urine. These findings have been referred to as "the classic triad" for diagnosis of primary hyperparathyroidism. They are caused by an increase in circulating parathyroid hormone (PTH).

Parathyroid hormone has three basic actions, each of which tends to elevate the serum calcium concentration. It causes bone resorption through osteocytic osteolysis and by increasing osteoclastic activity; it acts on the gut to increase the absorption of calcium; and finally, it acts on the kidneys to increase the tubular resorption of the filtered calcium ion and to decrease the proximal renal tubular resorption of phosphate ion. Thus more calcium and less phosphate is conserved.

The cellular action of parathyroid hormone appears to be mediated through the adenyl cyclase, cyclic adenosine monophosphate system. The importance of vitamin D has also been recognized. It has clearly been shown that the action of parathyroid hormone on the gut, kidney, and bone requires the presence of vitamin D metabolites.

Clues to the Diagnosis of Primary Hyperparathyroidism

Although most of the original patients with primary hyperparathyroidism had severe bone disease, it was soon realized that more patients had renal stones (Table 5-4). Peptic ulcer, pancreatitis, fatigue, hypertension, arthritis, arthralgias, and mental disorders were later recognized to be associated with this condition. Today, because of the common use of multiphasic chemical screening, many patients are diagnosed who are relatively asymptomatic.

Hypercalcemia per se causes a number of clinical manifestations. Among them are nausea, vomiting, and constipation; polyuria and polydipsia; fatigue, weakness, stupor, and coma; depression and psychosis; and finally, metastatic calcification.

Changes in Calcium Homeostasis During Normal Pregnancy

During pregnancy 25 to 30 g of calcium is transferred from the mother to the human fetus through the placenta. Total serum calcium levels as well as ionized calcium values decrease in the mother during pregnancy (Fig. 5-2). This change is accompanied by progressive increases in the values of circulating PTH and the active metabolite of vitamin D [$1,25(OH)_2D$] in the maternal circulation. Calcium ion freely crosses the placenta to the fetus, but PTH and calcitonin do not.[10]

Fetal concentrations of both total and ionized calcium are higher than the maternal levels, which has led some investigators to postulate an active calcium pump. Our studies in sheep demonstrate PTH levels of the fetus to be lower than maternal levels during the later part of pregnancy, whereas calcitonin levels are higher in the fetus than in

TABLE 5-4. Signs and symptoms of hyperparathyroidism.

Sign or symptom	Percent of patients
Hypercalcemia on screening test (routine physical examination)	40[a]
Renal stones	28
Symptoms of hypercalcemia (weakness, constipation, lethargy, polyuria, polydipsia)	
Peptic ulcer disease	5
Hypertension	4
Bone disease	4
Diabetes mellitus	3
MEN-I	3
Thyroid disease	3
Pancreatitis	2
Lump in neck	1

[a]Eight percent of the patients were completely asymptomatic.
(From Clark OH, Way LW. The hypercalcemic syndrome: Hyperparathyroidism. In Friesen SR, Bolinger RE (eds): *Surgical Endocrinology*. Philadelphia, Lippincott, 1978.)

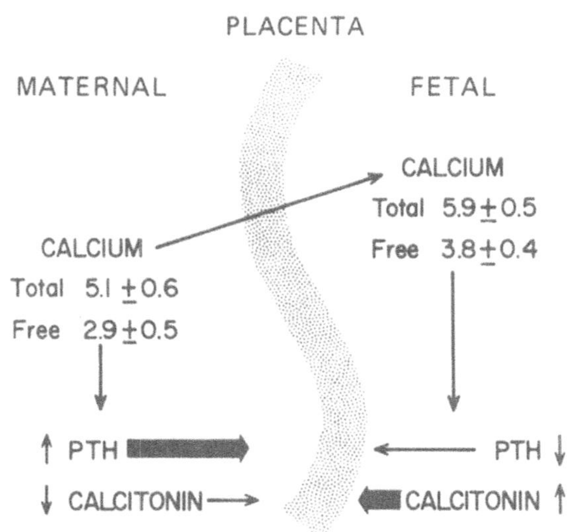

FIGURE 5-2. Relations of calcium (free and total), calcitonin, and parathyroid hormone (PTH) in mother and fetus and the placental transfer of calcium. (From reference 7 with permission.)

the mother. This peptide might aid in mineralization of bone. Vitamin D metabolites are higher in fetal than maternal circulations.[11]

After birth, the neonatal serum calcium level normally drops about 1 mEq/liter during the first 24 to 48 hours of life. As is outlined later, this physiologic decrease in serum calcium is greatly accentuated if the mother has primary hyperparathyroidism.

Maternal Hyperparathyroidism

When primary hyperparathyroidism occurs in the mother, the normal physiologic changes are exaggerated. Increased concentrations of serum calcium, PTH, and $1,25(OH)_2D$ in the mother result in even greater hypercalcemia in the fetus. The further elevation of fetal serum calcium concentration is apparently toxic to the fetus and leads to fetal wastage. In addition, fetal parathyroid development may be retarded or at least suppressed. Hence after birth the expected fall in serum calcium concentration is exaggerated and prolonged, and severe neonatal tetany may result.

Complications

Hyperparathyroidism during pregnancy is associated with an increased rate of complications for both mother and child. Maternal complications, reported in 33 to 48% of such pregnancies, include nephrolithiasis, metabolic bone disease, generalized weakness, hyperemesis gravidarum, pancreatitis, and psychiatric disorders. Fetal complications, reported in up to 80% of pregnancies associated with hyperparathyroidism, include decreased birth weight and an increased rate of spontaneous abortions, stillbirths, and neonatal deaths. Neonatal tetany is common for reasons that have been explained, and permanent hypoparathyroidism of the neonate has also occurred. There is evidence that early detection and treatment reduce complications of the mother's hyperparathyroidism.

Diagnosis

The diagnosis of primary hyperparathyroidism is difficult during pregnancy. Nausea, emesis, fatigue, muscle weakness, mental symptoms, and hypertension are symptoms of hyperparathyroidism and are similar to the symptoms of normal pregnancy or those of other diseases of pregnancy, e.g., toxemia. Skeletal discomfort, weight loss, insomnia, anorexia, constipation, abdominal or back pain, polydipsia, and polyuria may occur normally in pregnancy as well. Frequently hyperparathyroidism is asymptomatic during pregnancy. Thus the diagnosis is often not made until the mother suffers a serious complication, e.g., nephrolithiasis. Unfortunately, the diagnosis of hyperparathyroidism is sometimes not made until a history of unexplained fetal loss has been prominent or until neonatal tetany has occurred.

The diagnosis of hyperparathyroidism during pregnancy is confirmed by laboratory values. An elevated serum calcium concentration, a reduced serum phosphate level, and an increased PTH value are diagnostic. Thus any patient displaying any of the symptoms or complications of hyperparathyroidism during pregnancy must have a serum calcium determination. Furthermore, the frequency of asymptomatic hyperparathyroidism during pregnancy leads many authors to advise measurement of the serum calcium concentration in all pregnancies.

Treatment

The primary treatment of primary hyperparathyroidism during pregnancy is parathyroidectomy, usually during second trimester, when surgery is safest. In recent years, parathyroidectomy has been performed in 16 patients. A graphic representation of one of our patients is shown in Figure 5-3.[12] After operation both fetal deaths and neonatal hypocalcemia were almost totally eliminated. Immediately after parathyroidectomy, maternal serum calcium levels must be maintained by calcium and vitamin

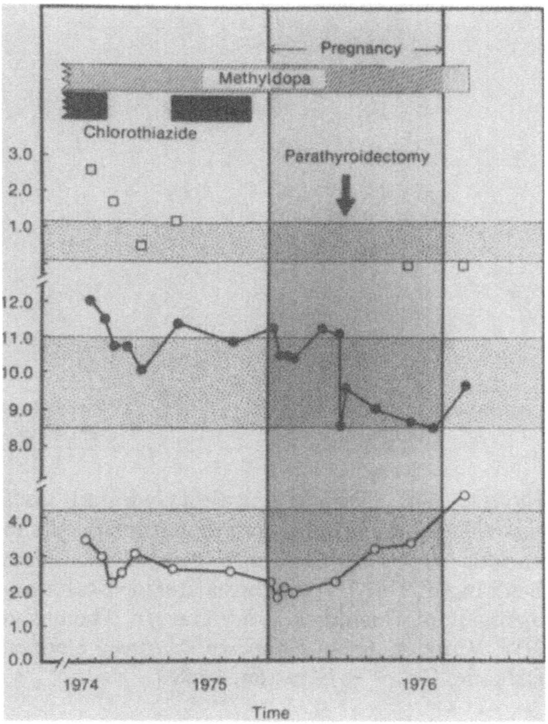

FIGURE 5-3. Clinical course of maternal primary hyperparathyroidism of pregnancy. Serum parathyroid hormone (□), calcium (●), and phosphorus (○) concentrations are depicted before, during, and after pregnancy. Normal values of serum determinations are designated by shaded areas. Antihypertensive therapy time is designated by the cross-hatched bar. Parathyroidectomy (arrow) was performed 21 January 1976. (From reference 12 with permission. Copyright 1977, American Medical Association.)

D replacement, as necessary, to prevent hypocalcemia.

Not all asymptomatic patients have been operated on during pregnancy, however. Some have been treated by oral phosphate therapy, which exposes the mother to the risk of deterioration of renal function. Furthermore, the possibility of "parathyroid crisis" in the mother with severe hypercalcemia is increased after the birth of the child because at that time calcium shunting to the fetus is stopped. Finally, there may be a greater risk of neonatal hypocalcemia and tetany when the mother is treated in this way. A neonatologist must examine and observe all such infants. Cow's milk is avoided because of its high phosphate content. Breast milk or a formula containing low phosphate content and magnesium supplementation is appropriate.

Adrenal Tumors

Adrenal tumors rarely occur during pregnancy, but each is associated with high fetal and maternal mortality if it is not diagnosed and treated appropriately. This discussion includes Cushing syndrome, pheochromocytomas, aldosteronomas, and virilizing tumors of the adrenal.

The normal adrenal gland contains two types of tissue: adrenal cortex and adrenal medulla. The cortex is divided into three portions. The outermost area, or *zona glomerulosa,* is the site of mineralocorticoid production. The central area, or *zona fasciculata,* is the site of glucocorticoid production. The innermost layer, or *zona reticularis,* is the site of sex hormone production. The adrenal medulla is a completely different organ; it is derived from neuroectoderm and is part of the APUD series of cells. This area is a site of epinephrine and norepinephrine production. Thus tumors that arise from different sites of the adrenal gland have different signs and symptoms.

Cushing Syndrome

Cushing syndrome is a collection of signs and symptoms that are due primarily to overactivity of glucocorticoid production (Table 5-5). The same syndrome can be reproduced by exogenous administration of excessive cortisol. Patients typically have a characteristic appearance: moon facies, buffalo hump, central obesity, and thin extremities. Purple striae and easy bruisability are common. Women with this disease manifest loss of scalp hair, acne, facial plethora, and increased body and facial hair. Hyperten-

TABLE 5-5. Signs and symptoms of Cushing syndrome.

Moon facies	Hirsutism
Truncal obesity	Easy bruising
Hypertension	Acne
Muscle wasting	Fungal infections
Buffalo hump	Back pain
Diabetes (latent or overt)	Headache
Osteoporosis	Emotional lability
Plethora	Pathologic fractures
Amenorrhea	Renal calculi
Impotence	Atherosclerosis
Weakness and fatigue	Polyuria
Striae	Hypocalcemic alkalosis

sion, cardiomegaly, and renal calculi may also occur.

Chemistry changes include normal or elevated serum sodium and decreased serum potassium levels owing to the effects of mineralocorticoids on the kidney. There is also a decrease in lymphoid tissue, lymphocytes, and eosinophils and an increase in polymorphonuclear leukocytes in the blood.

Pathophysiology

Normal control of the adrenal cortex is through a servomechanism, whereby ACTH from the pituitary stimulates cortisol secretion from the adrenal cortex and an elevation of circulating cortisol turns off ACTH secretion from the pituitary (Fig. 5-4).

When a *pituitary basophilic adenoma* is present, it secretes an excessive amount of ACTH into the circulation, resulting in stimulation of the adrenal cortex, which becomes hyperplastic, and in a rise of cortisol secretion from the adrenal gland. Thus under these circumstances serum levels of both ACTH and cortisol are elevated (Fig. 5-4).

The *ectopic ACTH syndrome* is caused by autonomous secretion of an ACTH-like molecule from a carcinoma of the lung or elsewhere. In this instance, circulating ACTH is often greatly elevated, more so than with a pituitary microadenoma. In response, serum cortisol levels are also elevated, and the adrenal cortex is stimulated to become hyperplastic.

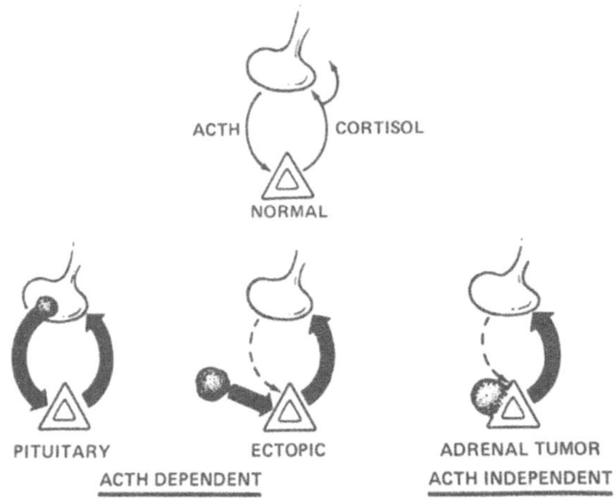

FIGURE 5-4. Normal pituitary-adrenal axis (above) and abnormalities in the pituitary axis in various types of Cushing's syndrome. (From Schteingart DE. The diagnosis and medical management of Cushing's syndrome. In Thompson NW, Vinik AI (eds): *Endocrine Surgery Update.* Orlando, Grune & Stratton, 1983.)

An adrenocortical adenoma or carcinoma secretes cortisol autonomously. In these instances, because the pituitary is normal, ACTH secretion is suppressed. Hence under these circumstances serum ACTH levels are low, and circulating cortisol values are elevated.

Cushing Syndrome During Pregnancy

Pregnancy associated with Cushing syndrome is uncommon because of the associated amenorrhea, suppressed ovulation, and resultant depression of fertility that occurs in 50 to 75% of women. However, some patients with active Cushing syndrome are fertile, and the risk to mother and fetus is high. Spontaneous abortion, premature delivery, or stillbirth occurs in 50 to 60% of these pregnancies. Only 40% of women deliver term infants. In one review 45 pregnancies were reported in 41 women.[13] Fetal mortality was 9% in 11 patients whose Cushing syndrome was diagnosed and treated during pregnancy versus 30% in undiagnosed cases. Premature labor occurred in 25% of the 8 treated patients and in close to 50% of the untreated patients who ultimately had living babies. Maternal mortality was low, but significant morbidity is related to hypertension, emotional lability, diabetes mellitus, weakness, and myopathy. Thus the diagnosis and treatment of Cushing syndrome must be accomplished as early as possible during pregnancy.

Diagnosis

The diagnosis of Cushing syndrome is more difficult during pregnancy because cortisol secretion from the adrenal increases during this period. Circulating ACTH values increase mildly during pregnancy, but it results in a two- to threefold increase in adrenal corticosteroid output. Thus values of serum cortisol, as well as urinary 17-hydroxycorticosteroid and free cortisol, are expected to be higher during pregnancy than in the nonpregnant state.

In general, however, the diagnosis of Cushing syndrome during pregnancy is made by following the same principles as in the nonpregnant state. In normal individuals both ACTH and cortisol are secreted in a circadian rhythm with a diurnal variation, which means that both serum cortisol and ACTH levels are higher during sleep and in the early morning and are lower in the afternoon. Multiple blood levels are assayed to determine if a normal diurnal variation is present. Patients with Cushing syndrome generally exhibit abnormally high plasma cortisol values at early morning and late afternoon as well as a loss of circadian rhythm. These patients also excrete high levels of the metabolic breakdown products of adrenocortical steroids (17-hydroxycorticoids and free cortisol) in their urine.

As described earlier, the plasma ACTH level depends on the underlying pathogenesis of the Cushing syndrome. Plasma ACTH values are elevated when either a pituitary microadenoma or ectopic ACTH secretion from a tumor is present, although the highest values are observed in the presence of the ectopic ACTH syndrome. On the other hand, when an adrenal tumor is present, the plasma ACTH levels are low.

Dexamethasone Suppression Test

Dexamethasone is a potent adrenocortical steroid. When given in a small dose (e.g., 1 or 2 mg daily) it suppresses normal pituitary ACTH secretion and hence suppresses cortisol secretion from the adrenal gland. If a pituitary adenoma is present, on the other hand, ACTH suppression occurs with this drug only at higher doses, e.g., 8 mg daily. Finally, if the ectopic ACTH syndrome or if an adrenal tumor is present, dexamethasone administration does not result in suppression of adrenocortical steroid production. Thus when low doses are used, this test helps to differentiate patients with true Cushing syndrome from others who do not have this condition. When high doses are utilized, pituitary-dependent Cushing syndrome (Cushing Disease) can be differentiated from that due to adrenal tumors. It must be remembered, however, that pituitary ACTH secretion is somewhat more resistant to suppression during pregnancy than in the nonpregnant state.

Localization Studies

After biochemical studies, it is common to perform localizing studies. Computed tomography (CT) scanning or magnetic resonance imaging of the sella turcica is usually done to identify a pituitary adenoma. Although it is possible during pregnancy, CT scanning of the adrenal glands and isotope scanning with ^{131}I-cholesterol to identify adrenal pathology must not be performed because of the risk of radiation damage to the fetus. Ultrasonography may be helpful for identifying large adrenal masses but is not as valuable as the CT scan. Magnetic resonance imaging (MRI scanning) shows great promise, as there is no fetal irradiation. This procedure was used during pregnancy with great efficacy in one of our patients with a pheochromocytoma.[14]

Pathogenesis

By far the most common cause of Cushing syndrome in the *nonpregnant state* is bilateral adrenocortical hyperplasia, which occurs in close to 80% of cases. It is usually the result of a basophilic adenoma of the anterior pituitary gland, which secretes an excessive amount of ACTH. Usually these lesions are small when the disease is first diagnosed. They do not enlarge the sella turcica and are called microadenomas; however, large basophilic and chromophobe adenomas may be encountered. In a few cases, it is postulated that the cause of the Cushing syndrome is hypothalamic-pituitary dysfunction.

Adrenocortical hyperplasia can also be the result of ectopic production of ACTH from a group of tumors found throughout the body (10%). Oat cell and bronchogenic carcinoma of the lung, carcinoid tumor of the lung, appendix, small bowel, and elsewhere, pancreatic islet cell tumors, medullary carcinoma of the thyroid gland, and other neoplasms have been shown to secrete an ACTH-like substance at times. Although many of these tumors are widely metastatic when patients manifest Cushing syndrome, in other instances (e.g., carcinoid or islet cell tumors) the lesions may be resectable and even curable.

In about 20% of cases, Cushing syndrome is caused by a unilateral adrenal tumor. Usually these tumors are benign adenomas, but some adrenocortical carcinomas also result in this syndrome.

Of interest is the fact that of the 41 women who were diagnosed as having Cushing syndrome during pregnancy, 23 (56%) had an adrenal tumor. Five of these lesions (12% of all cases) were adrenal carcinomas, although the malignant diagnosis of one tumor has been questioned. Thus there appears to be a preponderance of adrenal tumors and carcinomas of the adrenal in these pregnant women.[13] The exact reasons for these differences remain to be clarified.

Treatment

For nonpregnant women the following general rules apply: If an adenoma or carcinoma of the adrenal cortex is the cause of the Cushing syndrome, this tumor is operated on and removed. Similarly, if the cause of the Cushing syndrome is ectopic production of ACTH, that tumor (e.g., cancer of the lung) is removed if possible; otherwise, the overactive adrenocortical state can be treated medically.

Controversy still exists as to what is the proper treatment for hyperplasia due to a pituitary microadenoma. The usual treatment in the past was bilateral total adrenalectomy. This therapy necessitates a lifetime of adrenocortical steroid replacement, however. Irradiation to the pituitary has resulted in some remissions of the disease. Today, however, most endocrinologists favor transsphenoidal resection of the microadenoma if a neurosurgeon skilled in this technique is available. In the best of hands, about 80 to 90% of patients are cured of Cushing syndrome by this technique. Long-term results, including the incidence of recurrence, are not known. The advantages of this technique are the low morbidity and mortality associated with the operative procedure in these sick patients and the fact that no long-term medical therapy is required if the microadenoma can be removed and the rest of the anterior pituitary left intact.

Is operation safe during pregnancy? Eleven patients have been treated: Unilateral adrenalectomy for adenoma was performed in five patients, bilateral adrenalectomy in four, pituitary irradiation in one, and medical therapy with metyrapone (a cortisol inhibitor) in one.[13] Fetal loss and premature labor seemed to be lessened by these procedures, and no maternal mortality occurred. No patient was treated by transsphenoidal resection of the basophilic microadenoma; however, this approach seems reasonable for bilateral hyperplasia.

Preoperative and Postoperative Management

Patients with an adrenal adenoma, carcinoma, or hyperplasia that has resulted in Cushing syndrome require treatment with adrenocortical steroids at the time of an adrenal operation. If bilateral total adrenalectomy is performed, the patient requires steroid therapy for the rest of her life. After removal of an adrenal adenoma or carcinoma, the opposite adrenal gland is suppressed, but it regains its function with time in most instances. Although such patients require maintenance steroids for varying periods, eventually most require no replacement therapy.

On the day of operation, cortisol or cortisone 300 mg is given, one-half intramuscularly and one-half intravenously. Over the next few days these amounts are rapidly decreased. Oral cortisone acetate can be given as soon as oral feedings are resumed. A maintenance dose is usually between 37.5 and 50.0 mg of cortisone per day, given as 25.0 mg in the morning and 12.5 mg in the evening. Many patients with Cushing syndrome have to be weaned to this level gradually, however, or else they have severe symptoms. We also add 9α-fluorocortisone (Florinef) at about 0.1 mg per day and adjust it as needed. This agent is a potent mineralocorticoid and results in the retention of sodium and excretion of potassium from the body. Always instruct the patient to increase the oral dose of cortisone during times of stress, infections, or trauma. The cortisone is given intramuscularly if vomiting occurs. It is important to realize, however, that patients with bilateral adrenalectomy lead a long and perfectly healthy life so long as they take their medication.

It is important as well to monitor the newborn infant for any signs of an addisonian crisis or adrenocortical insufficiency. It is especially true if the Cushing syndrome was untreated in the mother. If any signs of trouble occur, glucocorticoid replacement is given immediately to the neonate.[15]

Aldosteronoma

During the mid-1950s, Conn described the syndrome of primary hyperaldosteronism, which bears his name. Patients with an aldosteronoma demonstrate elevated blood pressure, absence of edema, and low serum potassium concentration; frequently they manifest polyuria, polydipsia, muscle weakness, and occasionally tetany due to low calcium and magnesium levels secondary to a metabolic alkalosis. They also have a diabetic glucose tolerance curve in many instances. The removal of an aldosteronoma leads to correction of the hypertension in most instances.

Aldosteronomas are almost always benign. They are much smaller than most of the other adrenocortical tumors and are generally 2 cm in diameter or less. More than 90% of the tumors are unilateral.

Aldosteronoma During Pregnancy

Primary aldosteronism during pregnancy is rare. Pregnancy is normally associated with an increase in both aldosterone and renin. When an aldosteronoma is present, with or without pregnancy, elevated aldosterone but decreased renin levels are found, thereby facilitating diagnosis. Primary hyperaldosteronism during pregnancy is associated with severe hypertension and hypokalemia. This tumor must be included in the differential diagnosis of eclampsia and other hypertensive states. The hypertension can be life-threatening and can result in congestive heart failure. Although adrenalectomy with removal of the tumor is curative, the consensus is that aggressive medical management must be performed during pregnancy, followed by definitive surgical treatment after delivery. Medical treatment includes aldosterone blockade and aggressive potassium replacement. A peripheral vasodilator and α- and β-blockades are also sometimes needed. Care must be taken to closely monitor serum potassium levels in the mother to prevent fetal hyperkalemia.[16]

Pheochromocytoma

Pheochromocytomas are tumors of the sympathetic nervous system. Most arise from the adrenal medulla, but 10 to 15% originate from the sympathetic nervous chain anywhere from the neck to the pelvis. The term pheochromocytoma, meaning "gray tumor", is appropriate, as this tumor is not yellowish-tan like the adrenocortical tumors. Pheochromocytomas are potentially dangerous tumors, as striking elevations of blood pressure can occur that sometimes lead to a stroke and death if the patient is not diagnosed early and treated appropriately.

Incidence

Pheochromocytoma is a rather rare tumor. The exact incidence of this disease is uncertain. Studies performed on large autopsy series indicate a possible incidence of 0.1%. Moreover, in 0.05 to 0.10% of patients with diastolic hypertension, a pheochromocytoma is thought to be the cause of the altered blood pressure. There is no predilection for sex. This tumor occurs at all the ages, but there is an increased incidence during the fourth and fifth decades.

Clinical Signs and Symptoms

Pheochromocytomas are clinically dangerous tumors because of the catecholamines—epinephrine and norepinephrine—they secrete. Patients with pheochromocytomas exhibit hypertension, either constant (49%), paroxysmal (42%), or both. Attacks of hypertension are often accompanied by excessive perspiration, headaches, palpitations, tachycardia, pallor, nausea, trembling, nervousness, anxiety, chest pain, or epigastric pain. Orthostatic hypotension is often prominent. These symptoms are due to excessive circulating catecholamines. Norepinephrine is an α-adrenergic agonist, and its primary manifestations are vasoconstriction and hypertension. Epinephrine acts primarily as a β-adrenergic stimulator that causes vasodilation, an increase in the heart rate, an increase in ventricular conductivity, dilation of the bronchial musculature, and an increase in blood glucose levels. Thus the symptoms may differ in each patient according to which catecholamine is excessively secreted. A listing of symptoms and signs of pheochromocytoma is presented in Table 5-6.

Associated Diseases

In about 10% of pheochromocytomas a familial history can be identified. In such cases medullary cancer of the thyroid (or C-cell hyperplasia) as well as primary hyperparathyroidism due to parathyroid hyperplasia are associated with the pheochromocytomas—an entity called multiple endocrine neoplasia, type IIa (MEN-IIa) (Table 5-3). In these cases the pheochromocytomas are usually bilateral; often they are multiple, and usually adrenal medullary hyperplasia is present as well. In some of these cases adrenal medullary hyperplasia is the cause of hypertension, and no true tumors are present. In the MEN-II syndrome with hypertension a bilateral adrenalectomy is necessary because if a unilateral resection is done hypertension and pheochromocytomas occur later in the other adrenal gland. Some investigators favor a unilateral adrenalectomy in children, with the total adrenalectomy being completed at a later date.

The coexistence of pheochromocytoma, medullary cancer of the thyroid, mucosal neuromas, ganglioneuromas, and marfanoid habitus establishes another familial entity, called MEN-IIb. These individuals have a specific appearance (Fig. 5-1). Their bilateral pheochromocytomas and adrenal medullary hyperplasia also require a bilateral adrenalectomy for cure. With these MEN syndromes, the pheochromocytomas are always operated on first, if they occur synchronously, because of the risk of these lesions.

An association with neurofibromatosis is present in 5% of pheochromocytomas. In such a patient a somatostatin-secreting tumor of the ampulla of Vater has occasionally

TABLE 5-6. Signs and symptoms of pheochromocytoma.

Signs and symptoms	% of patients (approx.)
Symptoms	
Persistent hypertension	65
Paroxysmal hypertension	30
Headache	80
Sweating	70
Palpitation, nervousness	60
Pallor of face	40
Tremor	40
Nausea	30
Weakness, fatigue	25
Weight loss	15
Abdominal or chest pain	15
Dyspnea	15
Visual changes	10
Constipation	5
Raynaud's phenomenon	5
Convulsions	3
Polydipsia, polyuria	
Puffy, red, cyanotic hands	
Signs	
BMR > +20%	50
Fasting blood glucose >120 mg/dl	40
Glycosuria	10
Eye ground changes	30

From Hume DM. Pheochromocytoma. In Astwood EB, Cassidy CE (eds): *Clinical Endocrinology,* Vol. II. Orlando, Grune & Stratton, 1968; 519.

been described. Another rare association is with von Hippel-Lindau disease (cerebellar hemangioblastoma and retinal angioma).

Pheochromocytoma During Pregnancy

Pheochromocytoma is an uncommon, but potentially lethal, complication of pregnancy. Only about 150 cases have been reported to date, but most were undiagnosed until a severe complication occurred. When unrecognized prior to the onset of labor and delivery, this condition carries a risk of mortality to both mother and fetus of more than 50%. Only 48 patients with this tumor have been diagnosed prior to delivery; these cases have reduced maternal mortality (18%), but fetal mortality remains high at 50%. Death is caused by hypertensive crisis, shock, and convulsions. Thus early diagnosis is imperative.[17]

The differential diagnosis during pregnancy includes essential hypertension and eclampsia. Differentiation from eclampsia is complicated by the fact that 25% of patients with pheochromocytoma are also proteinuric. Thus the history assumes an important role, as recurrent flushing, sweating, anxiety, and faintness often precede or accompany hypertension and serve as clues to the clinician. Visual complaints and convulsions are stated to be more common during pregnancy than in nonpregnant patients with pheochromocytoma.

Diagnosis

The most important element in the diagnosis of a pheochromocytoma is for the clinician to think of this possibility. The diagnosis is made with certainty in the pregnant as well as the nonpregnant woman by the demonstration of elevated levels of plasma and urinary catecholamines or their metabolites. The easiest and safest diagnostic tool is to demonstrate elevated catecholamine, metanephrine, and vanillylmandelic acid (VMA) levels in the urine. It must be remembered that some drugs interfere with measurement of the catecholamines and thus cause falsely elevated values.

Pharmacologic tests have been investigated over the years that inhibit or cause the release of catecholamines into the circulation by the tumor. Patients with pheochromocytomas are sensitive to α-adrenergic blockade. An inhibitory test with 1 mg of phentolamine (Regitine) may result in a rapid decrease of blood pressure and even shock if a pheochromocytoma is present. On the other hand, provocative tests using glucagon, histamine, or tyramine may result in severe hypertension, which can be dangerous. Hence these pharmacologic tests are performed only under controlled situations, with a physician in attendance and with the availability of proper treatment for hypotension or hypertension.

Blood catechol levels can be helpful but

sometimes are high in a normal person because of fear or stress. In such instances a *clonidine suppression test* has proved useful for distinguishing neurogenic hypertension from that due to a pheochromocytoma. Clonidine acts centrally and suppresses the release of neurologically mediated catecholamines, i.e., those released because of fear, stress, anxiety, and so on. Because the release of catecholamines in pheochromocytoma is not mediated through activation of the sympathetic nervous system, clonidine does not suppress the release of epinephrine and norepinephrine into the circulation in patients who have a pheochromocytoma. Thus when a pheochromocytoma is present, the plasma catecholamine levels remain elevated after clonidine is given.

Other laboratory findings include hyperglycemia, an effect of elevated circulating epinephrine values. The hematocrit and blood calcium levels are sometimes elevated because of hypovolemia and hemoconcentration, which are usually present. In cases of MEN-IIa, elevated circulating calcium, calcitonin, and parathyroid hormone values may be present.

Two of the most useful localization tests are CT scanning and isotope scanning with ^{131}I-metaiodobenzylguanidine (MIBG scanning). CT scanning can identify the precise location of the tumor in 95% of cases. MIBG is taken up selectively by adrenergic cells and is useful for localizing many pheochromocytomas that are extraadrenal and difficult to find. Unfortunately, neither of these studies can be utilized during pregnancy because of the harmful effects of radiation on the fetus. Utrasonography can localize a large adrenal mass at times.

The greatest advance for diagnosis of this tumor during pregnancy as well as in the nonpregnant state, however, is magnetic resonance imaging (MRI). This technique was helpful for finding an extraadrenal pheochromocytoma located at the aortic bifurcation in a pregnant woman we recently treated at the University of Chicago Medical Center (Fig. 5-5).

Management

Once a diagnosis has been made, the patient must be hospitalized. In the past most pheochromocytomas were removed soon after they were diagnosed because of the severe risk to mother and fetus. All such individuals are prepared for 7 to 10 days or longer with α-adrenergic blockade. β-Blockade is used if tachycardia or arrhythmias are present. Early operation is still appropriate therapy. However, we and others have successfully managed several pregnant women without immediate removal of the tumor, using both α-blockade (phenoxybenzamine) and β-blockade (propranolol) to control the effects of circulating catecholamines. The morbidity and mortality of the fetus are directly related to those of the mother. Labor and vaginal delivery can cause a surge in catecholamines and must be avoided. Near term, when the fetus is large enough, cesarean section and tumor extirpation are performed.[18]

An excellent anesthesiologist is imperative, as the patient must be monitored extensively during operation. An arterial catheter for measuring blood pressure is used, as well as a Swan-Ganz or central venous pressure catheter, a Foley catheter, and electrocardiography leads. Transesophageal echocardiography to evaluate the state of the heart has been used with success in the authors' institution.[19]

The anesthesiologist must be ready to control the dramatic changes in blood pressure that might occur. Elevations in blood pressure are usually treated by phentolamine (Regitine) or nitroprusside. A severe fall in pressure following clamping of the adrenal veins is treated by rapid volume expansion, norepinephrine, or both.

Once the adrenal tumor or paraganglioma is located, the surgeon must handle the tumor gently to avoid a massive release of catecholamines into the bloodstream, as might happen when the tumor is pulled or treated roughly. Immediate control of the vascular supply of the pheochromocytoma is desirable

FIGURE 5-5. Extraadrenal pheochromocytoma at the aortic bifurcation (crosses) visualized by MRI in a patient during her seventh month of pregnancy. The gravid uterus is seen above the mass. The vena cava is compressed below the tumor. (A) Sagittal view. (B) Axial view. (From reference 14 with permission of the Radiological Society of North America.)

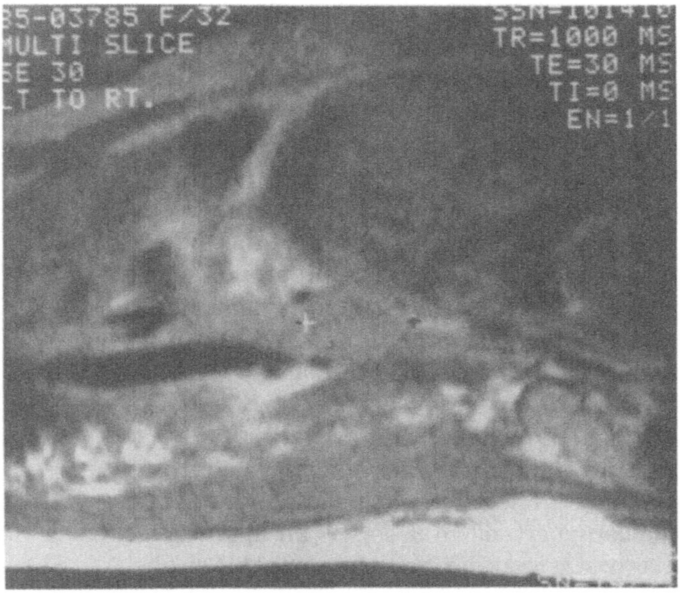

A

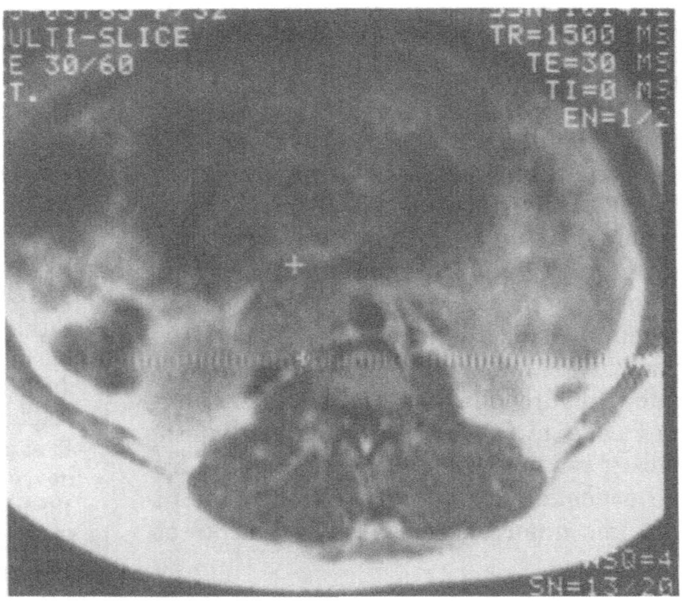

B

if possible. After removing an adrenal tumor, a careful exploration of the other adrenal gland and all the sites of extraadrenal pheochromocytoma is performed. Bilateral sites and extraadrenal sites of pheochromocytoma occur in 10 to 15% of adult patients. About 10% of adrenal tumors are malignant, whereas about 40% of extraadrenal lesions are cancerous.

Follow-up

If the fetus survives the pregnancy, a normal neonatal course can be expected. There have been no reports of birth abnormalities related to α-adrenergic blockade.

More than 75% of patients who are operated on for a pheochromocytoma are cured of hypertension. When malignant lesions are present, debulking operations are performed, and as much tumor as possible is removed. Neither chemotherapy nor external irradiation is useful for metastases. It is hopeful that high-dose ^{131}I-MIBG may be useful in the future; however, preliminary results have not been totally favorable. All patients with metastases that cannot be removed must be treated for a lifetime by adrenergic blockade with phenoxybenzamine and possibly with propranolol to control the hypertension and arrhythmias. α-Methyltyrosine blocks the synthesis of catecholamines and may be useful as well. Finally, a patient who has been normotensive after removal of a pheochromocytoma and later becomes hypertensive deserves an aggressive diagnostic work-up, as sometimes this situation is caused by the presence of a second benign tumor that can be easily removed, rather than by a metastatic lesion.

Virilizing Tumors

Several benign virilizing tumors of the adrenal have been described during pregnancy. In one case virilization was associated with hyperaldosteronism. Severe virilization with female pseudohermaphroditism of the offspring occurred in several cases but only minimal virilization of the mother occurred. The virilization appears to be caused by elevated dihydroepiandrosterone, which is augmented by hCG. Successful treatment of the mother can be achieved by excising the adrenal adenoma, but the virilization of the infant has unfortunately already had its effect.[20]

Acknowledgments. The authors wish to thank Ms. Joey Czerwonka for her help in typing this manuscript. The work reported was supported in part by a generous grant from the Nathan and Frances Goldblatt Society for Cancer Research.

References

1. Malkasian GD, Mayberry WE. Serum total and free thyroxine and thyrotropin in normal and pregnant women, neonates and women receiving progesterones. Am J Obstet Gynecol 1970;108:1234–1238.
2. Mestman JH. Management of thyroid diseases in pregnancy. Clin Perinatol 1980; 7:371–385.
3. Stoutler SS, Hamburger JT. Inadvertent ^{131}I therapy for hyperthyroidism in the first trimester of pregnancy. J Nucl Med 1976; 17:146–149.
4. Mujtaba Q, Burrow GN. Treatment of hyperthyroidism in pregnancy with propylthiouracil and methimazole. Obstet Gynecol 1975; 46:282–286.
5. Talbert LM, Thomas CG, Holt WA, et al. Hyperthyroidism during pregnancy. Obstet Gynecol 1970;36:779–785.
6. Fisher DA. Thyroid function in the fetus and newborn. Med Clin North Am 1975;59:1099–1107.
7. Fisher DA. Fetal endocrinology: endocrine disease in pregnancy. In Degroot LJ, Cahill GF, Martini L, et al (eds): Endocrinology. New York, Grune & Stratton, 1979;1649–1663.
8. Kristoffersson A, Dahlgren S, Lithner F, et al. Primary hyperparathyroidism in pregnancy. Surgery 1985;97:326–330.
9. Parham GP, Orr JW. Hyperparathyroidism secondary to parathyroid carcinoma in pregnancy: a case report. J Reprod Med 1987; 32:123–125.
10. Pitkin RM. Calcium metabolism in pregnancy and the perinatal period: a review. Am J Obstet Gynecol 1985;151:99–109.
11. Kaplan EL, Burrington JD, Klementschitsch P, et al. Primary hyperparathyroidism, pregnancy and neonatal hypocalcemia. Surgery 1984;96:717–722.
12. Gaeke RF, Kaplan EL, Lindheimer MD, et al. Maternal primary hyperparathyroidism of pregnancy: successful treatment by parathyroidectomy. JAMA 1977;238:508–509.
13. Bevan JS, Gough MH, Gilmer MDG, et al. Cushing's syndrome in pregnancy: the timing of definitive treatment. Clin Endocrinol (Oxf) 1987;27:225–233.
14. Greenberg M, Moawad A, Wieties BM, et al. Extraadrenal pheochromocytoma: detection during pregnancy using MR imaging. Radiology 1986;161:475–476.
15. Kreines K, Devaux WD. Neonatal adrenal insufficiency associated with maternal Cushing's syndrome. Pediatrics 1971;47:516–519.

16. Lotgering FK, Derkx FM, Wallenburg HC. Primary aldosteronism in pregnancy. Am J Obstet Gynecol 1986;155:986–988.
17. Mulcahy D, O'Dwyer WF, Carmody M, et al. Pheochromocytoma presenting in pregnancy. Ir J Med Sci 1984;153:389–391.
18. Venuto R, Burstein P, Schneider R. Pheochromocytoma: antepartum diagnosis and management with tumor resection in the puerperium. Am J Obstet Gynecol 1984;150:431–432.
19. Roizen MF, Schreider BD, Hassan SZ. Anesthesia for patients with pheochromocytoma. Anesthesiol Clin North Am 1987;5:269–275.
20. O'Leary TJ, Ooi TC, Miller JD. Virilization of two siblings by maternal androgen-secreting adrenal adenoma. J Pediatr 1986;109:840–842.

6 Gastroduodenal, Hepatobiliary, and Pancreatic Emergencies During Pregnancy

ANDREA PIETRABISSA and FABRIZIO MICHELASSI

Less than 1% of pregnant women undergo major nonobstetric surgery.[1] However, in these patients the diagnosis of acute abdomen can be difficult owing to the physiologic changes induced by pregnancy. As pregnancy progresses, the enlarged uterus becomes an obstacle to physical examination, and the abdominal wall becomes stretched and gradually less reactive. As a consequence, the usual diagnostic signs of tenderness, guarding, and rigidity become more difficult to elicit. The location where signs are expected is changed by the enlarging uterus. Nausea is so common during the first trimester that it loses its significance as an early symptom of the acute abdomen. The relative leukocytosis of pregnancy makes the interpretation of a somewhat increased white blood cell count unreliable. All these factors make the diagnosis of an acute abdomen in the pregnant patient a diagnostic challenge.

Peptic Ulcer Disease

The incidence and severity of peptic ulcer disease is markedly reduced during pregnancy, probably owing to the effect of estrogen. In addition, although research results are conflicting, it appears that acid output is reduced.

When ulcers do arise, the diagnostic and therapeutic alternatives are limited, and the complications can be major. Signs and symptoms may be modified or confused by physiologic changes of pregnancy. Increased levels of progesterone cause delayed gastric emptying and a decrease in the tone of the lower esophageal sphincter. Symptoms of peptic ulcer disease may therefore be dismissed as secondary to common esophageal reflux. To avoid the risks of radiation to the fetus, upper gastrointestinal endoscopy becomes the diagnostic procedure of choice in the gravid woman.

Most ulcers in the pregnant patient heal uneventfully. In fact, fewer than 100 pregnant patients have been reported with complications of peptic ulcerations, e.g., perforation, hemorrhage, and gastric outlet obstruction.[2–5] Most of these complications have occurred during the third trimester of pregnancy. The pathophysiologic changes underlying the predilection for the third trimester have not been elucidated. It has been postulated that the systemic complications of late pregnancy, e.g., toxemia, may predispose the pregnant woman to ulcer disease.

Hemorrhagic peptic ulcers pose a more serious threat to the pregnant patient than to the nongravid patient. During pregnancy more than 30% of total blood volume loss in 12 hours or less often leads to fetal death and maternal morbidity. Hence significant bleeding secondary to peptic ulcer disease requires urgent treatment, and prompt surgical intervention is recommended. At operation, ligation of the bleeding vessel with a nonabsorbable suture plus pyloroplasty with truncal vagotomy is indicated instead of resection, which causes higher morbidity and mortality for the patient and the fetus.

The treatment of established gastric outlet obstruction is surgical, but it must be preceded by decompression of the dilated stomach, correction of fluid and electrolyte imbalance, and establishment of parenteral alimentation. It has been suggested that the dilated gastric musculature as a result of the pyloric obstruction is more susceptible to the atonic effect of vagotomy, and prolonged gastric stasis has followed this operation. Gastric atony may also be made worse by the elevated levels of progesterone. Therefore the choice of operation lies between a subtotal gastrectomy without vagotomy and a vagotomy with drainage associated with a temporary gastrostomy to permit gastric decompression until adequate tone is regained. We prefer the first option to limit the time the patient may require parenteral alimentation.

Gastroduodenal perforation during pregnancy has a high mortality rate, partially because of the difficulty of establishing the proper diagnosis. In 1955 Durst and Klieger[6] reported 17 deaths from perforated peptic ulcers during pregnancy. The diagnosis was entertained antemortem in only 3 of the 17 patients treated. By 1974 there had been 31 perforations reported, and most of the patients had expired.[7] Single closure of the perforation is employed when dealing with an acute duodenal ulcer without previous ulcer history. Definitive surgery with resection or vagotomy and drainage procedure are otherwise performed in the presence of gastric or chronic ulcers.

In patients requiring surgery for complicated peptic ulcers during their third trimester, cesarean section is considered. The purpose of cesarean section in this situation is to prevent fetal death due to potential maternal hypoxemia and to help provide an adequate view of the upper abdomen. Infrequent as it may be, complicated peptic ulcer disease requires prompt and definitive intervention.

Acute Cholecystitis

Following acute appendicitis, acute cholecystitis is the second most common nonobstetric cause of an acute abdomen during pregnancy, with cholecystectomy being performed in about 3 to 8 of every 10,000 pregnant women.[8–10] Its relatively high frequency is not surprising considering that about 2.5% of women have cholelithiasis by their fourth decade of life[11] and that at least 3.5% of pregnant women harbor gallstones.[12] The rate of cholelithiasis is even higher in women who have taken oral contraceptives in the past, approaching twice that of those who have not used contraceptives.[13,14] Thus gallbladder disease represents one of the most common underlying medical conditions likely to be present during pregnancy.

In addition, research has indicated that there are physiologic changes during pregnancy that lead to an increased incidence of gallstone formation. Kern et al.[15] demonstrated that during early pregnancy there is increased synthesis of bile acids leading to an increase in both the cholate and chenodeoxycholate bile acid pools. During the second and third trimesters of pregnancy, however, there is a progressive decrease in the chenodeoxycholate pool secondary to a decrease in the rate of its synthesis, but there is no decrease in the cholic acid pool. The total rate of cholesterol secretion remains constant; but relative to the now altered chenodeoxycholate/cholate ratio its concentration in bile increases, resulting in the creation of lithogenic bile.

In addition, during pregnancy there are significant changes in gallbladder function.[16] During the second and third trimesters, the volume of the gallbladder increases almost twofold. Moreover, after stimulation the rate and percentage of bile discharged are reduced. The combination of increased residual volume and lithogenic bile leads to an increased opportunity for cholesterol crystals to form, allowing gallstone formation to begin.

The symptoms and signs of cholecystitis during pregnancy are similar to those in the nonpregnant individual. The major diagnostic difficulty that pregnancy imposes is differentiating between cholecystitis and appendicitis, as in the pregnant woman the appendix may also occupy the right upper quadrant. Pain, usually localized to the right upper quadrant, is constant and accompanied by nausea and fever. Physical examination often reveals right upper quadrant tenderness and severe pain under the costal margin with deep inspiration (Murphy's sign). Laboratory examination usually reveals an increased white blood cell count with an increase in immature forms.

The diagnosis of cholelithiasis in the pregnant woman has been greatly facilitated by the application of ultrasonography, which can now be performed without fear for the patient or the fetus. With acute cholecystitis, ultrasonography confirms the presence of stones in the gallbladder and sometimes reveals a thickened gallbladder wall suggestive of acute inflammation. Nevertheless, in case of diagnostic uncertainty, a technetium 99m-iminodiacetic acid (Tc-IDA) scan can be performed with little risk for the fetus because of the minimal radiation dose associated with it.

Severe preeclampsia, with associated right upper quadrant pain and abnormal liver function tests can be confused with acute cholecystitis. This diagnosis is entertained whenever acute cholecystitis is considered during pregnancy. The distinguishing clinical features include the presence of proteinuria, pedal edema, and severe hypertension as well as disproportionate and sustained increases in the SGOT and SGPT levels compared to the alkaline phosphatase level.[17]

Once the diagnosis of cholecystitis is made, the general approach to therapy in the pregnant patient is based on the fact that surgery may be complicated by spontaneous abortion, especially during the first trimester[18] or by preterm labor and delivery during the third trimester. As a consequence, it appears prudent to attempt to manage patients medically. Failure to respond to medical therapy, suspected perforation with peritonitis, obstructive jaundice, and cases in which a perforated peptic ulcer, appendicitis, or intraabdominal emergency cannot be excluded are indications for surgery. The late second trimester is the most favorable time for a cholecystectomy because the peak period for spontaneous abortion has passed, organogenesis is by and large complete, and the uterus is not large enough to impinge on the operating field.

Hepatic Rupture

Spontaneous rupture of the liver is a rare complication of pregnancy that is associated with preeclampsia/eclampsia in 75 to 85% of patients. Preeclampsia induces histologic changes in the liver that characteristically appear as periportal hemorrhages, scattered fibrin deposition, and subcapsular hemorrhage. The fibrin deposits or fibrin thrombi occur in hepatic sinusoids and are near areas of liver cell necrosis and hemorrhage. Presumably, these fibrin deposits occlude the sinusoids, rendering a patch of parenchyma ischemic, and hepatocyte necrosis ensues. If liver cell necrosis is somewhat more severe and extends to involve the structure of the sinusoid itself, the process results in collections of extravasated blood, which are recognized as hemorrhages.

Seventy percent of patients with preeclampsia/eclampsia and spontaneous rupture of the liver are multiparas.[19–21] It has been postulated that preeclampsia in previous pregnancies, regardless of whether recognized, impairs removal of fibrin deposits by the reticuloendothelial system. Therefore subsequent preeclampsia leads to larger fibrin deposits, more vascular occlusion, larger hemorrhages, and ultimately sizable subcapsular hematomas with hepatic rupture.

Because this potentially disastrous complication is associated with a maternal mortality and fetal death rate reported to be higher than 50%, multiparous pregnant patients with preeclampsia/eclampsia must be monitored closely. Routine periodic ultrasound examinations of the liver are indicated to demonstrate possible intrahepatic hematomas. Location of one or more of these hematomas close to the liver surface could be evidence enough to proceed with surgical evacuation of the uterus if the patient is in the third trimester in order to stop the pathophysiologic processes of preeclampsia/eclampsia. The decision to continue or interrupt the pregnancy during the first or second trimester must be individualized.

Most cases of spontaneous rupture occur close to term or immediately postpartum. Rupture involves the right lobe in 90% of reported cases. Multiple ruptures can occur, and the size and extent of the rents in the liver vary from a few millimeters to several centimeters. The initial symptoms are epigastric or right upper quadrant pain, right shoulder pain, and possibly nausea. Patients who develop large ruptures associated with rapid, even massive blood loss into the peritoneal cavity present with hypotension and shock. On physical examination, the abdomen can show a sudden increase in distention, beyond that seen with pregnancy. The differential diagnosis includes uterine rupture and ectopic pregnancy, but in these cases immediate surgical intervention must be carried out. Therapy of hepatic rupture is based on evacuation of the uterus and hepatorrhaphy.

Acute Pancreatitis

The incidence of pancreatitis during pregnancy is unknown, but probably acute pancreatitis is uncommon[22,23] with over 40% of cases during the third trimester.[22–25] It may be due in part to the known tendency of serum triglyceride levels to rise sharply during this period or to the increased chance that cholesterol microcrystals have formed by the end of the pregnancy in atonic gallbladders with increased lithogenic bile. Most cases occur in multigravidas.[22–25]

Cholelithiasis is the most commonly identified cause, occurring in one-third to over one-half of cases of pancreatitis during pregnancy. Other causes of acute pancreatitis include hyperlipidemia (types I, V, and occasionally IV), hyperparathyroidism, and drug and alcohol intake. The incidence of tetracycline- and thiazide-induced pancreatitis is decreasing because of the decreased use of these drugs during pregnancy. Moreover, during pregnancy there have been only a few reports of alcohol-induced pancreatitis, presumably because alcoholism is uncommon in women of childbearing age and voluntary abstinence of alcohol during pregnancy is common.

Regardless of the etiology of the pancreatitis, the clinical presentation is the same. The patient presents with steady epigastric pain, at times severe, often radiating to the back. Physical examination may reveal a tachycardic and sometimes hypotensive patient with epigastric tenderness and a silent abdomen. In severe cases of hemorrhagic pancreatitis there may be bloody discoloration of the umbilicus (Cullen sign) or retroperitoneal dissection of the hemorrhage to the flank (Grey Turner sign). Chest examination might show evidence of a pleural effusion or signs of respiratory distress syndrome.

The clinical diagnosis of acute pancreatitis is confirmed by determination of the serum amylase level, which is above normal in 95% of cases. Elevated serum amylase levels may be occasionally associated with other causes of abdominal pain, such as perforated peptic ulcer, ruptured ectopic pregnancy, small bowel obstruction, and acute cholecystitis. The amylase level does not reflect the severity of the disease process. In cases where the serum amylase level is normal but the clinical suspicion of pancreatitis is strong, a 24-hour urine collection with determination of urinary amylase may be helpful for confirming the clinical diagnosis.

The differential diagnosis includes preeclampsia, which may be associated with epigastric or more characteristically right upper quadrant pain. Features of preeclampsia such as hypertension, edema, and proteinuria are not characteristic of pancreatitis. Ruptured ectopic pregnancy and pancreatitis may present with abdominal pain and elevated serum amylase.[26] The location of the abdominal pain is epigastric with acute pancreatitis and usually hypogastric with ectopic pregnancy. Ultrasound examination of the pancreas and the pelvis is safe and may be helpful for demonstrating an ectopic pregnancy or an enlarged head of the pancreas. Radiologic tests may be necessary to exclude other etiologies of an acute abdomen, including acute cholecystitis and perforated peptic ulcer when the diagnosis of acute pancreatitis is not clear. A Tc-IDA scan can be performed safely with respect to the fetus because a minimal radiation dose is associated with it, and an upright chest radiograph for detection of free air under the diaphragm can be obtained after lead shielding of the patient's abdomen.

There is little reason to believe that pancreatitis is a more serious condition during pregnancy than in the nonpregnant state. Medical management of acute pancreatitis is basically supportive.[26–36] Intravascular volume must be restored vigorously and monitored. The use of parenteral analgesics for pain relief is often necessary; meperidine is the drug of choice because it has the least effect on the sphincter of Oddi. The patient is kept fasting initially until all pain has resolved. Despite numerous control trials, no specific therapy has been shown to be of benefit for acute pancreatitis. Prophylactic antibiotics are clearly inappropriate,[37,38] pancreatic enzyme inhibitors have not been successful,[39] and the variety of strategies designed to reduce the physiologic activity of the pancreas, including cimetidine, glucagon, and anticholinergics,[40,41] have not been shown to be effective. Several small studies have failed to demonstrate an advantage for the use of nasogastric suctioning of patients with mild alcoholic pancreatitis,[42,43] but it is not clear if this finding can be generalized to other etiologies or to more severe cases. Furthermore, clinical experience suggests that nasogastric suctioning increases patient comfort.

Timing of surgery for hyperparathyroidism and gallstone pancreatitis deserves special consideration. With hyperparathyroidism, where pancreatitis tends to be severe, neck exploration during pregnancy is often indicated.[44,45] In gallstone pancreatitis symptoms and signs usually subside with supportive management. In the nonpregnant patient optimal timing for surgical treatment is thought to be after the acute attack has resolved but on the same hospital admission. During pregnancy it is clear that postponing surgery until after delivery is preferable in general,[46] but management must be individualized.

To protect the fetus from protracted fasting, total parenteral nutrition must be considered in the pregnant patient with pancreatitis of any cause when the illness is severe or protracted or whenever feeding causes an exacerbation of symptoms.

All clinical syndromes producing peritoneal irritation (e.g., perforated ulcer, acute cholecystitis, pancreatitis) trigger uterine contractions and hence the risk of either spontaneous abortion or premature labor. The current recommended treatment (in addition to attacking the provoking syndrome) consists in the administration of uterine relaxant medications. The family of β-mimetic agents (e.g., ritodrine, terbutaline, orciprenaline) have been successful in abating uterine contractions long enough for the irritating process to be brought under control by either medical or surgical means. In the latter circumstance it is highly desirable to use anesthetic agents (unless otherwise contraindicated) that are uterine relaxants, as discussed in Chapter 1. Even under ideal circumstances, one sometimes faces the case of premature labor refractory to pharmacologic control. It may be the consequence of either the cardiovascular system's intolerance to the dose of medication needed to produce uterine relaxation or to premature rupture of the membranes, a common consequence of premature labor and early cervical changes. If premature delivery occurs, it is managed with appropriate analgesia-anesthesia, with the neonatologist in attendance.

References

1. Shnider SM, Webster GM. Maternal and fetal hazards of surgery during pregnancy. Am J Obstet Gynecol 1965;92:891.
2. Paul M, Tes WL, Holliday RL. Perforated peptic ulcer in pregnancy with survival of mother and child: case report and review of the literature. Can J Surg 1976;19:427.
3. Honiotes G, Clark PH, Cavanagh D. Gastric ulcer perforation during pregnancy. Am J Obstet Gynecol 1970;106:691.
4. Dilts PV, Coopersmith B, Sweitzer C. Perforated peptic ulcer in pregnancy. Am J Obstet Gynecol 1967;99:293.
5. Jones PF, McEwan AB, Bernard RM. Hemorrhage and perforation complicating peptic ulcer in pregnancy. 1969;Lancet 2:350.
6. Durst JB, Klieger JA. A report of fatal hemorrhage due to peptic ulcer in pregnancy. Am J Obstet Gynecol 1955;70:448.
7. Parry GK. Perforated duodenal ulcer in the puerperium. NZ Med J 1974;80:448.
8. Hill LM, Johnson CE, Lee RA. Cholecystectomy in pregnancy. Obstet Gynecol 1975; 46:291.
9. Sparkman RS. Gallstones in young women. Ann Surg 1957;145:813.
10. Aufes AH Jr. Biliary tract disease. In Rovinsky JJ, Guttmacher AF (eds): Medical, Surgical and Gynecologic Complications of Pregnancy, 2nd ed. Baltimore, Williams & Wilkins, 1965;251.
11. Zahor Z, Sternby NH, Kagan A, et al. Frequency of cholelithiasis in Prague and Malmo: an autopsy study. Scand J Gastroenterol 1974;9:3.
12. Honre LH. Cholesterol cholelithiasis in adolescent females. Arch Surg 1980;115:62.
13. Evron S, Frankel M, Diamant Y. Biliary disease in young women and its association with pregnancy or oral contraceptives. Int Surg 1982;67:448.
14. The Boston Collaborative Drug Surveillance Program: Surgically confirmed gallbladder disease, venous thromboembolism, and breast tumors in relation to postmenopausal estrogen therapy. N Engl J Med 1974;290:15.
15. Kern F, Everson GT, Demark B, et al. Biliary lipids, bile acids, and gallbladder function in the human female: effects of pregnancy and the ovulatory cycle. J Clin Invest 1981; 68:1229.
16. Braverman DZ, Johnson ML, Kern F. Effects of pregnancy and contraceptive steroids on gallbladder function. N Engl J Med 1980; 302:362.
17. Friedenberg WR, Burrill RE, Young J. Severe preeclampsia presenting as hepatobiliary disease. Wis Med J 1978;77:117.
18. Kammerer WS. Nonobstetric surgery during pregnancy. Med Clin North Am 1979; 63:1157.
19. Bis KA, Waxman B. Rupture of the liver associated with pregnancy. Obstet Gynecol Surv 1976;31:763.
20. Golan A, White RG. Spontaneous rupture of the liver associated with pregnancy. S Afr Med J 1979;56:133.
21. Westergaard L. Spontaneous rupture of the liver in pregnancy. Acta Obstet Gynecol Scand 1980;59:559.
22. Corlett RC Jr, Mishell DR Jr. Pancreatitis in pregnancy. Am J Obstet Gynecol 1972; 133:281.
23. Wilkinson EJ. Acute pancreatitis in pregnancy: a review of 98 cases and a report of 8 new cases. Obstet Gynecol Surv 1973;28:281.
24. Langmade CF, Edmondson HA. Acute pancreatitis during pregnancy and the post partum period: a report of nine cases. Surg Gynecol Obstet 1951;92:43.
25. Herfort K, Fialova V, Srp B. Acute pancreatitis in pregnancy. Mater Med Pol 1981;13:15.
26. Flege JB. Ruptured tubal pregnancy with elevated serum amylase levels. Arch Surg 1966;92:397.
27. Orda R, Roda S, Baron J, et al. Lipase turbidimetric assay and acute pancreatitis. Dig Dis Sci 1984;29:294.
28. Van Gossum A, Seferian V, Rodzynek JJ, et al. Early detection of biliary pancreatitis. Dig Dis Sci 1984;29:97.
29. Kelly ML Jr. Elevated serum amylase level associated with ruptured ectopic pregnancy. JAMA 1957;164:406.
30. Lunt RL. Acute pancreatitis simulating intrapelvic bleeding in pregnancy. Br J Obstet Gynaecol 1957;64:437.
31. Hochbert CJ. Tubal amylase. Obstet Gynecol 1974;43:129.
32. Banks PA. Pancreatitis. New York, Plenum, 1979, Ch. 7.

33. Chakravorty AP. Acute pancreatitis during pregnancy: review of literature with a case report. Br J Clin Pract 1963;17:131.
34. Boyle JM, McLeod ME. Pancreatic cancer presenting as pancreatitis of pregnancy. Am J Gastroenterol 1979;70:371.
35. Regan PT. Medical treatment of acute pancreatitis. Mayo Clin Proc 1979;54:432.
36. McPhee M. Treatment of acute pancreatitis. Hosp Pract 1985;20:83.
37. Howes R, Zuimeda GD, Cameron JL. Evaluation of prophylactic antibiotics in acute pancreatitis. J Surg Res 1975;18:197.
38. Finch WT, Sawyers JL, Schenker S. A prospective study to determine the efficacy of antibiotics in acute pancreatitis. Ann Surg 1976;83:667.
39. MRC Working Party. Death from acute pancreatitis; MRC multicentre trial of glucagon and aprotinin. Lancet 1977;2:632.
40. Gilsanz V, Oteyza CD, Rebollar JL. Glucagon vs. anticholinergics in the treatment of acute pancreatitis, a double-blind controlled trial. Arch Intern Med 1978;138:535.
41. Regan PT, Malaagelada JR, Go VLW, et al. A prospective study of the antisecretory and therapeutic effects on cimetidine and glucagon in human acute pancreatitis. Mayo Clin Proc 1981;56:499.
42. Levant JA, Secrist DM, Resin H, et al. Nasogastric suction in the treatment of alcoholic pancreatitis, a controlled study. JAMA 1974; 229:51.
43. Field BE, Hepner GW, Shabot MM, et al. Nasogastric suction in alcoholic pancreatitis. Dig Dis Sci 1979;24:339.
44. Thomason JL, Sampson MB, Garb HF, et al. Pregnancy complicated by concurrent primary hyperparathyroidism and pancreatitis. Obstet Gynecol 1981;57(suppl 6): 345.
45. Gaeke RF, Kaplan EL, Lindheimer MD, et al. Maternal primary hyperparathyroidism of pregnancy: successful treatment by parathyroidectomy. JAMA 1977;238:508.
46. McKay AJ, O'Neill J, Imrie CW. Pancreatitis, pregnancy and gallstones. Br J Obstet Gynaecol 1980;87:47.

7
Acute Appendicitis During Pregnancy

A.H. MOAWAD

Acute appendicitis during pregnancy is an important complication that deserves careful consideration. It is perhaps the most common indication for performing a laparotomy for extrauterine lesions. Various anatomic and physiologic changes that occur during pregnancy contribute to the complexity of making the diagnosis of this disorder. The symptoms and signs of acute appendicitis during pregnancy may, however, mimic those commonly encountered in normal nongravid women. This chapter reviews these two principal aspects of this condition and discusses points that may be useful in its prompt diagnosis and treatment.

Anatomic Changes

In the nonpregnant state the vermiform appendix is located in the right lower quadrant 65% of the time. In 30% of cases it is situated in the pelvic region, and in only 5% is it located retrocecally. It is in this last position, or when it is anchored by adhesions, that it may be fixed deeply within the pelvis.[1,2] In the pregnant state the progressive increase in the size of the uterus displaces the base of the appendix upward. We owe much of our knowledge in this regard to the classic work of Baer and colleagues in 1932.[1] These authors traced the anatomy of the appendix using barium studies performed on healthy gravid women with no history of appendicitis; the subjects were evaluated in the supine position. The authors reported on the progressive displacement of the appendix by the enlarging uterus, upward and outward, as well as a counterclockwise rotation of the tip into the right upper quadrant; the latter effect is so marked that during the last 2 weeks of pregnancy the tip overlies the right kidney (Fig. 7-1). The positional changes described by Baer et al.[1] may be hampered by the presence of adhesions, which restrict the free movement of the appendix. The retrocecal appendix is also limited in its mobility.

The enlarging uterus pushes the parietal peritoneum outward and the greater omentum upward, making it difficult for the latter to reach the inflamed appendix. These alterations accelerate the spread of peritonitis when appendicitis occurs during late pregnancy. Intermittent uterine contractions may further contribute to the inflammatory progression of the disease if labor is in the active phase.

Pathology

Acute appendicitis usually begins as an obstructive phenomenon. The obstruction may be caused by a fecolith, a calculus, a contracture at the base of the organ, a previous fibrosis at the proximal end, or even by a kink due to the presence of a congenital fold or adhesions. Within a matter of hours after

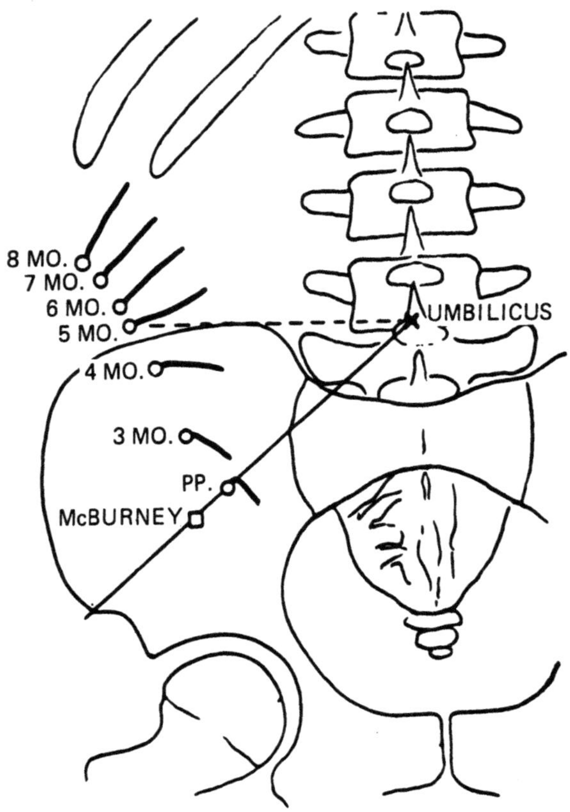

FIGURE 7-1. Changes in the position and direction of the appendix during pregnancy. (From Baer et al.,[1] reprinted by permission of the Journal of the American Medical Association 98:1359. Copyright 1932, American Medical Association).

the obstruction occurs, luminal bacteria invade the devitalized mucosa, setting off a rapidly advancing inflammatory process that results in swelling of the appendix. Because of the tight peritoneal boundaries, it leads to necrosis and finally perforation.[3] If there is enough time for the greater omentum and intestines to come to the rescue, the site is walled off and a local abscess is formed. If this protective action does not occur, generalized peritonitis may set in.

Although mild cases of obstruction are sometimes relieved spontaneously, the attacks may be repeated later leading to the gradual development of fibrosis. Recurrent attacks indicate, therefore, that the obstruction occurred without progression to inflammation.

The pathology of the disease described above applies equally to the pregnant state with one added variable to the anatomic changes that occur within the abdominal cavity because of the enlarging gravid uterus. During the second and third trimesters of pregnancy the upward displacement of the greater omentum makes it difficult to isolate a perforated appendix; the mechanisms that normally act to contain the spread of infection into the peritoneal cavity may be overwhelmed, allowing rapid progression of the disease.

Incidence

One would expect appendicitis to occur more frequently during pregnancy than in the nongravid state because the elevated progesterone levels seen during pregnancy can lead to hypomotility of the gut. The incidence of the disease, however, is the same during pregnancy as it is in the nonpregnant state. Acute appendicitis occurs more than 70% of the time in women under the age of 35 years. Therefore it is more prevalent in the reproductive age population. During pregnancy it accounts for two-thirds of all gastrointestinal-related surgical cases.[4] The incidence is on the average 0.69 per 1000 deliveries and ranges from 1.2 per 1000 in tertiary referral centers (24 cases in 19,187 deliveries)[5] to 0.87 per 1000 (29 cases in 33,300 deliveries),[6] 0.67 per 1000 (503,000 pregnancies),[4] 5.5 per 1000 (25 cases in 50,089 deliveries) in a study that included the entire city of Aberdeen and suburbs,[7] and 0.37 per 1000.[8] It is important to note that, based on the results of two large study series,[3,9] the incidence of appendicitis in pregnant women is practically the same during all three trimesters. In our own institution over the last 7 years, we have seen 12 cases of appendicitis during pregnancy among 22,000 deliveries, an incidence of 0.55 per 1000.

Symptomatology

During the initial stages of appendicitis, minimal changes are seen in the patient's vital signs (temperature and pulse); even tenderness may be absent. In 20% of cases the bladder is involved, producing dysuria. Twelve percent of the time irritation of the rectal wall is responsible for an increase in bowel activity.[10]

Although pain is the most common presenting symptom in cases of appendicitis (90% of cases), patients may also present with nausea (86%), vomiting (62%), and anorexia (41% of cases).[6] Diarrhea is a rare symptom identified only 5 to 10% of the time. The presence of dysuria is also relatively rare. The hallmark of all of the symptoms and signs of acute appendicitis during pregnancy is that their manifestation is blunted.[2,3,8,9–15]

Pain

Abdominal pain is the most constant symptom in acute appendicitis. It must be emphasized that almost all patients with this disease report having vague pain in the right side of the abdomen. In the early stages of appendicitis, following obstruction of the lumen, visceral pain results from the inflammation and distention of the appendiceal capsule. The appendix is embryologically a midline structure, so the location of the initial pain is usually centered around the umbilicus. This pain is referred to the periumbilical and epigastric areas regardless of the rotation of the appendix.[5,8] Because the initial pain of acute appendicitis is due to appendiceal smooth muscle contractions trying to overcome an obstruction, it is often colicky in nature. Therefore, as pregnancy progresses beyond the first trimester, pain becomes a less definite characteristic and may be located in the right upper quadrant, the right side, or the right flank. Cunningham and McCubbin[8] identified the location of the pain to be the right lower quadrant in 90%, 75%, and 37% of the cases during the first, second, and third trimesters, respectively. In the remaining cases the pain was in the right upper quadrant or diffused to other areas of the abdomen. Also, because this pain is mediated by the autonomic nervous system it is usually experienced as dull and diffuse.

With the onset of acute suppuration and the progression to perforation or gangrene, radiated pain is maximal at the point of contact between the appendix and the peritoneum. Because the retrocecal appendix is farthest removed from peritoneal contact, it is in this location that the incidence of perforation of the appendix is highest, as the acuity of the pain is blunted, thus delaying the diagnosis.

Anorexia

Although it is considered a constant symptom of appendicitis during pregnancy, anorexia is absent in 25 to 60% of all cases in the nonpregnant state.[6,8,16]

Nausea

The presence of nausea may be confusing, as it is so commonly seen during normal pregnancy. However, it cannot be ignored when it is accompanied by pain. The cause of nausea and vomiting must be vigorously pursued when they occur later than the first trimester of pregnancy.

Dysuria

In 20% of cases of appendicitis during pregnancy the bladder is involved, producing dysuria as a presenting symptom. Urinary tract infection, a relatively common condition during pregnancy, must be ruled out. This symptom seems to become more common as pregnancy advances. In one study series[8] symptoms suggestive of lower urinary tract infection were present in 25% of cases of acute appendicitis. The incidence in this study was 10% during the first trimester, 19% the second trimester, and 50% the third trimester.

Diarrhea

In 12% of the cases in one study, irritation of the rectal wall was responsible for the increased frequency of bowel motion.[10]

Physical Findings

Temperature

Fever is usually low grade (in 75% of cases it is 100.8°F or less), occurring in most patients at the time of their presentation. The presence of fever is not a useful sign for reaching a diagnosis, but it may predict perforation. Weingold[5] found that more than half of his pregnant patients with appendicitis were afebrile during the preoperative period.

Tachycardia

Tachycardia is not a useful diagnostic sign because of the physiologic tachycardia that accompanies pregnancy. Only 25% of patients show a pulse of more than 100 beats per minute.

Tenderness

Tenderness is the most constant sign of acute appendicitis and is present in most cases. Cunningham and McCubbin[8] found that direct abdominal tenderness was present in 94% of the patients in their study. The localization of tenderness in the right lower quadrant is common during the first two trimesters, but this sign is frequently absent during the third trimester because of the growing uterus (three of eight cases during the third trimester).[8] The upper quadrant or the entire right side of the abdomen are more common sites for tenderness during the latter part of gestation.[16] Therefore it is important to regard tenderness as suggestive of appendicitis if elicited anywhere on the right side of the abdomen and not just in the right lower quadrant. Tenderness may also be elicited at other sites. Rectal tenderness may be found in 33%[3] to 80%[8] of all cases examined. This sign is negative more frequently during the third trimester than in early pregnancy. Sites where localized tenderness is elicited less frequently are the right costovertebral angle, the left iliac fossa, and the right lumbar region.

Adler's Sign

Adler's sign may be useful for differentiating uterine from appendiceal sources of pain. To elicit this sign, one localizes the pain with the patient in the supine position. Without changing the position or pressure of the fingers the patient is turned to her left side, allowing the uterus to shift with her to the left. If the intensity of the pain remains the same and its location is still in the right lower quadrant, appendicitis is suspected. It is known as the "sign of fixed tenderness." If the pain shifts to the patient's left, it may be uterine in origin.[17]

Bryan's sign[16,18] is positive if right lower quadrant pain is aggravated when the uterus is shifted to the right by turning the patient to her right side. This sign is useful for ruling out uterine causes of abdominal pain.

Rebound tenderness and abdominal muscle spasm, valuable findings in the nonpregnant patient, are found less frequently in the pregnant patient because of the fact that the abdominal wall muscles are more relaxed. In one study, these signs occurred in only one-half of the cases of appendicitis during the third trimester.[8]

Progression of the disease process of appendicitis can lead rapidly to the development of various degrees of peritonitis. This progression is marked by the presence of hypoactive bowel sounds, abdominal distention, and generalized tenderness and rigidity. The localization process can result in a walling-off of the perforated appendix and consequently the development of a right lower quadrant mass. If it occurs, the spread and progression of appendicitis to peritonitis may be halted.

Laboratory Findings

Leukocytosis is commonly found in cases of gestational appendicitis but may be confused with the physiologic leukocytosis of pregnancy. Cunningham[8] defined leukocytosis as a white blood cell count of more than 11,000/cu mm. Using this definition he found that most patients with appendicitis during pregnancy had leukocytosis and immature forms (a shift to the left).

Urinalysis is usually normal. Concomitant urinary tract infection may be coexistent and confuse the diagnosis. Involvement of the ureters or renal pelvis may lead to the appearance of a small number of white blood cells in the urine. Early in an attack of pyelonephritis the urine may be normal or show only a few pus cells. If the urine does not show a large number of cells when the examination is repeated in a few hours, it is unsafe to diagnose pyelonephritis.[3]

Ultrasonography is helpful for evaluating the status of the pregnancy (fetus and placenta). The exact gestational age, the presence of fetal anomalies, and the insertion and normality of the placenta are important information to be gathered. Uterine abnormalities, such as the presence of fibroids, must also be noted. These considerations are important for the differential diagnosis. Ultrasonography is also useful for the diagnosis of adnexal masses, either solid or cystic, which may be the cause of acute abdominal distress.

Differential Diagnosis

A frequent cause of lower abdominal pain and tenderness during pregnancy is urinary tract infection. Thus when fever and bacteriuria accompany abdominal or flank pain, appendicitis may be overlooked. It is important to note that a positive urine culture does not exclude appendicitis.

The diagnostic error in men of matched ages with appendicitis is reported to lie between 5 and 10%, in contrast to much higher figures for women (25–40%).[5] Pregnancy is a contributing factor in delayed and mistaken diagnosis in women. Conditions to be considered in a differential diagnosis are ruptured corpus luteal cyst, torsion of an adnexal mass, adnexal torsion, ectopic pregnancy, abruptio placentae, labor, round ligament syndrome, chorioamnionitis, degenerating leiomyoma, salpingitis, pyelonephritis, cholangitis, and mesenteric adenitis.[5] Ultrasonography, especially the application of vaginal probes, has immeasurably improved our ability to diagnose these adnexal and pelvic conditions, which warrant serious consideration in the differential diagnosis.

Acute appendicitis during pregnancy is a medical condition that has not benefited from modern advances in diagnostic or therapeutic techniques. Diagnosis still depends to a large extent on clinical acumen.

Treatment

Acute appendicitis during pregnancy is a surgical problem, and the only treatment available is a prompt operation. If the diagnosis is certain, it is the opinion of the author that the incision should be over the point of maximum tenderness. This point may be high late in pregnancy, and a high lateral oblique muscle-splitting incision may be required. On the other hand, if the diagnosis is uncertain, a transverse muscle-cutting incision at the level of the umbilicus allows better visualization than a paramedian incision. It also carries the advantage of avoiding vigorous manipulation of the uterus when attempting to retract it laterally. Manipulation of the uterus may precipitate labor and is avoided as much as possible.

Some investigators advocate a low vertical midline incision. The advantages claimed for this incision are as follows: (1) it is adequate for appendiceal exposure; (2) it can be used for treatment of other abdominal conditions that simulate acute appendicitis; and (3) it is convenient for uterine evacuation by cesarean section should it become necessary.[8,19] Tilting the patient approximately 30 degrees to the left side enhances the exposure and pulls the uterus to a more physiologic position relative to the vena cava.

Cesarean delivery during an appendectomy is reserved for obstetric indications.[16] McComb and Laimon[20] have suggested that with modern neonatal care a cesarean section may decrease fetal mortality in cases of appendicitis with perforation close to term. However, the maternal risk of abdominal delivery is increased in the presence of gross infection.[16] Immediate delivery by section may also be necessary in cases of general peritonitis to reduce the fetal risk from impending septic shock.[2] Access to the appendix can be difficult with a full-term gravid uterus. In such situations, one may have to resort to delivery of the fetus by cesarean section to decrease the size of the uterus sufficiently to proceed with the appendiceal surgery.

Perforated or Ruptured Appendix and Appendiceal Abscesses

Spitzer and Kaiser,[19] in a review of the literature, found 43 cases of perforated appendix with sufficient details to warrant analysis. Thirty-three of these patients delivered within several days of the perforation. The fetal mortality rate in these cases was 37% (16 of 43); maternal mortality was 4.65% (2 of 43).

If the infection is still localized in the appendix, removal is the recommended surgery. If peritonitis is evident, antibiotics and drains must also be used. An abscess is treated by drainage and perhaps cecostomy if the pregnancy is in the third trimester, with the appendectomy deferred to a later date.[5]

Anesthesia

If regional anesthesia is to be used it is perhaps more appropriate during the first trimester when teratologic side effects are a concern. Later in pregnancy, general intubation anesthesia is the treatment of choice because of the added advantages of oxygenation, control of hypotension, and abdominal and uterine relaxation when necessary. The method of anesthesia is secondary in importance to the presence of a skillful anesthesiologist who is familiar with the problems specific to the gravid patient.

Premature Labor

Prevention

Fetal morbidity and mortality rise sharply when perforation of the appendix occurs. The best way to decrease or prevent premature labor and delivery is to act promptly while the acute inflammation is limited to the appendix. Several series have demonstrated minimal or no fetal losses when the patient is treated prior to perforation.[4,16,20,21] It is also important to exercise caution and gentleness when manipulating the uterus intraoperatively.

Tocolytic agents have been used with some success for the treatment of existing premature contractions. The value of the prophylactic use of this class of drug in the prevention of premature labor, however, has not been established. It must be stressed that premature labor may be the presenting symptom for a case of acute appendicitis. In recent years so much emphasis has been placed on the symptomatic treatment of premature labor that the possibility of a nonobstetric etiologic factor cannot be overemphasized. A single-minded focus on stopping uterine contractions may delay the diagnosis and treatment of an acute appendiceal emergency.[22]

Case Reports

A 19-year-old woman (T.T), para 1 (via cesarean section), 34 weeks' gestation, experienced intermittent, crampy pain for 2 days prior to admission that developed into more right-sided abdominal pain with nausea on the day of admission.

The patient denied experiencing anorexia, fever or chills, or diarrhea. Physical examination revealed a healthy woman intermittently in moderate distress with diffuse tenderness of the abdomen that was not limited to the uterus and slightly more right than left. Bowel sounds were within normal limits. The fundal height was 34 cm. Temperature was 96.8°F (36°C). Irregular uterine contractions were noted on the external fetal monitor. Fetal heart tones were 150 beats per minute. Laboratory results throughout the first hospital day showed a white blood cell count of 13.8 to 15.6/cu mim with a left shift and hematocrit of 38.1% initially, falling to 30.8% within a few hours.

During the patient's first hospital day the pain progressed from mild and diffuse to more severe and localized with rebound tenderness in the right middle quadrant. She remained afebrile.

An exploratory laparotomy was performed to evaluate possible appendicitis or adnexal pathology. An enlarged indurated appendix was found and removed. The uterine scar was found to be intact, and there was no adnexal pathology. Postoperative uterine contractions were treated with intravenous ritodrine therapy (Fig. 7-2). The patient was discharged on postoperative day 5.

C.B., a 32-year-old woman (gestation 6, para 3) was admitted to the hospital at 27 weeks' gestation with vomiting after the onset of sharp right side abdominal pain. Uterine contractions started 2 weeks prior to, and became more intense during, this episode. The patient's vital signs were normal, and her temperature was 98.4°F (36.9°C). Abdominal examination showed a uterine fundal height of 27 cm and a soft, nontender uterus. There was only one point of tenderness on the right side at the level of the umbilicus, with no guarding or rebound. Initial laboratory findings of a white blood cell count of 15,500/cu mm rose to 16,000/cu mm with a definite left shift over a few hours, along with increased pain and vomiting. The patient underwent an exploratory laparotomy and was found to have an unruptured retrocecal appendicitis, but

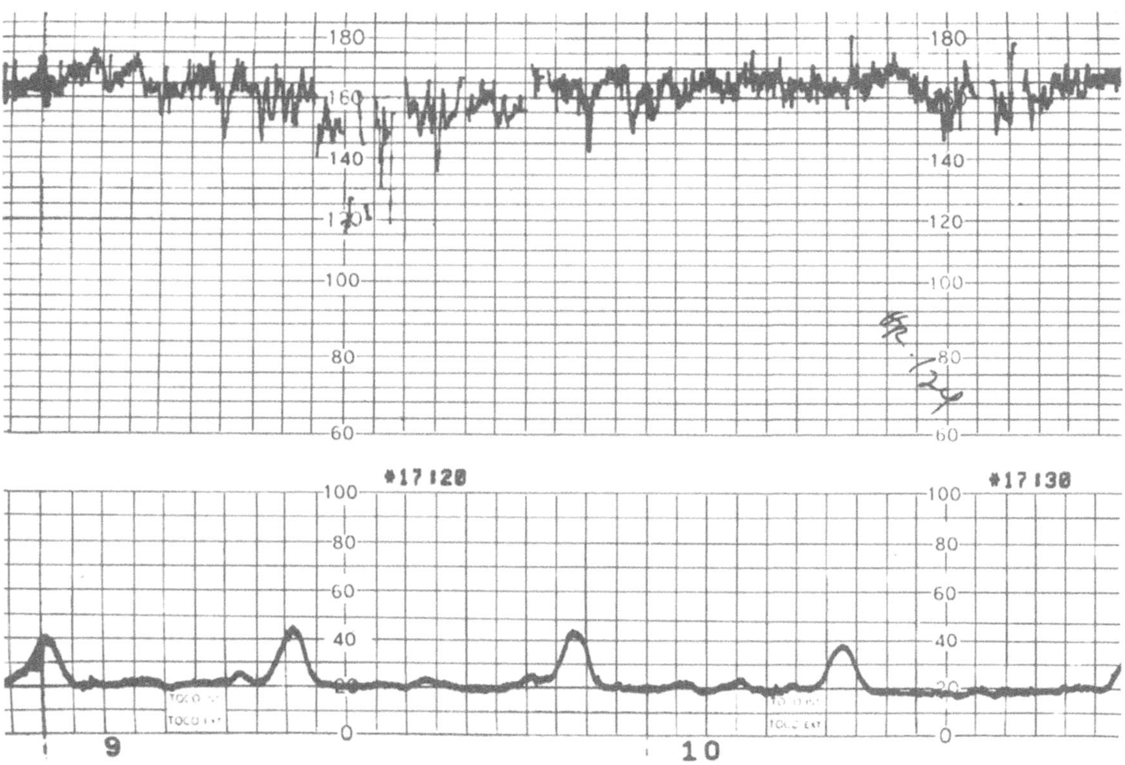

FIGURE 7-2. Fetal heart rate and uterine activity (external monitor applied) tracing of case 1 showing intermittent, regular contractions that responded to ritodrine tocolysis.

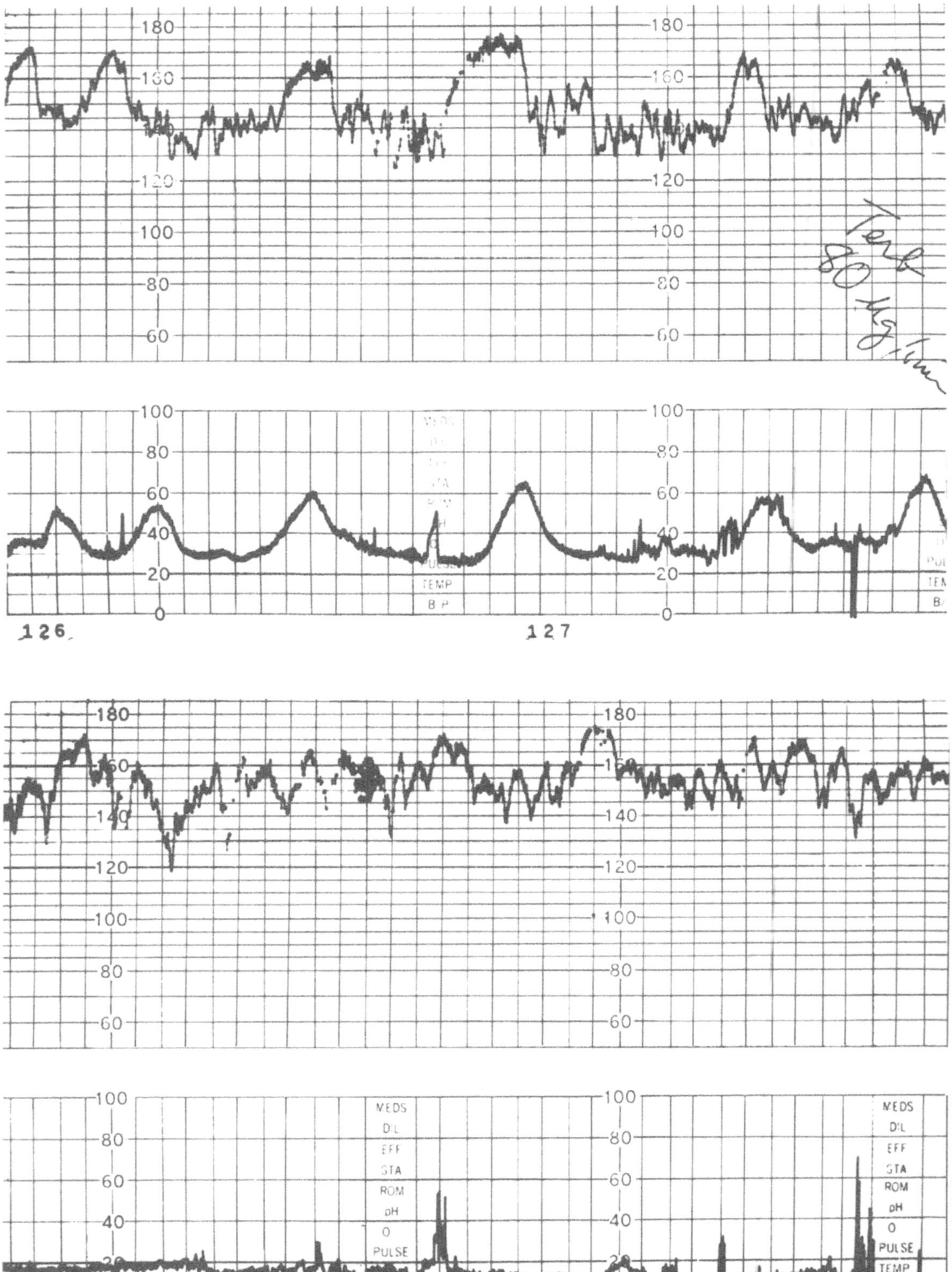

FIGURE 7-3. (A) Premature labor with normal fetal heart tracing following acute appendicitis at 27 weeks' gestation (case 2). (B) The patient responded well to ritodrine therapy. Note the development of fetal tachycardia as ritodrine is infused.

surrounding tissues were the site of a fibrinous discharge. The appendix was removed (Fig. 7-3). The patient had a benign postoperative course and was discharged after 1 week. She returned 1 week later with a fever of 101.3°F (38.5°C) and incisional tenderness. This condition was surgically explored, and an incisional abscess was evacuated and drained. Despite tocolytic therapy, premature labor ensued after this procedure (Fig. 7-4) and progressed to spontaneous rupture of membranes and delivery at 29.5 weeks' gestation. Both mother and child did well.

These two case reports illustrate the lack of the classic symptoms and signs of acute appendicitis in the pregnant patient, particularly the absence of elevated temperature; and they demonstrate the need for prompt surgical intervention. Concomitant premature labor can usually be controlled with tocolytic agents. The presence of purulent material adjacent to the uterus, as in the second case, resulting from delayed intervention or a postoperative complication may make premature delivery impossible to avoid.

Conclusion

Appendicitis during pregnancy presents a clinical dilemma. The anatomic changes described above, the result of the enlarging gravid uterus, and the increased vascularity of the pelvis hinder the natural mechanisms for the containment of infection. Clinical and laboratory findings may be deceptive in that they do not accurately reflect the gravity or the extent of the infection. Conservative treatment is never a primary choice and can prove disastrous. It is safer to err on the side of surgical exploration with negative findings than to wait for a progression of symptoms and signs.

Prompt surgery is the treatment of choice, and results are excellent at any gestational

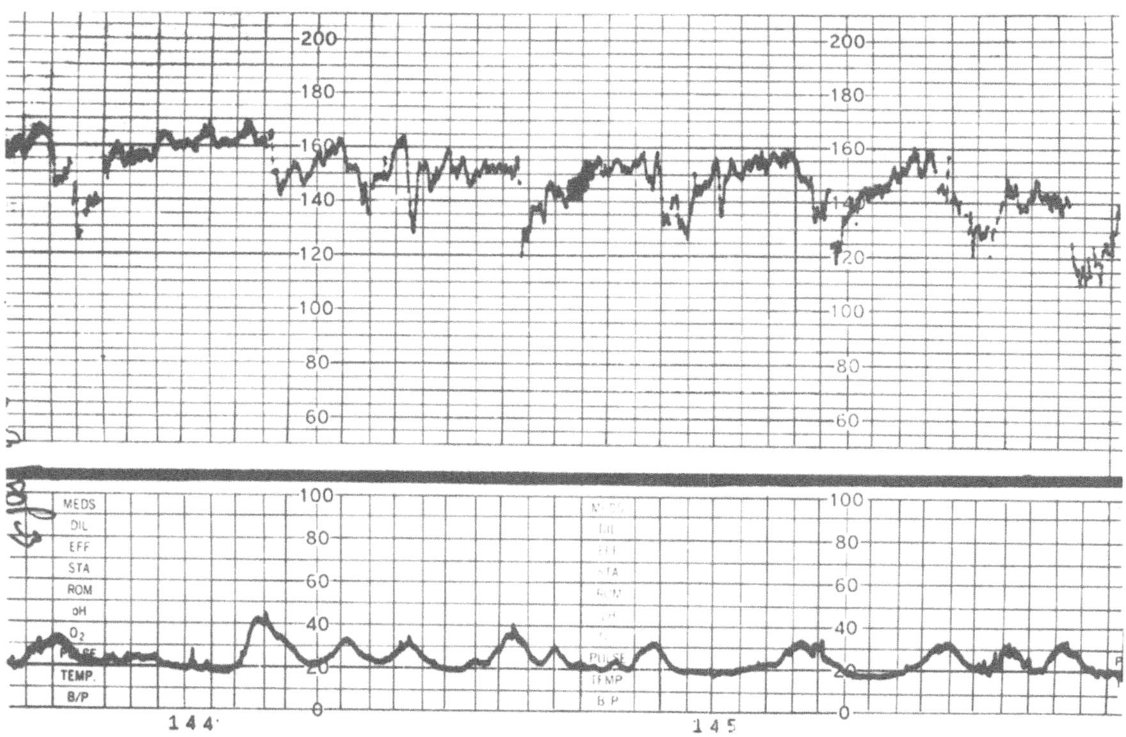

FIGURE 7-4. With the development of an incisional abscess (case 2), premature labor ensued that was not responsive to tocolytic management.

age if performed prior to rupture or perforation. Administration of anesthesia by personnel who are experienced in caring for the pregnant patient is more important than the specific anesthetic method used. Gentle handling of the uterus intraoperatively is important if obstetric sequelae are to be avoided.

Proper drainage for purulent material must be used during closure to minimize postoperative suppurative complications. Adjuvant antibiotic therapy is necessary in most cases. There is a reasonably wide choice of drugs that are relatively safe during pregnancy; agents such as tetracycline and chloramphenicol are to be strictly avoided. Postoperative tocolysis is necessary only when there are indications that premature labor is evolving.

Rupture or perforation of the appendix abscess formation with a localized mass adjacent to the uterus, and generalized peritonitis carry a poor prognosis for the pregnancy. In general, however, with prompt intervention and in the absence of suppurative complications, the prognosis for the continuing health of mother and fetus following appendicitis during pregnancy is excellent.

Acknowledgment. The author gratefully acknowledges the editorial assistance of Mr. L. Philip River in the preparation of this manuscript.

References

1. Baer JL, Reis RA, Arens RA. Appendicitis in pregnancy with changes in position and axis of the normal appendix in pregnancy. JAMA 1932;98:1359.
2. Sarson EL, Bauman S. Acute appendicitis in pregnancy: difficulties in diagnosis. Obstet Gynecol 1963;22:388.
3. Brant HA. Acute appendicitis in pregnancy. Obstet Gynecol 1967;29:130.
4. Babaknia A, Parsa H, Woodruff JD. Appendicitis during pregnancy. Obstet Gynecol 1977;50:40.
5. Weingold AB. Appendicitis in pregnancy. Clin Obstet Gynecol 1983;26:801.
6. Townsend JM, Greiss FC. Appendicitis in pregnancy. South Med J 1976;69:1161.
7. Farquharson RG. Acute appendicitis in pregnancy. Scott Med J 1980;25:36.
8. Cunningham FG, McCubbin JH. Appendicitis complicating pregnancy. Obstet Gynecol 1975;45:415.
9. Black WP. Acute appendicitis in pregnancy. Br Med J 1960;1:1938.
10. Bruce J. Acute appendicitis. Practitioner 1964;192:731.
11. Hoffman ES, Suzuki M. Acute appendicitis in pregnancy, a ten year survey. Am J Obstet Gynecol 1954;67:1338.
12. Parker RB. Acute appendicitis in late pregnancy. Lancet 1954;1:1252.
13. Schelpert JW. Acute appendicitis in the obstetric patient. NY J Med 1961;61:4032.
14. Priddle HD, Hesseltine HC. Acute appendicitis in the obstetric patient. Am J Obstet Gynecol 1951;62:150.
15. Hinshaw WR. The acute abdomen complicating pregnancy. J Okla Med Assoc 1963;56:4.
16. Horowitz MD, Gomez GA, Santiesteban R, Burnett G. Acute appendicitis during pregnancy. Arch Surg 1985;120:1362.
17. Adler N. A sign for differentiating uterine from extrauterine complications of pregnancy. Br Med J 1951;2:1194.
18. Kurtz GR, Davis RS, Sproul JD. Acute appendicitis in pregnancy and labor, a report of 41 cases. Obstet Gynecol 1964;23:528.
19. Spitzer M, Kaiser IH. Perforative appendicitis in the third trimester of pregnancy. NY State J Med 1984;132.
20. McComb P, Laimon H. Appendicitis complicating pregnancy. Can J Surg 1980;23:92–94.
21. Lee RA, Johnson CE, Symmonds RE. Appendicitis during pregnancy. JAMA 1965; 193:966.
22. Herczeg J, Kovacs L, Keseru T. Premature labor and coincident acute appendicitis not resolved by beta-mimetic but surgical treatment. Acta Obstet Gynecol Scand 1983; 62:373–375.

8 Inflammatory Bowel Disease During Pregnancy

WOLFGANG H. SCHRAUT

Inflammatory bowel disease (IBD) is a common, serious gastrointestinal disorder that frequently afflicts women of childbearing age. Thus there is concern about the effects of the disease on fertility and pregnancy and, conversely, about the effects of pregnancy on the course of the disease. It is incumbent on the obstetrician to recognize this disorder and to be prepared, with the help of appropriate consultants, to provide the pregnant IBD patient with counseling and any medical and surgical treatment that may be required.

By common usage, the term "inflammatory bowel disease" encompasses Crohn's disease and chronic ulcerative colitis. The two diseases differ in their prognosis and their clinical manifestations, and they may affect different segments of the alimentary tract. Both diseases are considered medical rather than primarily surgical diseases; however, most of the patients eventually require surgery after the benefits of medical management have been exhausted or when special problems evolve.

Crohn's disease usually recurs, and operations are palliative in nature inasmuch as the surgeon removes only the diseased segment of the intestine from which a complication (abscess, fistula, obstruction, hemorrhage) arises. In contrast, ulcerative colitis, which can pose some special problems, such as evolution into colon cancer, is cured by removing the colon and rectum.

Because of the differences in presentation, treatment, and prognosis, the two diseases are considered separately, when appropriate, with regard to fertility and to the special circumstance of their occurrence during pregnancy and postpartum. In 1951 Abramson et al.[1] classified pregnant women with IBD into several groups: those with antecedent IBD that was quiescent at the beginning of pregnancy; those with active antecedent IBD; and those who developed IBD either during pregnancy or postpartum. This classification, originally employed for pregnant women with ulcerative colitis, is equally useful and applicable for pregnant patients with Crohn's disease and therefore is used in this chapter.

Fertility of Patients with Inflammatory Bowel Disease

The likelihood that IBD and pregnancy will coincide is a function of the fertility of the female patient with IBD, unless the disease arises during the course of a pregnancy. Fertility in women with IBD who are of childbearing age has been evaluated repeatedly, but it is difficult to draw statistically valid conclusions from the data. Patients with IBD are subject to the same physiologic, psychological, and environmental factors and variables that influence the fertility rate in women without IBD. Thus it is difficult to attribute changes in fertility to a single factor such as the presence of IBD.

Most authors agree that ulcerative colitis is not a cause of impaired fertility.[2–6] With regard to Crohn's disease, the findings of deDombal et al.[7] indicate a lowering of the fertility rate to 53.5%. In their opinion, women with Crohn's disease are rendered temporarily subfertile, but not infertile, by the activity of their disease. It is reasonable to assume that the systemic impact of active Crohn's disease, as manifested by fever, abdominal pain, diarrhea, and inanition, interferes with conception. Furthermore, the proximity of the ovaries and fallopian tubes to the inflamed bowel may lead to impaired ovulation[5] and extensive perineal infection and inflammation, which interferes with conception. This point of view is supported by a European report[8] in which the authors concluded that Crohn's disease was primarily responsible for the subfertility of these patients. Fielding and Cooke,[9] Homan and Thorbjarnarson,[10] and Morl et al.[5] found that the extent to which fertility in various subgroups of Crohn's disease patients was reduced depended on the bowel segments involved. However, their reports are at variance with each other and with reports on other series; thus broad statements cannot be made as to a possible association between infertility or subfertility and a specific form of Crohn's disease.

The data in the reports just cited suggested that fertility may be impaired more severely in patients with Crohn's disease (conception rate 42–68%) than in patients with ulcerative colitis (conception rate 81–90%).

Pregnancy in Patients with Quiescent Inflammatory Bowel Disease

If the IBD, whether ulcerative colitis or Crohn's disease, is quiescent at the start of pregnancy, there is a high probability that it will remain in remission throughout the pregnancy and the puerperium. Medical treatment designed to induce and maintain remission before and during pregnancy must be continued and can be employed without hesitation. If a relapse should occur in patients with ulcerative colitis, it is most commonly seen during the first trimester of pregnancy.[2,6,11,12] Even in case of relapse, the risk to the mother is minimal, provided the colitis is treated vigorously. Although the risk of spontaneous abortion and premature delivery is increased under this circumstance,[13] a therapeutic abortion is not indicated or required for the patient who has a severe attack of ulcerative colitis during the first trimester.[6]

The abortion rate in women with quiescent ulcerative colitis is the same as that in women without IBD.[14,15] Investigators have demonstrated this fact using historical controls from the general population; more directly, Maddix[16] compared the outcome of pregnancies (spontaneous abortion, stillbirth) in the same women before and after the onset of ulcerative colitis. Thus ulcerative colitis does not represent a contraindication to pregnancy in regard to either the course of the bowel disease or the course and outcome of the pregnancy.

Ever since a woman who had Crohn's disease during pregnancy was first described by Babson in 1946,[17] many additional reports on such cases have been published.[7–10,14,15] The site of disease (ileocecal, colorectal, small bowel, diffuse) had no influence on the course of pregnancy, the number of children born, or the number of miscarriages.[8] Overall, the early report by Crohn et al.,[14] together with other, subsequent clinical studies,[18] has led to the view that Crohn's disease, particularly if inactive at the time of conception, does not significantly affect the course of pregnancy or result in premature birth. A European survey of fertility and pregnancy in women with Crohn's disease[8] showed an increased incidence of prematurity (16% compared to 5% in other reports[9,14] and to 8% in the general population). A conclusive explanation for this difference is not possible. The European survey failed to differentiate between patients with active and those with inactive disease. Perhaps the increased prematurity rate mentioned above reflects that most patients had severe Crohn's disease.

Summarizing the various reports, one can conclude that Crohn's disease, like ulcerative colitis, does not seriously affect the course of a pregnancy. Especially those patients in whom the disease is quiescent can expect a normal pregnancy, as the inactive bowel disease does not expose the fetus to any risk.

Pregnancy in Patients with Active, Severe Inflammatory Bowel Disease

It is to be expected that a pregnancy can be affected adversely by any severe systemic illness such as active IBD, be it ulcerative colitis or Crohn's disease. Numerous reports indicate that this situation is the case.[13,14,18–23] For pregnant patients with severe active ulcerative colitis that requires aggressive medical management, the likelihood that a normal, live baby will be born is reduced, according to Willoughby and Truelove.[6] In 1976 Fielding[20] reported a 15% maternal mortality rate for such patients, but all of the more recent studies on this subject indicate a favorable prognosis. In fact, no maternal deaths due to a complication of pregnancy in 416 patients with active, severe ulcerative colitis have been reported in three publications.[6,11,12] The risk of spontaneous abortion and premature delivery is slightly but significantly higher in women who have active ulcerative colitis at the start of the pregnancy than is the risk in women in whom the disease is inactive.

Women who conceive while their bowel disease is active are examined at frequent intervals. Furthermore, they are treated for their disease according to the same principles applied to any other patient with IBD, i.e., with dietary measures, administration of drugs (prednisone, sulfasalazine, and other antibiotics as indicated), total parenteral nutrition, and if deemed absolutely necessary surgical intervention.

Considering the benign course for the pregnant patient with active ulcerative colitis, the coexistence of active colitis and pregnancy is not necessarily an indication for a therapeutic abortion, although this measure was suggested in the past by Crohn et al.[13] In one review, Willoughby and Truelove[6] expressed the currently predominating viewpoint on this subject. They reported that in several of their patients who underwent a therapeutic abortion the course of the disease and the difficulty of its control were the same as encountered in nonpregnant patients. Yet, however manageable and usually successful a pregnancy may be in patients with active ulcerative colitis, an increased risk for fetal complications exists. Therefore it is recommended that, whenever feasible, a woman with ulcerative colitis plan her family in such a way that she conceives when the disease is quiescent.

Patients with active or exacerbated Crohn's disease must expect an incidence of complications similar to that for active ulcerative colitis. According to Khosla et al.,[23] pregnancy is twice as likely to end in spontaneous abortion in patients with active Crohn's disease as it is in patients with quiescent disease. Baiocco and Korelitz[24] found that, among their patients with Crohn's disease whose pregnancies were associated with fetal complications, 62.5% had active disease. They also reported that disease activity rather than drug treatment was likely to be responsible for fetal complications, supporting earlier studies on the safety of corticosteroids and sulfasalazine given during pregnancy.[25]

In women with active Crohn's disease at conception, women whose disease is reactivated or develops de novo during pregnancy, or those who require bowel resection during pregnancy, the risks of premature delivery and spontaneous abortion are increased.[8,19,23,26] Thus women with Crohn's disease, just like women of childbearing age who have ulcerative colitis, must try to avoid conception when their bowel disease is active and severe.

Effect of Pregnancy on Established Inflammatory Bowel Disease

The notion that the normal function of the reproductive system may adversely affect the course of IBD has not been substantiated. Crohn et al.[14] reported a 16% rate of relapse of ileitis during pregnancy in patients in whom the disease was quiescent at conception. Among patients who had active disease, exacerbation occurred in 13%. For patients with ulcerative colitis, there appears to be no significant difference between quiescent and active disease in terms of disease progression, with reported relapse rates and exacerbation rates of 25% and 21%, respectively. deDombal et al.[2] found that the rates of exacerbation and relapse were the same for pregnant and nonpregnant patients.

Mogadam et al.[11] undertook an extensive study in which they found that the course of IBD during pregnancy was affected directly by the disease activity present before conception: in 75% of those patients whose disease had been quiescent to mild, it remained so (the remaining 25% experienced a flare-up), whereas 51% with active IBD continued to have moderate to severe disease; the other 49% improved. Specifically, among patients with active Crohn's disease, 65% continued to have signs and symptoms of active disease during pregnancy; for patients with ulcerative colitis the rate was 44%. For patients with quiescent to mild disease, however, there was no difference between Crohn's disease and ulcerative colitis; 78% and 72%, respectively, had no exacerbation during their pregnancy. Only 30% of the patients who had quiescent to mild IBD at term experienced moderate to severe disease postpartum, in contrast to 53% of those with active disease at term. The data presented by Mogadam et al. provided no link between a physiologically functioning reproductive system, i.e., pregnancy, and the induction or exacerbation of IBD.

The various reports, including in particular the report by Mogadam et al.,[11] can serve as a rational basis for the following conclusions: (1) The course of inflammatory bowel disease is independent of and not adversely affected by concurrent pregnancy or by the postpartum period. (2) Exacerbations during or after pregnancy are more frequent in patients with active or uncontrolled disease than in those whose disease is inactive. (3) Because the risk of relapse or exacerbation during pregnancy and postpartum is significantly lower in patients whose disease is inactive, treatments designed to induce and maintain remission during the course of pregnancy are likely to minimize reactivation or worsening of IBD.

Newly Diagnosed Inflammatory Bowel Disease in the Pregnant Patient

Previously undetected IBD may emerge for the first time during pregnancy. Early reports indicated a poor prognosis for these patients.[1,13,14] More recently, however, it was demonstrated that the outcome of pregnancy and the course of ulcerative colitis for these patients are not necessarily worse than in patients in whom the colitis was diagnosed previously.[6] In regard to Crohn's disease, the effect of new-onset disease during pregnancy is not as clearly defined. The incidence of newly diagnosed Crohn's disease during pregnancy is much lower than that for ulcerative colitis; up to 1975, a total of ten cases had been reported in the English-language literature.[7,9,14,19] Of these ten patients, six required surgery during pregnancy or shortly thereafter, and two more required operative treatment subsequently, not in association with a pregnancy. Of the 11 fetuses (one pair of twins), six died in conjunction with the operative treatment or shortly after birth. This figure indicates that the new onset of Crohn's disease during pregnancy carries a poor prognosis for the fetus. These observations have not been further substantiated.

Khosla et al.[23] reported on seven patients with Crohn's disease that occurred for the first time during the puerperium. They found the disease here not to be necessarily associated with a more severe course. Only two of these seven patients subsequently developed severe disease.

Obviously, the development of ulcerative colitis or Crohn's disease during pregnancy is a rare event; nevertheless, the obstetrician may be faced with this possibility, and a correct diagnosis must be made. It is beyond the scope of this review to elaborate on the differential diagnosis of gastrointestinal diseases that arise during pregnancy. As to the specific aspects of and approaches to the diagnosis of IBDs and their differential diagnosis, the reader is referred to specialized texts.[27,28]

Medical Management of Inflammatory Bowel Disease During Pregnancy

With regard to medical management, ulcerative colitis and Crohn's disease are treated in a similar manner. Corticosteroids, sulfasalazine, and dietary measures are the mainstays of therapy. Often patients with quiescent disease are already receiving maintenance therapy with corticosteroids and sulfasalazine; thus it is highly likely that these medications will be given during pregnancy, and their possible effects on the fetus must be considered.

Sulfonamides are known to displace bilirubin from serum albumin and thus have the potential to cause kernicterus. Sulfasalazine, although a sulfonamide, displaces bilirubin only weakly; and so it engenders a minimal risk, if any, for kernicterus in the newborn and can safely be given to patients during pregnancy and postpartum.[29,30] Sulfasalazine, however, reduces intestinal absorption of folate, a vitamin that is important for fetal development.[30,31] It follows that the administration of supplemental folic acid, a common practice in the care of IBD patients, is particularly important in the pregnant patient with IBD.

Corticosteroids have been found to increase the rates of abortion and stillbirths and to cause a variety of congenital defects in the offspring of rats and mice.[32] Studies in human subjects failed to substantiate similar complications and demonstrated no significant harmful effects.[33]

The effects of sulfasalazine and corticosteroids on fetal outcome were reviewed by Mogadam et al.[25] in 174 patients receiving sulfasalazine either alone or in combination with corticosteroids. They found that, compared to reported rates of fetal complications in the general population of pregnant women, the frequency of such complications was lower in their pregnant patients treated with either or both drugs or with neither drug. The authors concluded that, for the management of IBD in pregnant women, either or both medications may be used with just as much confidence as in nonpregnant patients. Although most authors have not reported adverse effects on the fetus, Schade et al.[22] noted a decrease in the birth weight (but no other effects on the fetus) in women treated with prednisone and sulfasalazine. They stated that the administration of these medications might be a contributing factor, in addition to the possible but unspecified effects of IBD on the nutritional status of the fetus.

The administration of 6-mercaptopurine or azathioprine has been associated with fetal complications in animals.[34,35] The use of these cytotoxic drugs in pregnant women has not been associated with an increased rate of fetal complications.[36] Several instances of successful pregnancy in patients with IBD who received cytotoxic immunosuppressive drugs have been reported.[24,37] Nevertheless, the use of cytotoxic drugs, especially during the first trimester or at the time of conception, is of concern. Until further information becomes available, it appears advisable to discontinue these medications when a patient with IBD (male or female) plans to have a child.

Metronidazole is frequently given to patients with IBD, especially if fistulae are present. Its safety during pregnancy has been demonstrated;[38] however, adequate and well controlled studies in pregnant women are not available. Thus metronidazole, like any other drug, is used during pregnancy only if the therapeutic benefits outweigh the potential risks to the fetus.

In certain situations, pregnant patients with IBD may require and benefit from the institution of total parenteral nutrition (TPN).[39] The activity of IBD may lead to chronic starvation and weight loss, which are associated with increased perinatal mortality.[40] TPN is instituted for such patients so their health and the pregnancy can be sustained and so an operative procedure can be postponed until some time after delivery. The specific guidelines for instituting and maintaining TPN are the same for pregnant as for nonpregnant patients.[41]

Surgical Treatment of Inflammatory Bowel Disease During Pregnancy

If possible, operative therapy of IBD is avoided during pregnancy. In general, however, the indications for surgery for IBD are the same regardless of whether the patient is pregnant. There are absolute indications for operation, including failure of medical management to relieve the patient's debilitating symptoms and to halt a course of deterioration. They also include the presence of peritonitis, gastrointestinal hemorrhage. complete and persistent bowel obstruction, abdominal abscess and perforation, carcinoma, and enterocutaneous fistulas. Obviously, some of these indications do not require instant operation during pregnancy, although under certain circumstances surgical intervention during pregnancy may be unavoidable.

Diffuse peritonitis, gastrointestinal hemorrhage, and complete bowel obstruction do not allow postponement of an operative procedure and therefore must be dealt with quickly. All of the other conditions that may arise during pregnancy and that require surgical treatment eventually, but not instantly and emergently, can be managed by medical means, including bowel rest, TPN, and, if indicated, administration of antibiotics. Another special circumstance requiring surgical therapy is the presence of an intraabdominal abscess that cannot be drained safely with a needle and catheter under sonographic or computed tomographic (CT) guidance. In such a situation, open surgical drainage may have to be performed. All other conditions that would normally require and receive operative treatment when they develop can be dealt with temporarily in pregnant patients by aggressive medical management, the mainstay of which is total bowel rest with TPN. Often a patient with extensive IBD but a patent gastrointestinal tract is nutritionally supported or balanced by enteral alimentation with elemental diets. This approach is also useful in patients who have internal or external fistulae with low output.

The pregnant patient who has an acute exacerbation of ulcerative colitis associated with sepsis and bleeding is likely to require operative treatment. An operation must be scheduled if aggressive medical management, including TPN and bowel rest, fails to ameliorate a worsening course. The conduct of the operation includes active participation of the obstetrician for fetal monitoring so that a spontaneous abortion can be avoided or if it occurs can be attended to appropriately.

The surgical procedure for a pregnant patient with fulminant colitis is conducted in the same manner as it would be in any other patient undergoing an emergency abdominal colectomy. After resection an end-ileostomy is constructed, and either a distal mucous fistula is placed or Hartmann's closure of the rectal stump is employed. If the pregnancy is in the second or third trimester, a Hartmann's closure below the peritoneal reflection, which is preferable to an intraabdominal closure, may be difficult to achieve; in such a case, a mucous fistula to the anterior abdominal wall may be necessary. However, this procedure may require retention of a rather lengthy sigmoid rectal stump, i.e., a diseased segment of intestine. The correct approach must be determined for the individual patient and is guided by the severity of the colitis, the presence or absence of intraabdominal contamination and sepsis, the status of the pregnancy, the overall condition of the patient, and the experience of the surgeon. However, it must be emphasized that the distal, diseased colorectal stump is kept as short as possible, so that the persistence of incapacitating colitis in the stump, with all its risks and complications, can be minimized. The ileostomy is fashioned in the right lower quadrant as in any other patient, the techniques applied being the same as those for nonpregnant patients who require an ileostomy. The postoperative care of the pregnant patient after an abdominal colectomy must be supervised closely and must involve the surgical and obstetric teams.

A patient with a complication of Crohn's disease that requires operative treatment during pregnancy is approached in the same way as is a pregnant patient with ulcerative colitis who is in need of a surgical procedure. Again, the surgical and obstetric teams must join forces to resolve the problem and, at the same time, preserve the pregnancy. A patient with diffuse peritonitis resulting from a free perforation requires an immediate operative procedure after being resuscitated with the administration of intravenous fluids, blood, and antibiotics. The site of perforation is resected in these patients, and both bowel ends are exteriorized and matured as ileostomy and mucous fistula. Peritoneal débridement and extensive lavage for removal of all infected, purulent material are essential.[42] It is advisable to leave the skin open and to permit healing by secondary intention. As to the choice of the incision for these patients, a transverse incision has been recommended because of its low potential for disruption and exenteration if deep wound sepsis should develop.[43]

If a patient requires laparotomy because of acute complete small-bowel obstruction that does not resolve with medical management, or if the patient has an acute gastrointestinal hemorrhage, the surgical principles that apply to nonpregnant patients are employed. Identifying the site of obstruction and determining if it must be resected with either a primary intestinal anastomosis or exteriorization of the proximal and distal intestinal stumps as enterostomas are intraoperative decisions made in accordance with the individual circumstances.

In patients with acute hemorrhage (a rare occurrence in Crohn's disease), preoperative identification of the site of hemorrhage, which may require endoscopy or angiography, is of the utmost importance. Bleeding into the upper gastrointestinal tract from a duodenal ulcer or bleeding from the colon must be ruled out prior to surgical treatment. Radiographic demonstration of the site of bleeding may not be a real option during the first and second trimesters of pregnancy but may be considered during the third trimester. If the exact site of bleeding cannot be verified preoperatively, laparotomy is an acceptable and almost invariably successful procedure by removing the intestinal segment involved with Crohn's disease.

Occasionally, patients with Crohn's disease develop rather large intraabdominal abscesses that require drainage. Drainage can be carried out either percutaneously under sonographic or CT guidance or, if these procedures are not possible, by means of an open surgical procedure. In the latter circumstance, the abscess is approached directly, incised, and drained. If possible, the free peritoneal cavity is not entered during the course of the drainage procedure. Soft drains are placed in the abscess cavity. Eventually, open drainage leads to establishment of an enterocutaneous fistula.

When recuperation is complete the patient continues to receive TPN; or, if possible, enteral nutrition is resumed with a regular or elemental diet.

The postoperative course is determined by the site of the fistula and the amount of its output. In most cases patients can be cared for easily at home. A definitive procedure is performed during the postpartum period after a complete evaluation of the status of the IBD by means of radiologic and endoscopic techniques.

For the patient with an established ileostomy following proctocolectomy, certain potential complications may arise on account of the stoma.[44] Of these complications, the most serious is intestinal obstruction, the incidence of which has been reported to be as high as 10%.[45] Aside from this seemingly specific complication of the pregnant patient with an ileostomy, the usual problems of a stoma (e.g., ostomy polyposis, bleeding, ulceration, retraction) can be seen during pregnancy and require the appropriate medical management. Rarely is a surgical revision of the enterostoma necessary during pregnancy.

The ileoanal anastomosis has become a real alternative to the standard proctocolectomy and ileostomy. Pregnancy and delivery in women with an ileoanal anastomosis can be expected to be uncomplicated.[46,47]

Conclusion

Concomitant pregnancy and IBD is not a rare event. The risk of exacerbation of the IBD during pregnancy is not greater than it would be in nonpregnant women. The risk to the fetus from the IBD is increased slightly in those patients who have active disease at the time of conception and in those who require operative therapy during the pregnancy. Direct surgical intervention becomes necessary if its postponement would jeopardize both mother and fetus. This situation is the case in the presence of acute diffuse peritonitis, complete small-bowel obstruction that does not resolve on medical management, and acute gastrointestinal hemorrhage. In all other situations in which operative therapy would be indicated for nonpregnant patients, a delaying strategy consisting of aggressive medical management with TPN is advisable. Definitive surgical therapy can then be performed during the postpartum period. Overall, the coincidence of pregnancy and acute IBD can be regarded as a complicating factor for either situation, but it is not a catastrophic event that precludes successful outcome of the pregnancy.

References

1. Abramson D, Jankelson IR, Milner LR. Pregnancy in idiopathic ulcerative colitis. Am J Obstet Gynecol 1951;61:121–129.
2. DeDombal FT, Watts JM, Watkinson G, et al. Ulcerative colitis and pregnancy. Lancet 1965;2:599–601.
3. Weeke E, Binder JH, Olsen PR, et al. The relationship between pregnancy and hemorrhagic proctocolitis. Acta Med Scand 1966; 180:179–185.
4. Webb MJ, Sedlack RE. Ulcerative colitis in pregnancy. Med Clin North Am 1974;58:823–837.
5. Morl M, Rottler H, Hartwich G. Pregnancy in ulcerative colitis and Crohn's disease. Dtsch Med Wochenschr 1977;102:387–391.
6. Willoughby CP, Truelove SC. Ulcerative colitis and pregnancy. Gut 1980;21:469–474.
7. DeDombal FT, Burton IL, Goligher JC. Crohn's disease and pregnancy. Br Med J 1972;3:550–553.
8. Mayberry JF, Weterman I. European survey of fertility and pregnancy in women with Crohn's disease: a case control study by European collaborative group. Gut 1986;27:821–825.
9. Fielding JF, Cooke WT. Pregnancy and Crohn's disease. Br Med J 1970;2:76–77.
10. Homan WP, Thorbjarnarson B. Crohn's disease and pregnancy. Arch Surg 1976;111: 545–547.
11. Mogadam M, Korelitz BI, Ahmed SW, et al. The course of inflammatory bowel disease

pregnancy and postpartum. Am J Gastroenterol 1981;75:265–269.
12. Nielsen OH, Andereasson B, Bondesen S, Jarnum S. Pregnancy in ulcerative colitis. Scand J Gastroenterol 1983;18:735–742.
13. Crohn BB, Yarnis H, Crohn EB, et al. Ulcerative colitis and pregnancy. Gastroenterology 1956;30:391–403.
14. Crohn BB, Yarnis H, Korelitz BI. Regional enteritis complicating pregnancy. Gastroenterology 1956;31:615–625.
15. Norton RA, Patterson JF. Pregnancy and regional enteritis. Obstet Gynecol 1972;40:711–712.
16. Maddix BL. Ulcerative colitis and pregnancy. Minn Med 1962;45:1097–1101.
17. Babson WW. Terminal ileitis with obstruction and abscess complicating pregnancy. N Engl J Med 1946;235:544–546.
18. Vender RJ, Spiro HM. Inflammatory bowel disease and pregnancy. J Clin Gastroenterol 1982;4:231–249.
19. Martimbeau PW, Welch JS, Weiland LH. Crohn's disease and pregnancy. Am J Obstet Gynecol 1975;122:746–749.
20. Fielding JF. Inflammatory bowel disease and pregnancy. Br J Hosp Med 1976;15:345–352.
21. Sorokin JJ, Levine SM. Pregnancy and inflammatory bowel disease: a review of the literature. Obstet Gynecol 1983;62:247–251.
22. Schade RR, VanThiel DH, Gavaler JS. Chronic idiopathic ulcerative colitis: pregnancy and fetal outcome. Dig Dis Sci 1984;29:614–619.
23. Khosla R, Willoughby, Jewell DP. Crohn's disease and pregnancy. Gut 1984;25:52–56.
24. Baiocco PJ, Korelitz BI. The influence of inflammatory bowel disease and its treatment on pregnancy and fetal outcome. J Clin Gastroenterol 1984;6:211–216.
25. Mogadam M, Dobbins WO III, Korelitz BI, Ahmed SW. Pregnancy in inflammatory bowel disease: the effects of sulfasalazine and corticosteroids on fetal outcome. Gastroenterology 1981;80:72–76.
26. Nielsen OH, Andreasson B, Bondesen S, et al. Pregnancy in Crohn's disease. Scand J Gastroenterol 1984;19:724–732.
27. Atlay RD, Weekes ARL. The treatment of gastrointestinal disease in pregnancy. Clin Obstet Gynecol 1986;13:335–347.
28. Kirsner JB, Shorter RG (eds). *Inflammatory Bowel Disease*. Philadelphia, Lea & Febiger, 1988.
29. Hensleigh PA, Kauffman RE. Maternal absorption and placental transfer of sulfasalazine. Am J Obstet Gynecol 1977;127:443–444.
30. Azad Khan AK, Truelove SC. Placental and mammary transfer of sulfasalazine. Br Med J 1979;2:1553.
31. Halstead CH, Gandhi G, Tamura T. Sulfasalazine inhibits the absorption of folates in ulcerative colitis. N Engl J Med 1984;305:1513–1517.
32. Fraser FC, Feinstat TD. Production of congenital defects in the offspring of pregnant mice treated with cortisone. Pediatrics 1951;8:527–533.
33. Bongiovanni AM, McPadden AJ. Steroids during pregnancy and possible fetal consequences. Fertil Steril 1959;11:181–186.
34. Scott JR. Fetal growth retardation associated with maternal administration of immunosuppressive drugs. Am J Obstet Gynecol 1977; 128:688–676.
35. Reimers TJ, Sluss PM. 6-Mercaptopurine treatment of pregnant mice: effects on second and third generations. Science 1978;201:65–67.
36. Blatt J, Mulvihill JJ, Ziegler JL, et al. Pregnancy outcome following cancer chemotherapy. Am J Med 1980;69:828–832.
37. Levy N, Roisman I, Teodor I. Ulcerative colitis in pregnancy in Israel. Dis Colon Rectum 1981;24:351–354.
38. Sutherland JM, Light IJ. The effect of drugs upon developing fetus. Pediatr Clin North Am 1965;12:781–806.
39. Rosenberg IH. Nutritional support in inflammatory bowel disease. Gastroenterology 1979;77:393–398 (editorial).
40. Chase HC. Infant mortality and weight at birth: 1960 United States cohort. Am J Public Health 1969;59:1618–1628.
41. Jacobson LB, Clapp DH. Total parenteral nutrition in pregnancy complicated by Crohn's disease. JPEN 1987;2:93–96.
42. Hudspeth AS. Radical surgical débridement in the treatment of advanced generalized bacterial peritonitis. Arch Surg 1975;110:1233–1236.
43. Block GE, Schraut WH. Complications of the surgical treatment of ulcerative colitis and Crohn's disease. In Kirsner JB, Shorter RG (eds): Inflammatory Bowel Disease. Philadelphia, Lea & Febiger, 1988;685–713.
44. Gopal KA, Amshel AL, Shonberg IL, et al. Os-

tomy and pregnancy. Dis Colon Rectum 1985;28:912–916.
45. Hudson CN. Ileostomy in pregnancy. Proc R Soc Med 1972;65:281–283.
46. Pezim ME. Successful childbirth after restorative proctocolectomy with pelvic ileal reservoir. Br J Surg 1984;71:292.
47. Metcalf A, Dozois RR, Beart RW, Wolff BG. Pregnancy following ileal pouch-anal anastomosis. Dis Colon Rectum 1985;28:859–861.

9
Renal Stones During Gestation

Patsy Maikranz, Fredric L. Coe, and
Marshall D. Lindheimer

Pregnant women with renal calculi present several special problems. They have an increased incidence of urinary tract infections,[1,2] and stone passage can cause premature labor.[3] They present both diagnostic and management problems; a stone obstructing the ureter may be difficult to diagnose because of anatomic changes that occur in the collecting system. Management of stones is more complicated than in the nongravid state because most of the therapeutic interventions commonly employed in stone disease can potentially adversely affect the developing fetus. This chapter reviews the incidence, etiology, diagnosis, natural history, and management of kidney stones during pregnancy.

Incidence

A nonselective survey by our group in 1978 revealed no increased incidence of kidney stones either developing or recurring during gestation; the actual incidence is reported to range between 0.03% and 0.53%.[1,4,5] Despite these and several other reports, the true incidence of kidney stones during pregnancy is still not clearly established because almost all of this literature is selective in nature: Patients are identified and reported because their stones have become symptomatic.

One's geographic location may influence the incidence of stone formation. Drach[6] has stated that mountainous and southern areas of the United States have the highest incidence of nephrolithiasis. If one includes the Western hemisphere, calcium nephrolithiasis appears to occur most frequently in the more industrialized countries.[7] Presumably the higher incidence of calcium stones in these countries is due to increased dietary intake of protein. Large amounts of protein in the diet lead to increased acid production and increased urinary calcium and uric acid excretion.[8] Augmented excretion of these ions increases the amount of substrate available for stone formation. Given these geographic variations, it is difficult to accurately compare patients' results with those in the literature, which come from diverse locations.

Etiology

Stone formation may be facilitated by anatomic and physiologic changes that occur normally during gestation. Dilatation of the calyces, ureters, and renal pelves—the renal collecting system—occurs during the first trimester and lasts until parturition. Ureteral peristaltic activity decreases as these anatomic changes evolve.[9] Hormonal and mechanical factors have each been implicated in causing the gestational changes in the collecting system, but a combination of these factors seems the most likely explanation. Urinary stasis may be the sequela of decreased ureteral peristalsis occurring in the dilated collecting system; this situation can lead to urinary tract infection, which can facilitate growth of preexisting stones or allow new stones to form.

A 25 to 50% increase in glomerular filtration rate (GFR) occurs early during the first trimester of normal pregnancy and persists until delivery.[10] This increase in GFR is accompanied by an increase in the filtered loads of calcium, sodium, uric acid, and other urinary ions. Urine must be severely oversaturated from dehydration, hyperuricosuric, or highly acidic for either uric acid crystalluria to develop or stones to form.[11] In the latter respect, the tendency of urinary pH to increase above nonpregnant means[12,13] affords some protection against uric acid stone formation during gestation.

The results of measurements by our laboratory of urinary ions during pregnancy resemble the observations of others[14,15] and reveal that urinary calcium excretion may increase by one- to twofold during normal gestation. We measured urinary supersaturation, an expression of concentration and solubility with respect to calcium salts, and noted that this parameter markedly increases during pregnancy.[13,14] Such an increase in urinary supersaturation would be expected to promote stone formation; but, interestingly, women with nephrolithiasis form no more stones when pregnant than when not. Perhaps mechanisms exist whereby these hypercalciuric gravidas with increased urinary supersaturation are protected against kidney stone formation. One possible mechanism is the increased urinary excretion of both magnesium and citrate (known inhibitors of calcium kidney stones) during normal pregnancy. Another possibility involves nephrocalcin, an acidic glycoprotein normally present in urine, which has been demonstrated to inhibit calcium oxalate stone formation in humans.[16] Excretion of nephrocalcin increases markedly during normal human pregnancy.[17] Thus the hypercalciuric and supersaturated urine of normal gravidas may be partially offset by the presence of increased amounts of inhibitory factors, especially nephrocalcin.

Why gravidas are hypercalciuric is of interest. One reason for gestational hypercalciuria may be high circulating levels of 1,25-dihydroxyvitamin D (calcitriol), reported by Kumar et al.[18] Calcitriol acts on the intestinal epithelium to increase calcium absorption, which can result in hypercalciuria in humans.[19] Hypercalciuria from calcitriol is due to decreased renal calcium reabsorption owing to calcitriol's inhibitory effect on parathyroid hormone (PTH) secretion[20–22] rather than to an increased filtered load of calcium. PTH normally stimulates renal tubular calcium reabsorption in the distal convoluted tubules and the connecting segments[23] as well as in the thick ascending limb of the loop of Henle.[24] If PTH were suppressed by calcitriol, renal calcium reabsorption could be decreased. Both increased serum calcium[25,26] and calcitriol[20] are capable of suppressing PTH secretion. PTH levels in gravidas have been evaluated by several groups, but unfortunately no consensus is available because of the widely varying results reported. Some authors have attributed the variations to differences in the assay methods used.[15] Gestational hypercalciuria may merely represent an overadaptation of calcitriol-mediated gastrointestinal calcium absorption in excess of what the fetus needs to create its skeleton, approximately 25 g. The precise mechanism(s) responsible for the elevated calcitriol levels during pregnancy is unknown but may be related to changes in other hormones that occur early in pregnancy.

Struvite kidney stones, the "infection stones," account for approximately 20% of all kidney stone disease in women but form only when the urinary tract is infected with organisms that possess the enzyme urease. These infection stones can form around a nidus of calcium, uric acid, or cystine or may be only magnesium ammonium phosphate (struvite) admixed with calcium carbonate, an inevitable calcium precipitate in the alkaline urine that characterizes this pathologic entity and gestation. Women who produce pure struvite stones present with infection rather than stone passage, are classically normocalciuric, have undergone numerous urologic surgical procedures, and have stone formation in the contralateral kidney.[27] Struvite stones can form in completely normal collecting systems, but the anatomic changes of the urinary tract during pregnancy may predispose to struvite stone formation. Little clinical information is available regarding struvite stone disease during pregnancy.

Types and relative percentages of kidney stones occurring in nongravid women studied at our institution are given in Table 9-1; shown for comparison are data from stone-forming men. Most of the available review articles on nephrolithiasis during gestation that provide case details have incomplete information on kidney stone crystallography; but from the limited data reported, it appears that the frequencies of occurrence are about the same as in the nongravid women listed in Table 9-1.

Diagnosis

Renal colic is amongst the most common nonobstetric cause of abdominal pain requiring hospitalization during pregnancy.[28] Pain remained the most common presenting symptom in the series of pregnant patients reported by Jones et al.,[29] Lattanzi and Cook,[30] Drago et al.,[3] and Perreault et al.[31] Other common presenting signs and symptoms are the same as those in nongravid stone-formers (Table 9-2).

TABLE 9-1. Types of stones formed and their frequency.

	Women		Men	
Stone composition	No.	%	No.	%
Calcium oxalate	244	58.0	737	73.0
Calcium phosphate	12	3.0	20	2.0
Calcium + uric acid	24	5.5	113	11.0
Uric acid	9	2.0	25	2.0
Cystine	8	2.0	4	1.0
Struvite	76	18.0	21	2.0
Unknown	49	11.5	89	9.0
Total	422		1009	

TABLE 9-2. Presentation of nephrolithiasis during pregnancy.

Symptoms	Signs
Fever	Abdominal pain
Nausea	Pyuria
Emesis	Crystalluria
Frank pain	Hematuria
Abdominal pain	Costovertebral angle tenderness
Dysuria	
Urgency	

Modified from Maikranz et al.,[35] with permission.

A complete history and physical examination form the basis of the approach to the gravida with suspected nephrolithiasis. Specific areas to be covered are age at onset of stone disease, family members with stone disease, diet and medications, prior urinary tract infections, results of previous abdominal radiographs, and other medical disorders or prior surgeries.[32]

Renal ultrasonography is well suited for evaluating stone disease during pregnancy because it does not expose the fetus to potentially harmful ionizing radiation. In a prospective comparison study with excretory urograms in nongravidas, renal ultrasonography had a specificity of 74% and a sensitivity of 98%.[33] Unfortunately, most symptomatic stones during pregnancy are located in the ureters,[5] which may make ultrasonography a less sensitive test in this clinical situation. Additionally, the physiologic dilatation of the collecting system in gravidas makes delineation of obstruction from kidney stones more difficult than in nongravidas.

Currently, most standard excretory urograms deliver less than 1.5 rad to the developing fetus. When larger doses of radiation (5–15 rad) are delivered to the maternal pelvis, the risk of congenital anomalies increases from 1 to 3%.[34] Effects on the fetus of amounts smaller than 5 rad exposure are not well documented. Thus radiographs are obtained of only the involved side, the maternal pelvis is shielded as much as possible, and the number of images obtained is limited. Current indications for obtaining an excretory urogram to evaluate stone disease during gestation include (1) symptoms of renal calculi unresponsive to conservative therapy; (2) a decline in renal function in association with symptoms of kidney stones; and (3) severely symptomatic pyelonephritis refractory to antibiotics when ultrasonography fails to identify a stone. These guidelines are especially valid for a gravida with a history of stone disease.[35] It is important to bear in mind that radiologic studies necessary for proper maternal health care delivery are not withheld for fear of fetal radiation exposure, especially after the midtrimester. Magnetic resonance imaging (MRI) does not expose the fetus to ionizing radiation and has been used to evaluate both fetal and maternal anatomy.[36] MRI has promise on theoretical grounds, but little information regarding its use for visualizing kidney stones during pregnancy is available.

Spontaneous passage of asymptomatic kidney stones occurred in 70% of the patients reported by Coe et al.,[1] Jones and Correa,[29] Cumming and Taylor,[37] Lattanzi and Cook,[30] Drago et al.,[3] Perreault et al.,[31] and Cass et al.[38] Those remaining required interventions during either pregnancy or the puerperium. The percentage of women requiring intervention was higher (70%) in the series of Horowitz and Schmidt[5] than in the other reported series. Progressive ureteral dilatation may account for the high rate of spontaneous passage during gestation. Loughlin and Bailey[39] have reported a series of eight gravidas who had stent catheters placed for ureteral obstruction by stones. All patients were successfully managed without induction of premature labor by the procedure; and the only complication was encrustation of the end of one catheter, necessitating its replacement.

Stones that occur during gestation are more frequently diagnosed during the second and third trimesters[30,40]; the increased rate of occurrence later in gestation may be due to passage of previously asymptomatic stones through the dilated ureter to the level of the pelvic brim. The ureters resume their normal caliber at this level and may halt stone migration and block urine flow, causing renal colic. Multiparas are more likely to develop symptomatic stones during pregnancy than are primiparas,[29,40] which may merely reflect the fact that stone disease increases with age in both sexes.[6]

Even though the right ureter is reportedly more dilated than the left during pregnancy, symptomatic stones occur without a difference in laterality. This equality may be attributable to the fact that both sides of the collecting system are dilated during normal pregnancy, and the difference between the dilated right and left sides may not be of clinical importance in stone disease. However, preexisting anatomic abnormalities may directly influence the location of migrating or forming stones.

Natural History

Conservative management alone results in spontaneous passage of at least 50% of symptomatic stones in most of the reported series of gravidas. Up to 30% require operative or nonoperative interventions during gestation (Table 9-3); in the remaining gravidas procedures can often be avoided until after delivery. If intervention is required, the first option is usually cystoscopy to remove the stone or relieve the obstruction by passing a stent catheter.[29,39] If cystoscopy is unsuccessful or undesirable, a percutaneous nephrostomy or an open procedure must usually be performed.[5] None of the surgical interventions referred to in Table 9-3 caused fetal demise or serious maternal complications. All interventions, operative and nonoperative, are delayed until after delivery when feasible. Specific indications for urologic intervention for stone disease during pregnancy are listed in Table 9-4.

Urinary tract infection is the most common nonobstetric complication of stone disease during gestation, reportedly occurring in as many as 10 to 20% of gravidas with

TABLE 9-3. Intervention for kidney stones during pregnancy.

Reference	No. stones/ No. pregnancies	Stone passed spontaneously (No.)	Nonoperative procedures[a] (No.)	Operative procedures (No.)	Other[b]
Cass et al.[38]	24/24	18	3	3 (1 PP)[c]	0
Coe et al.[1]	20/15	20	0	0	0
Cumming & Taylor[37]	13/13	5	3	2	3
Drago et al.[3]	12/9	9	1	2	0
Horowitz & Schmidt[5]	17/17	5	10	2	0
Jones & Correa[29]	21/20	12	2 (1 PP)[c]	5 (4 PP)[c]	2
Lattanzi & Cook[30]	11/11	7	2	0	2
Perreault et al.[31]	19/19	14	0	4 (1 PP)[c]	1

[a]Cystoscopy, percutaneous nephrostomy, basket extraction.
[b]Patients either lost to follow-up or without clinical evidence of stone passage.
[c]PP = number of procedures done postpartum.
(Modified from Maikranz et al.,[35] with permission.)

stone disease.[1,2] Oral antibiotic therapy is indicated for lower urinary tract infection. In pregnancy pyelonephritis especially when related to calculous disease requires hospitalization, parenteral antibiotic therapy, and perhaps intervention to remove an obstructing stone.

The most common obstetric complication of kidney stones is precipitation of premature labor by renal colic.[3] Standard Tocolytic therapy is usually beneficial in arresting premature labor in these circumstances. Risk of premature labor appears to cease completely with stone passage or removal. Armon[41] reported a patient with a large bladder stone that impeded labor and vaginal delivery, necessitating cesarean section for delivery. Honore[2] reported results of a retrospective study of stone disease and the incidence of spontaneous abortions in women from Newfoundland. He concluded that a history of stone disease predisposed gravidas to spontaneous abortion. He hypothesized that abnormalities in calcium homeostasis may have caused some abnormal hormone secretion by the corpus luteum or placenta, had teratogenic effects on the zygote, or cause myometrial hyperirritability that accounted for the increased number of spontaneous abortions he noted in this group of women.

TABLE 9-4. Indications for urologic interventions for stone disease during pregnancy.

Persistent infection proximal to an obstructing stone
Intractable pain
Renal colic inciting premature labor that is unresponsive to β-Tocolytic drug therapy
Worsening renal function with persistent ureteral obstruction
Obstruction of a solitary kidney

When counseling gravid stone-formers or stone-forming women who are contemplating pregnancy, one may be asked how pregnancy will effect their stone disease and vice versa. Data from our retrospective study of 1978 together with our continuing clinical experience suggests that pregnancy has no adverse effects on stone disease; that is, neither severity nor activity of stone disease appears to change during gestation. Gravidas with preexisting stone disease are at greater risk to develop urinary tract infections and require close surveillance to facilitate early detection and therapy. As referred to above, the incidence of spontaneous abortions may be increased in this group, but this observation needs more careful investigation before a true cause-and-effect relation can be established.

Management

Conservative therapy, with hydration, bed rest, and analgesics, is the way to start the management of gravidas with symptomatic stones. This approach alone is likely to result in spontaneous passage of at least 50% of symptomatic stones during gestation.[29] Meperidine and morphine are narcotic analgesics that are of benefit in alleviating renal colic. Both drugs have been used during pregnancy and reportedly cross the placenta rapidly.[42,43] Meperidine may cause less fetal respiratory depression than morphine[44]; the respiratory depression from meperidine usage during labor appears to be time- and dose-dependent and may be due to persistent unmetabolized drug.[45] Oral codeine-containing preparations have been associated with fetal defects in four large studies[46] and are avoided if possible during gestation. The nonsteroidal antiinflammatory agent indomethacin has been used successfully as prophylaxis against recurrent ureteral colic[47]; it has also been found to be as efficacious as meperidine in alleviating ureteral colic[48] in nongravid adults. Complications of indomethacin usage during pregnancy include early closure of the ductus arteriosus, resulting in primary pulmonary hypertension and other congenital anomalies. This drug has also been demonstrated to impair labor and is not ideal for use during gestation.[46]

Crystallography is performed on any retrieved stone, because valuable information may be gained that can influence long-term postpartum therapy. Metabolic evaluation of gravid stone patients is delayed until at least 3 months after parturition and lactation, as changes in metabolism of calcium, urate, and other substances during these periods produce urine and serum chemistry values that are not truly reflective of the patient's stone disease. Because routine testing can be prohibitively expensive and given the current emphasis on cost containment in medicine, it is important to individualize the diagnostic approach to each patient. Laboratory evaluation of the patient with kidney stones is often influenced by a careful history. A suggested initial postpartum evaluation of nephrolithiasis is outlined in Table 9-5. Attempting to delay nonoperative and operative urologic interventions until after delivery is the most conservative approach; unfortunately, it is not always possible for both maternal and fetal reasons. Intrapartum interventions have become less hazardous for both mother and fetus with advances in perinatal health care delivery.

Most of the frequently prescribed medications for patients with metabolic stone disease have unacceptable side effects for both gravida and fetus. Thiazides are commonly used to treat hypercalciuric states[49,50]; they work by increasing distal tubular calcium reabsorption so as to decrease calcium excretion and thus the substrate for stone formation.[7] This class of drugs may result in fetal congenital anomalies if used during the first trimester[46] and has anecdotally been reported to cause fetal thrombocytopenia,[51] hypoglycemia,[52] and hyponatremia.[52] These drugs may also impair the normal volume expansion of pregnancy, which is considered important for optimal fetal development. In-

TABLE 9-5. Postpartum evaluation of stone disease.

Serum levels
Calcium
Phosphorus
Uric acid
Creatinine
Spot urine
Urinalysis
Culture
Fasting pH
Urine chemistries (24-hour collection)
Creatinine
Uric acid
Calcium
Phosphorus
Oxalate
Citrate
Stone analysis (crystallography if available)
Radiology
Flat plate of abdomen
Excretory urogram (not always indicated)

(Modified from *Kidney International* (reference 32) with permission.)

stead of using thiazides to treat hypercalciuric disorders during pregnancy, patients are instructed to increase fluid intake and avoid excessive ingestion of calcium until after parturition, when thiazide therapy can safely be resumed or started. To keep the urine appropriately diluted, output must be at least 2.0 to 2.5 liters per day. For severe stone disease, a lower calcium diet may also be of value. Because most commonly prescribed prenatal vitamins contain only 100 to 200 mg of calcium they may be continued in the patient with mild stone disease; however, a non-calcium-containing vitamin may be preferable for those with more severe disease. Thiazides have been measured in human breast milk in low concentrations by Werthmann and Krees.[53] These authors suggested that the risk to the nursing infant is small, but the possibility of a pharmacologic effect does exist.

Purine metabolism is stopped at the level of hypoxanthine by xanthine oxidase inhibitors. These drugs are frequently used for the treatment of uric acid stone disease and decrease both serum and urinary uric acid. Effects of this drug on the human fetus are not established, but no adverse effects on fetal animals have been cited. Because definitive information is lacking, these drugs are also probably best avoided during pregnancy. Treatment of uric acid stone disease during pregnancy includes the following measures: increasing the urine volume, modestly limiting purines in the diet, and alkalinizing the urine to at least pH 6.0 using any oral alkali preparation. An appropriate dose of oral alkali is 0.5 to 2.0 mmol/kg/day in three or four divided doses.

Cystinuria is a rare hereditary disorder that can result in renal tubule transport defects and stone formation. These patients are often treated with D-penicillamine. Part of this drug's therapeutic action is to form the disulfide of cystine and penicillamine, which is much more soluble than cystine (cystine-cystine) and thus more readily excreted.

D-Penicillamine has been reported to be teratogenic in both animals and humans.[46] Solomon et al.[54] described cutaneous fetal anomalies in infants born to mothers who took this drug during gestation. Gregory and Mansell[55] reported their experience with a series of gravidas who had cystinuria. These investigators discontinued D-penicillamine therapy between the 6th and 20th weeks of gestation because of the teratogenic potential in the 22 patients they cared for during pregnancy. They reported no fetal anomalies related to maternal penicillamine ingestion prior to the 6th week or after the 20th week of gestation. Alternate ways to treat cystinuria in gravidas are to increase urine volume to between 2.0 and 2.5 liters/day and to mildly alkalinize the urine using oral alkali preparations, as stated above for uric acid nephrolithiasis.

Asymptomatic bacteriuria during pregnancy requires prompt treatment (based on the results of a urine culture and antibiotic sensitivity determination),[56] an approach that is even more fundamental when calculous disease is present. One must be aware of the risks inherent in prescribing drugs to pregnant women, and in this respect the *Physician's Desk Reference* (*PDR*) includes information concerning the safety during pregnancy of individual antibiotics, information that is updated yearly. When symptoms are present, treatment may have to be empiric, in which case we recommend ampicillin, sulfonamides (except near term), cephalosporins, or nitrofurantoin. Therapy is maintained for periods longer than the traditional 10- to 14-day course, as the stone represents a nidus of infection that may be resistant to sterilization. We routinely treat symptomatic disease in pregnant patients with renal stones for 3 to 5 weeks and suggest that this period be followed by antibiotic suppression for the duration of the pregnancy (i.e., nitrofurantoin 100 mg each evening).

When a stone obstructs the ureter, surgery may be necessary; in this respect, approximately 30% of patients reported in the literature have undergone an operative procedure such as open lithotomy and percutaneous nephrostomy.[3,37,57] Rigid uteroscopy (direct vision) performed under general anesthesia has also been utilized early in gestation.[57] However, a less invasive technique, the placement of a stent, has made it possible to manage gravidas through gestation, deferring more definitive surgical procedures to the postpartum period.[39,57] For example, Loughlin and Bailey[39] were able to place double J stents through a cystoscope into the ureters of eight pregnant women with stone disease using only local anesthesia. They recommended changing the stents at 8-week intervals and administering 1 g of ampicillin in divided doses daily for antibiotic prophylaxis.

Finally, performance of any surgical procedure during pregnancy requires that surgeon and anesthesiologist be familiar with the physiologic changes of normal pregnancy. Evaluation of the pregnant patient requiring non-obstetric-related surgery, detailed by Barron,[58] is beyond the scope of this chapter. Briefly, anesthesiologists must be aware that caval or aortic compression by the uterus when the patient is positioned supine may decrease cardiac output. This posture may also lower PO_2. Furthermore, some normal laboratory values during gestation differ from those of nonpregnant subjects; for example, PCO_2 and plasma HCO_3^- decrease 10 mm Hg and 4 to 5 mEq/liter, respectively, and the pH is 0.4 units higher.

Experience with extracorporeal shock wave lithotripsy in either animal or human gravidas is currently limited. Lithotripsy for the treatment of nephrolithiasis during pregnancy cannot be recommended until its safety during gestation is firmly established in animals and humans.

Our preliminary data on urinary mineral excretion and inhibitors of calcium kidney stones suggest that excretion of these inhibitors increases during the first trimester of normal pregnancy and remains elevated throughout gestation. After delivery the levels fall rapidly to values normal for nongravid non-stone-forming women. This observation is important because the hypercalciuria of pregnancy also develops during the first trimester, persists until delivery, and falls rapidly postpartum to levels at or below prepregnancy levels. This work[13,17] together with previous work on kidney stones and nephrocalcin[16] has led us to speculate that these glycoproteins may play a biologic role in preventing kidney stone formation during gestational hypercalciuria. Ongoing clinical work in this area on both normal and stone-forming gravidas is needed to firmly establish this relation.

Acknowledgment. Supported by grant AM 33949 from the National Institute of Health.

References

1. Coe FL, Parks JH, Lindheimer MD. Nephrolithiasis during pregnancy. N Engl J Med 1978;298:324–326.
2. Honore LH. The increased incidence of renal stones in women with spontaneous abortion: a retrospective study. Am J Obstet Gynecol 1980;137:145–146.
3. Drago JR, Rohner TJ Jr, Chez RA. Management of urinary calculi in pregnancy. Urology 1982;20:578–581.
4. Miller RD, Kakkis J. Prognosis, management and outcome of obstructive renal disease in pregnancy. J Reprod Med 1982;28:199–201.
5. Horowitz E, Schmidt JD. Renal calculi in pregnancy. Clin Obstet Gynecol 1985;28:324–338.
6. Drach GW. Urinary lithiasis. In Harrison JH, Gitter RF, Perlmutter AD, et al (eds): Campbell's Urology. Philadelphia, Saunders, 1978;779–878.
7. Lemann J Jr, Adams ND, Gray RW. Urinary calcium excretion in human beings. N Engl J Med 1979;301:535–541.
8. Robertson WG, Peacock M, Hodgkinson A. Dietary changes and the incidence of urinary calculi in the U.K. between 1958 and 1976. J Chronic Dis 1979;32:469–476.
9. Waltzer WC. The urinary tract in pregnancy. J Urol 1981;125:271–276.
10. Lindheimer MD, Katz AI. The kidney in pregnancy. In Brenner B, Rector F (eds): The Kidney. Philadelphia, Saunders, 1986;1253–1296.
11. Favus MJ, Coe FL. Disorders of stone formation. In Brenner B, Rector F (eds): The Kid-

ney. Philadelphia, Saunders, 1986;1403–1442.
12. Lim VS, Katz AI, Lindheimer MD. Acid-base regulation in pregnancy. Am J Physiol 1976;231:1764–1770.
13. Maikranz P, Parks JH, Holler JL, et al. Gestational hypercalciuria causes pathological urine calcium oxalate supersaturations. Kidney Int 1989;36:108–113.
14. Kristensen C, Abraham PA, Davis M, et al. Hypercalciuria and risk factors for calcium nephrolithiasis during pregnancy. Kidney Int 1984;327:144 (abstract).
15. Pederson EB, Johannesen P, Kristensen S, et al. Calcium, parathyroid hormone and calcitonin in normal pregnancy and preeclampsia. Gynecol Obstet Invest 1984;18:156–164.
16. Nakagawa Y, Abram V, Parks JH, et al. Urine glycoprotein crystal growth inhibitors. J Clin Invest 1985;76:1455–1462.
17. Wabner C, Sirivongs D, Maikranz P, et al. Evidence for increased excretion in pregnancy of nephrocalcin, a urinary inhibitor of crystal growth. Kidney Int 1987;31:359 (abstract).
18. Kumar R, Cohen WR, Silva P, et al. Elevated 1,25-dihydroxyvitamin D plasma levels in normal human pregnancy and lactation. J Clin Invest 1979;63:342–344.
19. Kaplan RA, Haussler MR, Deftos LJ, et al. The role of 1,25-dihydroxyvitamin D in the mediation of the intestinal hyperabsorption of calcium in primary hyperparathyroidism and absorptive hypercalciuria. J Clin Invest 1977;59:756–760.
20. Chertow BS, Baylink DJ, Wergedal JF, et al. Decrease in immunoreactive parathyroid hormone in rats and in parathyroid hormone secretion in vitro by 1,25-dihydroxycholecalciferol. J Clin Invest 1975;56:668–678.
21. Slatopolsky E, Weerts C, Thielan J, et al. Marked suppression of secondary hyperparathyroidism by intravenous administration of 1,25-dihydroxycholeciferol in uremic patients. J Clin Invest 1984;4:2136–2143.
22. Norris KC, Kraut JA, Andress DL, et al. Intravenous calcitriol for severe secondary hyperparathyroidism in dialysis patients. Kidney Int 1985;27:168 (abstract).
23. Shareghi GR, Stoner LC. Calcium transport across segments of the rabbit distal nephron in vitro. Am J Physiol 1978;235:F367–375.
24. Bourdeau JE, Burg MB. Effect of parathyroid hormone on calcium transport across the cortical thick ascending limb of Henle's loop. Am J Physiol 1980;239:F121–126.
25. Brown EM. Parathyroid hormone secretion in vivo and in vitro: regulation by calcium and other secretagogues. Miner Electrolyte Metab 1982;8:130–150.
26. Sherwood LM, Mayer GP, Ramberg CF, et al. Regulation of parathyroid hormone secretion: proportional control by calcium, lack of effect of phosphate. Endocrinology 1968;83:1043–1051.
27. Kristensen C, Parks JH, Lindheimer M, et al. Reduced glomerular filtration rate and hypercalciuria in primary struvite nephrolithiasis. Kidney Int 1987;32:749–753.
28. Folger GK. Pain and pregnancy. Obstet Gynecol 1955;5:513–518.
29. Jones WA, Correa RJ Jr, Ansell JS. Urolithiasis associated with pregnancy. J Urol 1978;122:333–335.
30. Lattanzi DR, Cook WA. Urinary calculi in pregnancy. Obstet Gynecol 1980;56:462–466.
31. Perreault JP, Paquin JM, Faucher R, et al. Urinary calculi in pregnancy. Can J Surg 1982;25:453–454.
32. Smith L. Calcium-containing renal stones. Kidney Int 1978;13:383–389.
33. Ellenbogen PH, Scheible FW, Talner LB, et al. Sensitivity of gray scale ultrasound in detecting urinary tract obstruction. Am J Roentgenol 1978;130:731–733.
34. Swartz HM, Reichling BA. Hazards of radiation exposure for pregnant women. JAMA 1978;239:1907–1909.
35. Maikranz P, Coe FL, Parks JH, et al. Nephrolithiasis in pregnancy. Am J Kidney Dis 1987;9:354–358.
36. Lowe TW, Weinreb J, Santos-Ramos R, et al. Magnetic resonance imaging in human pregnancy. Obstet Gynecol 1985;66:629–633.
37. Cumming DC, Taylor PG. Urologic and obstetric significance of urinary calculi in pregnancy. Obstet Gynecol 1979;53:505–508.
38. Cass AS, Smith CS, Gleich P. Management of urinary calculi in pregnancy. Urology 1986;28:370–372.
39. Loughlin KR, Bailey RB. Internal ureteral stents for conservative management of ureteral calculi during pregnancy. N Engl J Med 1986;315:1647–1649.
40. Broaddus SB, Catalano PM, Leadbetter GW, et al. Cessation of premature labor following removal of distal ureteral calculus. Am J Obstet Gynecol 1982;143:846–848.

41. Armon PJ. Obstructed labor due to a vesical calculus. Br Med J 1977;2:498–499.
42. Crawford JS, Rudofsky S. The placental transmission of pethidine. Br J Anaesth 1965;37:929–932.
43. Fisher DE, Paton JB. The effect of maternal anesthetic and analgesic drugs on the fetus and newborn. Clin Obstet Gynecol 1974; 17:275–287.
44. Way WL, Costley EC, Way EL. Respiratory sensitivity of newborn infant to meperidine and morphine. Clin Pharmacol Ther 1965; 6:454–461.
45. Belfrage P, Boreus LO, Hartvig P, et al. Neonatal depression after obstetrical analgesia with pethidine: the role of injection-delivery time interval and the plasma concentrations of pethidine and norpethidine. Acta Obstet Gynecol Scand 1981;60:43–49.
46. Briggs GG, Freeman RK, Yaffe SJ. Drugs in Pregnancy and Lactation. Baltimore, Williams & Wilkins, 1986;102/c–103/c.
47. Grenabo L, Holmlund D. Indomethacin prophylaxis against recurrent ureteral colic. Scand J Urol Nephrol 1984;18:325–327.
48. Lehtonen T, Kellokumpu, I, Permi J, et al. Intravenous indomethacin in the treatment of ureteric colic. Ann Clin Res 1983; 15:197–199.
49. Coe FL. Treated and untreated recurrent calcium nephrolithiasis in patients with idiopathic hypercalciuria, hyperuricosuria, or no metabolic disorder. Ann Intern Med 1977; 87:404–410.
50. Yendt ER, Cohanim M. Prevention of calcium stones with thiazides. Kidney Int 1978; 13:397–409.
51. Rodriguez SU, Leiken SL, Hiller MC. Neonatal thrombocytopenia associated with antepartum administration of thiazide drugs. N Engl J Med 1964;270:881–884.
52. Gray MJ. Use and abuse of thiazides in pregnancy. Clin Obstet Gynecol 1968;11:568–578.
53. Werthmann MW Jr, Krees SV. Excretion of chlorothiazide in human breast milk. J Pediatr 1972;81:781–783.
54. Solomon L, Abrams G, Dinner M, et al. Neonatal abnormalities associated with D-penicillamine treatment during pregnancy. N Engl J Med 1977;296:54–55.
55. Gregory MC, Mansell MA. Pregnancy and cystinuria. Lancet 1983;19:1958–1960.
56. Cunningham FG. Urinary tract infections. Clin Obstet Gynaecol 1987;1:891–908.
57. Rodriquez PN, Klein AS. Management of urolithiasis during pregnancy. Surg Gynecol Obstet 1988;166:103–106.
58. Barron WM. Medical evaluation of the pregnant patient requiring nonobstetric surgery. Clin Perinatol 1985;12:481–496.

10 Trauma During Pregnancy

JOHN BARRETT

Traumatic injuries are estimated to occur in 6 to 8% of all pregnancies[1–3] and are the most frequent cause of nonobstetric death during pregnancy.[4,5] Trauma is the leading cause of death in patients under the age of 40 years,[6] and the pregnant patient does not appear to be protected from traumatic insults.[7] In today's society the pregnant woman maintains her activity and frequently rides and drives in motor vehicles during her entire pregnancy. Motor vehicle collisions are the leading cause of trauma in this group[5] and are responsible for ten times the number of deaths than from all other causes combined.[8]

Penetrating trauma is the second most common cause of trauma in the pregnant patient[9] especially in the inner city where the incidence of penetrating trauma from gunshot and stab wounds is greater. Self-inflicted gunshot wounds in an attempt to induce abortion have been reported,[10] and it has been pointed out that pregnancy may actively precipitate domestic violence directed against the pregnant patient.[11]

Falls are also a common cause of injury in the pregnant patient,[12] possibly due to the alteration in gait and difficulty of maintaining balance, coupled with the fatigue and tilt due to lumbar lordosis in the pregnant patient. During late pregnancy the nesting syndrome may cause the patient to climb on ladders or chairs while decorating for the new baby. Generally such a patient lands on her back, not her abdomen, after falls.[13]

When confronted with the pregnant trauma patient, the treating physician is frequently uncomfortable. Not only are two lives at risk, but the emergency physician may be uncomfortable with management of the obstetric problem, just as the obstetrician may be uneasy with management of the trauma. In addition, the pregnant patient undergoes various physiologic alterations during the course of her pregnancy that can significantly alter trauma management.

Physiologic Changes During Pregnancy That Affect Trauma Management

Cardiovascular Changes

The heart rate increases during pregnancy by about 15 beats per minute.[14] Diastolic pressure falls, and there is a wide pulse pressure.[15,16] Cardiac output increases by 40%,[17] and the central venous pressure decreases to less than one-half normal levels.[18] Importantly, the gravid uterus during late pregnancy causes complete inferior vena caval obstruction in 90% of patients in the supine position, with resultant severe decreases in cardiac output; the patient then becomes profoundly hypotensive—the "supine hypotensive syndrome" of pregnancy.[15,19] It is therefore of vital importance to displace the gravid uterus off the inferior vena cava in any pregnant patient in shock. It can be done by turning the patient to the left lateral decubitus position[20] or elevating the right hip by placing towels or sandbags. In patients in whom such movements are contraindicated because of the fear of spinal injuries the patient can be immobilized on a long back board, and after proper fixation the board and patient are tilted to the patient's left.[21]

Electrocardiographic changes during pregnancy include flattening or inversion of the T wave in lead III and left axis deviation.[22,23] Another important cardiovascular alteration is the expansion of the blood volume in the pregnant patient, which may be as much as 50% by the 28th week of gestation.[24] The increase in red blood cells fails to keep pace with the increase in plasma volume, so a "physiologic anemia" of pregnancy results. The average hematocrit decreases from 41% to 33% during late pregnancy.[25,26] Because of the greatly expanded plasma volume, classic clinical signs of shock in the pregnant patient may not occur until significant volume losses have been incurred.

Hematologic Changes

There is an increase in the white blood cell count during pregnancy, and values as high as 20,000/cu mm during the last trimester and 25,000/cu mm during labor have been reported.[3,27,28] Differential white blood cell counts show an increase in leukocytes and a decrease in lymphocytes.[29,30]

Significant alterations occur in the coagulation profile during pregnancy. There is a normal elevation of factors VII, VIII, IX, and X; and elevated fibrinogen levels of 400 to 500 mg/dl are found at term,[31] so "normal" fibrinogen levels may in fact represent a severe deficit. Although these alterations may be of some benefit in the presence of trauma for decreasing bleeding, they also increase the risk of thromboembolic disease in the immobilized pregnant patient.

Respiratory Changes

There is an increase in tidal volume by about 40% and a reduction in residual capacity of 20%. Vital capacity increases, and alveolar ventilation and minute ventilation increase by 40 to 50%.[20] This situation creates physiologic hyperventilation and a consequent reduction in PCO_2. This respiratory alkalosis may cause dyspnea and dizziness. To maintain a pH of 7.4 there is a reduction in a serum bicarbonate to 21.2 MEq/liter at term.[20] Arterial blood gases in the pregnant patient are thus interpreted in light of these normal physiologic alterations.

Because of the increased blood volume, the rise in cardiac output, and the elevation of the diaphragm due to the pregnant uterus, increased lung markings are frequently seen on the chest radiograph.[32]

Gastrointestinal Tract Changes

There are significant decreases in gastric mobility and emptying times.[33] Emptying times as long as 50 minutes have been cited,[20] and these alterations increase the risk of aspiration.[34] Nasogastric tubes are freely used in the pregnant trauma patients to decrease the risk of aspiration. Transit time in the small intestine is also prolonged during pregnancy.[35]

Renal Changes

There is a 30 to 50% rise in the glomerular filtration rate[36] and a decrease in serum creatinine and blood urea nitrogen. Dilatation of the renal calyces, pelvis, and ureter is considered a normal finding during pregnancy.[20]

Hormonal Changes

Significant hemorrhage in a pregnant patient may cause necrosis of the hypertrophied pituitary gland because of the reduction in blood flow to the organ (Sheehan syndrome).[20] The serum calcium level is reduced and its turnover increased, but fracture management is not altered because of this phenomenon.[20,37]

Initial Management of the Pregnant Trauma Patient

Although the pregnant patient undergoes significant physiologic alterations, it is important to emphasize that the general principles of trauma management are the same in the pregnant as in the nonpregnant patient. The most common cause of fetal death following trauma is maternal death,[38] and therefore the initial management consists in resuscitation of the mother.[4] Keeping the mother alive is the best way to keep the fetus alive. The standard protocol of airway, breathing, and circulation (ABC) is therefore followed.[38]

Airway

Maintenance or restoration of an intact airway is the primary consideration in a pregnant trauma patient.[38–40] Airway obstruction due to the tongue falling back and blocking the airway can sometimes be overcome by manually repositioning the jaw by lifting the chin (chin lift) or thrusting the mandible forward (jaw thrust). During the performance of these and other maneuvers to obtain an airway, great care must be taken to protect the cervical spine. Although the incidence of clinically occult unstable injury to the cervical spine is unknown[41] and numbers based on autopsy studies overestimate the incidence in survivors, it seems prudent to ensure adequate cervical spine stabilization in all patients considered at risk. Such patients certainly include all victims of significant deceleration injuries and all patients with significant head or facial trauma. These patients are therefore transported on long back boards with hard cervical collars in place and with the head, neck, and collar taped to the back board.

Manual repositioning of the mandible, suctioning, and oxygen administration are the primary steps in airway management. Should these simple steps fail to restore an intact airway, the more definitive procedure of placing a tube in the trachea must be performed. In the patient who is still spontaneously breathing, the method of choice is blind nasotracheal intubation.[38] This measure reduces the risk of moving the cervical spine. It has been argued that nasotracheal intubation is contraindicated during pregnancy because of the danger of epistaxis.[42] Should oral endotracheal intubation be used, great care must be taken to maintain cervical spine stabilization during the intubation.

In patients in whom spontaneous breathing is absent or in whom massive midface trauma makes nasal or oral intubation impractical, a surgical approach to the upper airway is indicated. Surgical airways are also indicated as a last resort in patients who cannot be successfully intubated from above and who have acute airway compromise. The currently preferred method of surgical airway control under these circumstances is via the cricothyroid membrane, not via the trachea.[38,43] The fear of subglottic stenosis following this procedure voiced by Chevalier Jackson in the early part of this century[44,45] has been shown to be unfounded.[46–49] A surgical incision through the cricothyroid membrane, which lies just inferior to the thyroid cartilage ("Adam's apple") gives direct entry into the upper airway. A tracheostomy tube can then be passed through this opening and secured in position. Because of the small size of the membrane, a tube about one-half to one size smaller than the standard endotracheal tube is used. It is important to emphasize that control of the airway is of primary importance. The operator must not be distracted by other injuries at this critical time but must persist in an organized approach until the airway is secure (Fig. 10-1).

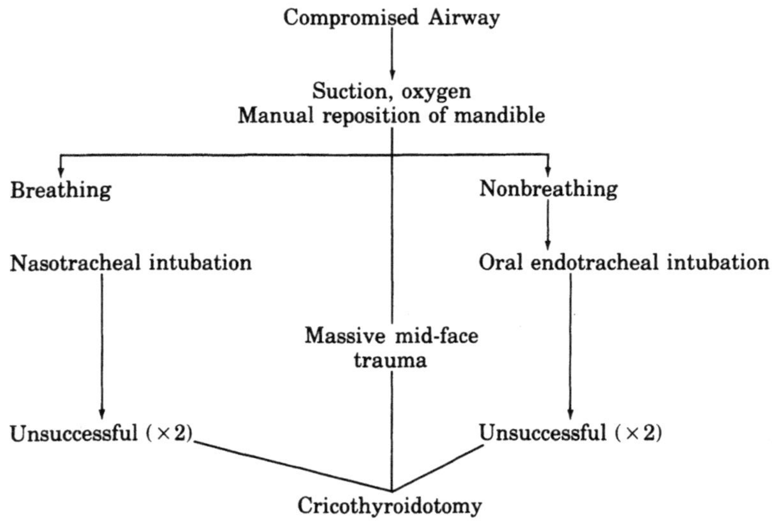

FIGURE 10-1. Airway management.

Breathing

After the upper airway is secured, attention is directed to the chest to assess the adequacy of ventilation.[38] Clothing is rapidly cut off to expose the chest. A number of immediately life-threatening conditions are then sought.

Tension Pneumothorax

The increased pressure within the thoracic cavity due to the accumulation of air causes displacement of the mediastinum and decreased venous return to the heart (Fig. 10-2). Physical findings under these circumstances include profound shock, absent breath sounds on the involved side, and distended neck veins. Tracheal deviation is not a consistent finding. The emergent nature of this injury does not allow time for confirmation by chest radiograph, and therefore the initial treatment consists in needle aspiration. Insertion of a needle in the midclavicular line in the second or third intercostal space confirms the diagnosis by the aspiration of air. Once the diagnosis is confirmed the definitive treatment is insertion of a large chest tube in the 4th or 5th intercostal space in the midaxillary line. Chest tube insertion under these conditions is performed by open tube thoracostomy technique. The operator inserts a finger into the thoracic cavity to ensure proper placement of the tube and to avoid injury to any viscus that may have herniated through the diaphragm.[50]

Open Pneumothorax ("Sucking" Chest Wounds)

With traumatic openings in the chest wall that approximate the size of the trachea, the resistance to airflow through the defect is less than the resistance to flow through the trachea. Under these conditions the air enters and exits preferentially via the chest wall defect and not through the trachea: Neither lung ventilates, and the patient dies from hypoxia. There are two methods of treatment[51]: The first and most frequently used is closing the defect in the chest wall by use of an occlusive dressing of impervious petrolatum gauze, followed by insertion of a chest tube to prevent the subsequent development of a tension pneumothorax. The second method is to intubate the patient and institute positive-pressure ventilation. This

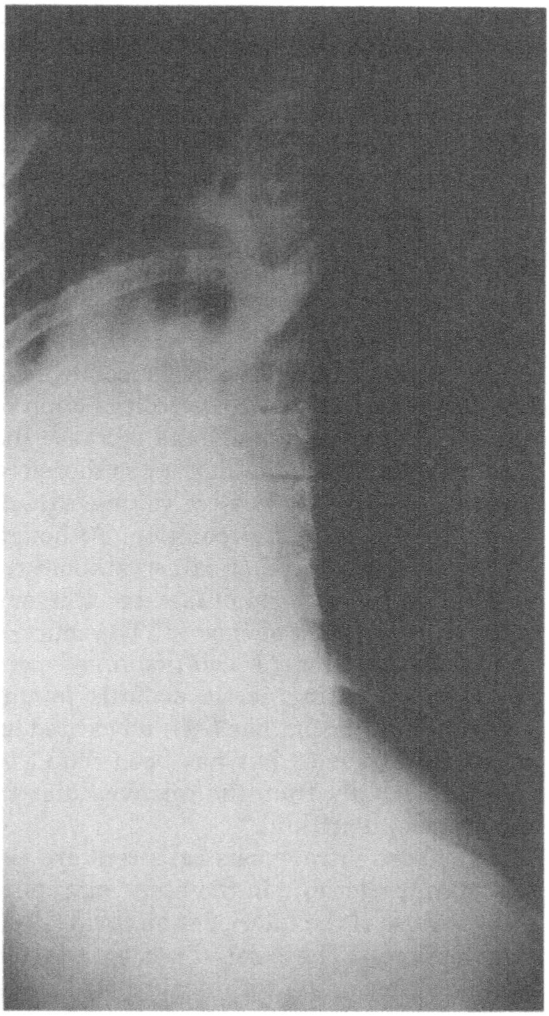

FIGURE 10-2. Tension pneumothorax (left side). It is an immediate life-threatening injury and is managed by needle decompression and chest-tube insertion. Normally chest radiographs *are not* obtained because of the delay in treatment. (Courtesy of the Cook County Hospital Trauma Unit.)

method restores normal oxygenation, and the defect in the chest wall can then be electively addressed.

Flail Chest

The presence of multiple rib fractures may cause a segment of chest wall to lose bony continuity with the remaining chest skeleton. This loose, or "flail," segment can be recognized clinically because of its paradoxical motion.[38] When the patient attempts to inspire, the flail segment is sucked into the chest, and during attempted expiration it is blown outward, creating the paradoxical movement. Although the mechanism of respiratory embarrassment under these conditions is disputed,[51] it appears to be related more to the underlying pulmonary damage than to the flail segment itself.[52–55] Initial treatment is therefore directed to improving oxygenation and relieving hypercarbia rather than to the flail segment. Intubation of the airway with volume cycled ventilation is therefore the treatment of choice under these conditions (Table 10-1).

Circulation

After the airway has been secured and oxygen is being delivered to the alveoli, attention is directed to the circulatory status of the patient. Initially, a brief examination for possible cardiac injury is carried out. In patients with frontal deceleration injuries to the chest, the possibility of myocardial contusion is entertained. Direct violence to the sternum may drive it forcefully backward against the heart (which in turn may be moving forward because of the deceleration). The resulting impact may bruise or contuse the cardiac muscle—most commonly the right ventricle because it lies directly behind the sternum.[56–59] Although the definitive diagnosis of this condition is complex, initial attention is directed to the occurrence

TABLE 10-1. Treatment of impaired breathing.

Diagnosis	Treatment
Tension pneumothorax	Needle decompression Chest tube insertion
Open pneumothorax	Close the hole Chest tube
Flail chest	Intubate airway Volume ventilate

of arrhythmias.[60–62] Electrocardiographic monitoring is therefore of importance in all patients with significant blunt chest trauma.

Pericardial Tamponade

After penetrating cardiac injury, blood escaping from the injury chamber may remain trapped within the pericardial sac. This problem is particularly likely to occur if the injury in the pericardial sac is small and prevents the escape of the blood. This situation is especially true with injuries to the heart due to long, thin implements, such as icepicks and stiletto-type knives.[63] The accumulation of blood within the pericardial sac results in an increase in pressure with subsequent interference to the systemic and pulmonary venous return to the heart; the patient may then expire because of inadequate venous return. Physical findings include distended neck veins, profound shock, and muffled heart tones. This presence of normal air entry facilitates distinction between this condition and tension pneumothorax.

When the presence of pericardial tamponade is suspected, prompt pericardiocentesis is performed. Utilizing a long needle and a subxiphoid approach, the needle is directed toward the inferior border of the left scapula.[38] The withdrawal of blood and clinical improvement in the patient confirms the diagnosis.[64,65]

Hemorrhage Control

External hemorrhage is controlled by direct pressure.[38] Blind clamping in the depths of the wound is hazardous because of the possibility of injury to associated nerve or venous structures. Tourniquets are not used unless a determination has been made that the limb is unsalvageable.

Hemorrhagic Shock

Because of the increased plasma volume in the pregnant patient significant blood loss may occur before any of the classic signs of hemorrhage are present in the mother. Maternal blood loss can be as much as 35% before any clinical manifestations occur.[66,67] Maternal hemorrhage decreases uterine perfusion[66]; and elaboration of maternal catecholamines causes placental vasoconstriction[40] and blood to be shunted away from the fetus.[1] Maternal hemorrhagic shock has been reported to be associated with a fetal mortality rate of 80%.[4] Use of vasopressins under these conditions is to be condemned because they lead to further constriction of the uteroplacental vessels and increase the hazard to the fetus.[68,69] Therapy is therefore directed to early, aggressive volume expansion of the maternal circulation. Although still somewhat controversial, crystalloid resuscitation in the form of lactated Ringer's solution is the fluid of choice.[38] The concern that lactated Ringer's solution may compound the existing lactic acidosis is unfounded.[70–73] Plasma has been advocated as a volume expander but has been found to dissipate rapidly from the intravascular to the extracellular fluid.[74]

Large-bore intravenous catheters are established, preferably in the upper extremities—because of the inferior vena caval (IVC) compression. The rate-limiting factor in fluid infusion via an intravenous line is the internal diameter of the catheter.[75] Large-bore catheters are therefore used for resuscitation of the pregnant patient (Table 10-2).

Pneumatic Anti-Shock Garment

Use of the pneumatic anti-shock garment (PASG) or military anti-shock trousers

TABLE 10-2. Treatment of impaired circulation.

Diagnosis	Treatment
Cardiac tamponade	Pericardiocentesis
External hemorrhage	Direct pressure
IV access needed	Prepare large-bore upper extremities

(MAST) has been advocated for treatment of hemorrhagic shock, although it has been argued that the time and effort involved is unwarranted.[40] It is generally agreed that in the pregnant patient during the last trimester the abdominal component must not be inflated,[20,38,40] but inflation of both of the lower extremity compartments can still be accomplished.

It cannot be emphasized too strongly that the presence of shock in the pregnant patient may be caused by IVC compression due to the gravid uterus. It is of the utmost importance to displace the gravid uterus off the IVC for management of these patients. Placing the patient in the left lateral decubitus position, elevating the right hip with sandbags or towels, or tilting the patient to the left on the backboard frequently achieves this goal.

At the conclusion of the ABCs of trauma resuscitation, it is important to "place a finger or a tube in every orifice." A nasogastric tube is placed to decompress the stomach and thus reduce the risk of aspiration associated with the decreased gastric motility and emptying. The presence of blood in the gastric aspirate may be an indication of injury to the esophagus or upper gastrointestinal tract, especially in the presence of penetrating trauma. In the presence of massive facial trauma or leakage of cerebrospinal fluid from the nose, a nasogastric tube is passed orally to avoid the possibility of passing the tube into the cranial vault via a fractured cribriform plate.

A Foley catheter is passed to empty the urinary bladder and obtain a urine sample to test for hematuria. Inability to pass a Foley catheter into the bladder suggests a urethral disruption, but such conditions are rare even in the presence of pelvic fractures.

A careful pelvic examination is carried out in all cases.[20] Fundal height, uterine irritability, fetal heart tones, and movement are estimated. The cervix is examined and any leaking fluid tested for pH and "ferning" in search of a possible amniotic fluid leak. Inspection of the cervix may show fluid leaking from the os.

Diagnostic Tests

Fetal Monitoring

Adequate fetal monitoring is started early and continued aggressively. The fetus may be in severe distress, even though the mother shows few if any changes[1]; fetal distress may be the earliest sign of maternal blood loss. The use of ultrasonography can confirm fetal movement and cardiac activity, as well as the location of the placenta. It may also be helpful for identification of retroplacental or pelvic hematomas.[1,20]

Radiography

Many physicians are reluctant to obtain needed radiographs in the pregnant trauma patient, and certainly radiography must not be used indiscriminantly. It must be pointed out, however, that radiation exposure in the order of 10,000 MRads is unlikely to cause significant problems.[76] A full series of intravenous pyelogram (IVP) exposures (ten films) is in the order of 140 MRads. It is clear therefore that whatever radiographs are needed to manage the mother must be obtained. If the fetus can be shielded during the x-ray exposure, it is done. Certainly, the cervical spine and chest x-ray films that are generally needed for management of the trauma patient can be obtained; and, in general, IVP and cystograms are obtained if needed.

Diagnostic Peritoneal Lavage

Diagnostic peritoneal lavage (DPL) is a highly accurate method of determining the need for exploratory laparotomy in a patient who has sustained blunt abdominal trauma.[77–79] The usual procedure consists in an infraumbilical incision with the catheter directed toward the pelvis.[38] During the latter stages of pregnancy an open supraumbilical approach can be used.[20] Prior to the lavage a Foley catheter is inserted to decompress the bladder, and a nasogastric tube is recommended as well. Once the lavage catheter is inserted, an attempt is made to aspirate blood. The withdrawal of nonclotting blood from within the abdominal cavity is considered positive, and exploratory laparotomy is recommended. If no blood can be aspirated, 1 liter of lactated Ringer's solution is infused and a sample of the return examined. Criteria for a positive lavage are (1) 100,000 red blood cells (RBCs)/cu mm; (2) 500 white blood cells (WBCs)/cu mm; and (3) the presence of bile, bacteria, or fecal matter in the lavage return.[39] Patients who have equivocal abdominal findings, who exhibit alterations in level of responsiveness, or who require operative intervention for major orthopedic injuries are candidates for DPL, as is any patient with unexplained hypotension or blood loss.[80] Culdocentesis has been advocated as an alternative method for diagnosing intraabdominal hemorrhage.[7] A positive culdocentesis is an indication for exploration, but a negative return cannot be relied on; DPL therefore appears to be a more specific and reliable diagnostic test.

Amniocentesis

Amniocentesis can be a useful test to determine fetal distress (meconium staining), intrauterine bleeding (RBCs), or infection (WBCs or bacteria).[1,20] Fetal maturity can also be estimated by means of the lecithin/sphingomyelin (L/S) ratio (which would be more than 2.0) or an amniotic fluid creatinine (which would be more than 2.0 mg/dl.[1]

Management of Specific Injuries

Blunt Abdominal Trauma

As the pregnancy progresses and the gravid uterus fills the abdominal cavity, the maternal abdominal wall becomes progressively more distended, thin, and lax and less able to protect the intraabdominal contents. These facts, coupled with the increasing size of the gravid uterus, make injuries to the uterus the most common intraabdominal injury in the pregnant patient.[20] In the nonpregnant abdomen the most frequently injured organs are the spleen and liver.[81] The most common cause of significant blunt trauma to the abdomen is motor vehicle trauma in both pregnant and nonpregnant patients.[5]

The presence of the gravid uterus frequently makes clinical detection of intraabdominal injuries difficult, and the increasing size of the gravid uterus displaces the intraabdominal contents from their usual positions.[1] Physical examination of the abdomen is frequently unreliable following blunt trauma, even in the nonpregnant abdomen,[82] especially in patients whose level of responsiveness during the physical examination is altered. Liberal use of additional diagnostic testing, either DPL or computed tomography (CT) scanning, is therefore indicated in these cases.

Of these two available modalities, CT scanning has the disadvantage of radiation exposure, although DPL appears to be the modality of choice especially during early pregnancy.[83] One of the disadvantages of DPL is that it affords little information with regard to injuries to the retroperitoneum.[38] The two diagnostic tests, however, can be viewed in a complementary light. DPL is used during early pregnancy and in all unstable patients, whereas CT scanning can be reserved for later pregnancy or cases in which the suspicion of retroperitoneal injury is high. CT scanning can also be used in a supplemental role in those cases where DPL may be unreliable. The patient with a scar on the abdomen has an increased risk of complication associated with DPL because of the numerous adhesions.[38] In addition, the presence of the adhesions may have compartmentalized the abdominal cavity, thereby reducing the accuracy of DPL. In patients with a pelvic fracture, DPL may be falsely positive if standard criteria are used. Although the aspiration of gross blood is still considered an indication for operation inpatients with pelvic fractures, standard criteria of 100,000 RBCs/cu mm, which is the usual criteria in blunt abdominal trauma, results in a high false positive rate for intraabdominal injuries. CT scanning can be used in these patients to reduce the negative laparotomy rate.[84,85]

Uterine Rupture

As the gravid uterus rises out of the pelvis it becomes more susceptible to direct violence, and uterine rupture due to major blunt trauma may occur (Fig. 10-3). Although uterine rupture has been reported as early as 2.5 to 3.0 months' gestation,[86] it is more common during later pregnancy and presents a more difficult diagnostic challenge than does rupture during labor.[3]

Traumatic rupture may present with massive hemorrhagic shock, or with few physical findings.[20] The most common site of rupture is the fundal portion of the uterus,[87] and radiographs may show extended fetal extremities, abnormal position, or free intraperitoneal air.[20] Fetal mortality is high under these conditions, and bleeding from ruptured uterus may be massive. Despite this fact, however, every effort must be made to preserve the uterus. With severe hemorrhage, internal iliac artery ligation may be utilized,[1] and subsequent pregnancies have been reported in uteri that have been subjected to extensive repair for rupture.[3] Frequently, the rent is isolated and can be easily repaired. If hysterectomy is needed, the cervix can be retained and the total blood loss reduced.[88]

Uterine rupture is a relatively rare entity. In one study it occurred in only two unrestrained victims among 208 pregnant victims of severe automobile collisions.[89]

Abruptio Placentae

Abruptio placentae is not an uncommon finding after severe blunt trauma: It has been reported in 3.4% of survivors of severe automotive collisions[89] and is the most common cause of fetal death if the mother survives the collision.[89] The exact mechanism of placental abruption is not clear, but shear forces between the placenta and the uterus have been implicated.[20] Small degrees of separation (up to 25%) present as vaginal bleeding and premature labor. Abruptio placentae must be considered in all patients who present with vaginal bleeding after severe blunt trauma.[88] If the separation is more than 50% of the placental surface, fetal death usually occurs.[7] The uterus may fill with blood, and hypovolemic shock may be present even in the absence of external hem-

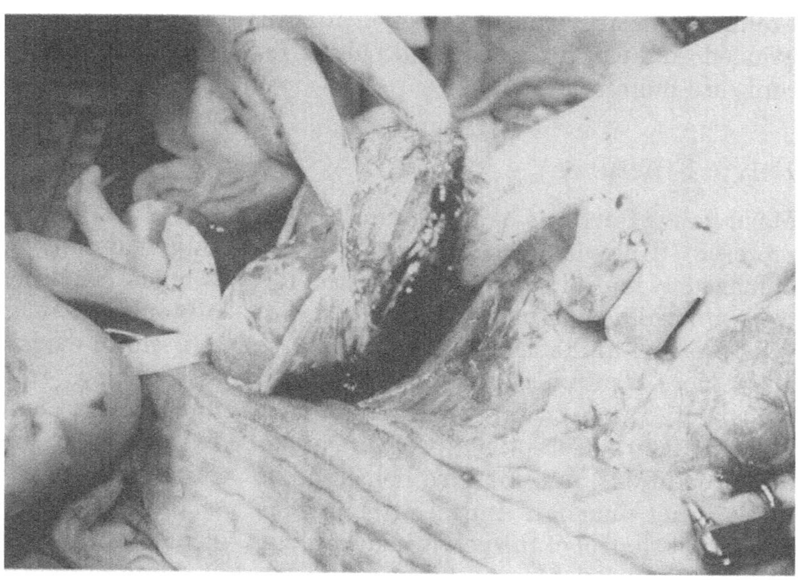

FIGURE 10-3. Spontaneous rupture of the uterus with extensive abruptio placentae due to an automotive collision. Fetal survival is unlikely under these conditions. (Courtesy of the Cook County Hospital Trauma Unit.)

orrhage.[88] Physical findings can include vaginal bleeding, shock, uterine irritability, and increasing fundal height. It is thought that the compression of the IVC by the gravid uterus may also induce placental separation,[20] and every effort is made to displace the gravid uterus off the IVC. In the presence of a mature fetus, cesarean section or labor induction is indicated for fetal distress, severe separation, or the appearance of clotting defects.[1,4] Fetal mortality is directly related to the degree of placental separation.[1,4] If the fetus is dead, normally spontaneous labor results in vaginal delivery. Disseminated intravascular coagulation (DIC) is more common under these conditions, however, and removal of the placenta may be needed to interrupt the condition.[88]

Premature Rupture of the Membranes

The management of this condition depends on the gestational age of the fetus and represents a difficult problem when the fetus is premature. Under these conditions every effort is made to avoid the occurrence of infection, and time is gained for fetal maturity. The occurrence of fetal distress mandates delivery; but unless this condition occurs, the patient is closely monitored and vaginal examinations are avoided.[20] Dehydration is avoided, and careful intake and output records are maintained.

Pelvic Fractures

Major pelvic fractures in a pregnant patient represent a significant challenge and are managed by multidisciplinary approach. Retroperitoneal hemorrhage, which is a significant problem in the nonpregnant patient, is much more severe because of the increased vascularity that occurs with pregnancy. Bleeding may be exceptionally severe under these conditions, and massive hemorrhagic shock is not unusual.[2] Initial attempts are directed at control of the pelvic hemorrhage. Associated intraabdominal sources of bleeding are ruled out by DPL or CT scanning in the stable patient. In the unstable patient in whom pelvic hematoma secondary to fracture has been confirmed as the source of the hemorrhage, diagnostic and therapeutic angiography may be indicated. Successful embolization of the bleeding vessels can frequently be achieved, and the reported surgical rate is in excess of 80%.[89–93]

Severe pelvic fractures may also distort the pelvic outlet and interfere with later vaginal delivery, necessitating cesarean section in 5 to 10%[94]. Most authorities agree, however, that even in the presence of significant pelvic fractures an attempt must be made to allow normal vaginal delivery[95–97] even in the presence of recent fractures.[98–100]

Penetrating Trauma

Gunshot Wounds During Early Pregnancy

During early pregnancy all gunshot wounds that penetrate the abdominal cavity must be explored (Fig. 10-4). In a series of 307 consecutive patients with gunshot wounds of the abdomen from the Cook County Trauma Unit, 97% of the 259 patients in whom the bullet entered the abdominal cavity had a significant injury that required repair.[101] In that same series there were 48 patients in whom the bullet struck the abdominal wall but did not enter the abdominal cavity. There were no intraabdominal injuries in any of the latter patients. It must be pointed out, however, that these wounds were due to relatively low velocity civilian hand gun shootings in an urban setting. Our policy for such tangential gunshot wounds has therefore been to avoid laparotomy.

Definite evidence of penetration includes shock, blood from an orifice, evisceration, and free air. Comparison of the entrance and exist sites, or the entrance site and the retained bullet, can also determine penetration. All patients in whom there is clear evidence of abdominal penetration must undergo exploratory laparotomy.

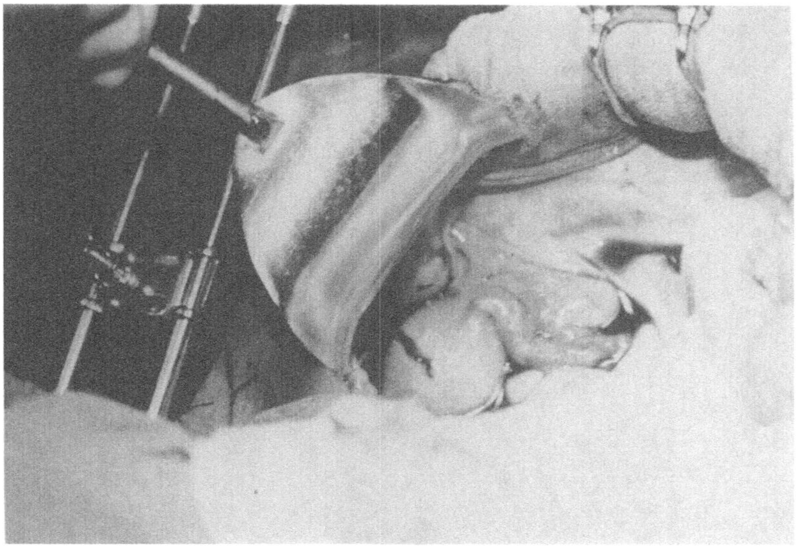

FIGURE 10-4. Gunshot wound of the uterus during early pregnancy. The patient was explored because of associated intraabdominal injuries. (Courtesy of the Cook County Hospital Trauma Unit.)

If it is not clear if the bullet entered the abdominal cavity, diagnostic peritoneal lavage may be useful. In a series of 235 such patients with penetrating trauma reported from our institution, a lavage count of 10,000 RBCs/cu mm was taken as an indication for exploratory laparotomy, with an overall accuracy rate of 96.6%[102] It appears therefore that lavage can be used to determine penetration provided a sufficiently sensitive (low) RBC count is used.

Gunshot Wounds During Late Pregnancy

During late pregnancy the gravid uterus fills the abdominal cavity and is the most frequently injured organ from gunshot wounds[7,103] (Fig. 10-5). Only 19% of women with gunshot wounds to the pregnant uterus sustain an additional injury[103,104] (Fig. 10-6). There has been no report of a maternal death associated with a gunshot wound of the gravid uterus since 1912.[103,104] During late pregnancy, therefore, it is generally accepted that certain gunshot wounds to the uterus can be observed. The following criteria need to be established: (1) fetal death; (2) bullet clearly within the uterus; and (3) stable maternal findings.[104] *Gunshot wounds to the upper abdomen require exploratory laparotomy even during late pregnancy.* The presence of a bullet within the uterine cavity is not in and of itself an indication for cesarean section. Infection is rare under these circumstances, and the need for cesarean section can be determined by the status of the fetus.[28]

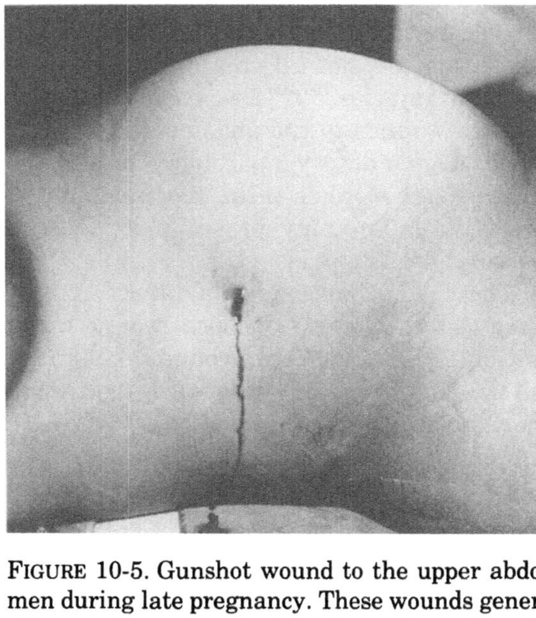

FIGURE 10-5. Gunshot wound to the upper abdomen during late pregnancy. These wounds generally require an exploratory laparotomy to rule out injuries to the intraabdominal contents, which are now compressed in the upper abdomen by the gravid uterus. (Courtesy of the Cook County Hospital Trauma Unit.)

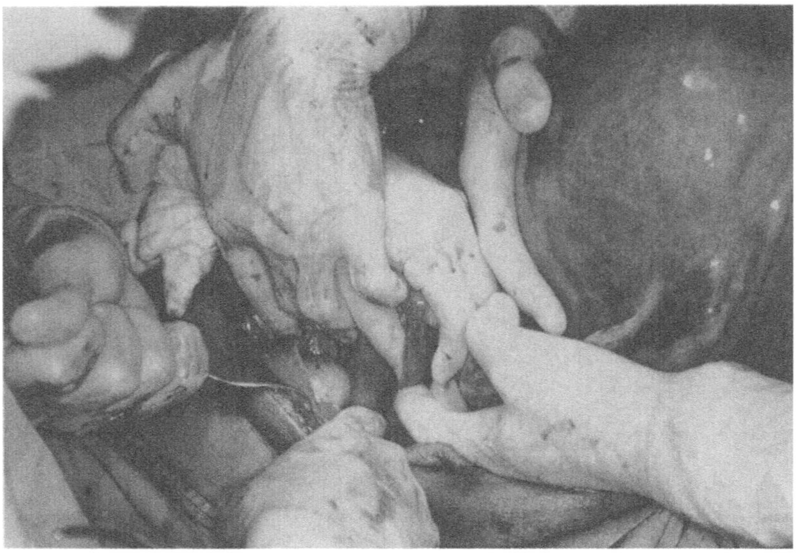

FIGURE 10-6. Exploratory laparotomy for penetrating trauma to the colon during late pregnancy. Just because the patient is undergoing an exploratory laparotomy is no reason for a cesarean section. The uterus can be carefully retracted and the associated intraabdominal injuries addressed. (Courtesy of the Cook County Hospital Trauma Unit.)

Stab Wounds to the Pregnant Abdomen

Unlike gunshot wounds, stab wounds result in relatively little energy dissipation into the tissues, and therefore their potential for significant organ injury is much less.[105,106] In excess of 25% of all stab wounds do not even penetrate the abdominal cavity; and of those that do, not all create significant abdominal injuries.[107,108] Two basic approaches to stab wounds to the abdomen have developed: (1) mandatory laparotomy for all penetrating stab wounds of the abdomen; and (2) selective laparotomy based on evidence of visceral organ injury.

Those who advocate mandatory laparotomy generally carry out local wound exploration. Then, if local wound exploration shows fascial penetration or is equivocal, peritoneal lavage is carried out.[109,113]

The alternate approach, which is our policy, is to perform exploratory laparotomy only if there is evidence of visceral organ injury. Evidence of visceral organ injury consists in shock or continued bleeding, peritonitis, bleeding from an orifice, evisceration, pneumoperitoneum, on abnormal radiographic results. The retained stabbing implement is also considered by us to be an indication for exploratory surgery.

During early pregnancy therefore the abdominal stab wound can be managed by either mandatory exploratory laparotomy for all penetrating injuries or selective management for evidence of visceral organ injury only. During late pregnancy most agree that isolated stab wounds to the pregnant uterus do not per se require exploratory laparotomy. Stab wounds to the upper abdomen during late pregnancy can be managed as for early pregnancy.

Remember that the abdomen extends upward to the fourth intercostal space (ICS) and even higher in pregnant patients. Thoracoabdominal stab wounds (below the nipples but above the costal margins) can therefore easily penetrate the abdominal cavity. Should penetration occur in these cases there must be an injury to the diaphragm. Our policy therefore has been to perform peritoneal lavage on all thoracoabdominal stab wounds. We have shown that lavage counts of 10,000 RBCs/cu mm or more call for laparotomy.[114]

Fetal Injury

During early pregnancy the embryo is well protected from the effects of maternal trauma. It lies deep within the maternal pelvic skeleton and is surrounded by a relatively large amount of amniotic fluid.[20] As the pregnancy progresses, however, the fetus becomes more exposed and the amount of protection provided by the amniotic fluid is reduced. Most significant direct fetal injuries occur during late pregnancy.[4] Intracranial hemorrhage and skull fractures are the most common direct fetal injury associated with blunt maternal trauma.[27,115] Parietal bone fractures are most common in these circumstances.[27,116] Extremity fractures of the fetus have also been reported, and spontaneous intrauterine healing with callous formation has been observed.[3,27] It must be emphasized that even direct fetal injury does not necessarily result in fetal death.[7]

Penetrating trauma from a variety of exotic implements has been reported to have caused fetal injury. Such implements have included knives, bullets, files, scythes, sickles, and even animal horns.[3,20] Among the gunshot wounds that strike the pregnant uterus, 60% injure the fetus, with a fetal mortality rate of 41 to 71%.[20,103] Cesarean section is not routinely required for penetrating injuries to the gravid uterus—even those caused by gunshot wounds—and indeed might even contribute to the mortality of premature infants. The decision to perform a cesarean section for penetrating uterine trauma must be based on the maturity of the fetus and the presence of fetal distress.

In the mature fetus cesarean section is indicated for signs of fetal distress or overt fetal bleeding. If there is vaginal bleeding, staining techniques can be performed on the sample to determine if fetal blood is present.[117,118] If the fetus is sufficiently mature, cesarean section can be performed and the mature fetus managed for its injuries.

The problem of a premature fetus in distress is a difficult one, and the risks of prematurity must be balanced against the possible benefits of surgical intervention in the immature fetus. Fetuses of 28 weeks or more are generally considered to be potentially survivable[118]; and depending on the particular institution, 36% survival rates have been reported at 24 weeks' gestation.[119] The problems posed by the individual patient must be answered by considering the degree of maturity and the expected survival rates for the particular fetus balanced against the degree of distress or bleeding and the possibility of surgical correction.

Cesarean Section in the Trauma Patient

Cesarean section is not indicated just because the mother has to undergo an exploratory laparotomy for trauma[4,7,76,96]; nor is it required because of fetal death following a traumatic episode.[7,120] Cesarean section is, however, indicated in the presence of a mature fetus in distress because of either maternal or fetal injury. It is also indicated if there is uterine rupture or extensive abruptio placentae. Occasionally the gravid uterus prevents repair of life-threatening injuries to the mother, and cesarean section may be indicated under these conditions.[1,118] In most cases, however, the maternal injury can be explored and repaired leaving the pregnancy undisturbed.

There is an extensive literature on the performance of postmortem cesarean section, and more than 150 successful cases have been reported[119,121] with survival rates of up to 40%. Continuous cardiopulmonary resuscitation (CPR) must be in progress on the mother,[20] and the fetus is ideally delivered within 5 minutes and certainly within 25 minutes.[3,122] An attempt must be made to salvage the fetus in any case beyond 28 weeks' gestation.[20,92]

Special Considerations

Abortion Associated with Trauma

The medical practitioner may be asked to make a judgment as to whether an abortion that is chronologically related to a traumatic episode is also causally related. These questions are of significant medicolegal importance, and great care must be taken to establish a cause-and-effect relation between the trauma and the abortion. Prior to the 20th week of gestation up to 20% of normal pregnancies end in a spontaneous abortion with no traumatic episode.[7] Fort and Harlin[123] reviewed the outcome of pregnancy after noncatastrophic maternal trauma and showed that the incidence of abortion is the same in both traumatized and nontraumatized mothers. Javert,[124] in a review of 2000 spontaneous abortions, could not implicate trauma in any, although it was antecedent in seven cases. This finding was supported by studies showing that even when a traumatic episode is followed by an abortion the products of pregnancy are usually defective.[7] It is also possible that the abortion may have been in progress at the time of the traumatic incident.[125] Because of these facts great care is taken to establish a cause-and-effect relation between the trauma and the abortion; in general, the temporal relation is hours (rather than days) in a normal pregnancy with the delivery of a normal fetus and placenta supporting the relation.[7,20]

Psychosocial and Rehabilitative Considerations

The health care team must at all times remember that the trauma to the pregnant patient has significant psychosocial aspects. These aspects must be recognized and dealt with as early as possible to facilitate the rehabilitative process.

Invasive tests, especially diagnostic peritoneal lavage, are often perceived by the mother as being harmful to her unborn baby.[126] In the pregnant trauma patient, therefore, adequate explanations, reassurance, and support are needed during all diagnostic testing. The same holds true for x-ray examinations. Repeated explanations of the reasons for the various procedures and reassurance as to their safety for the fetus help reduce maternal anxiety.[127–129]

It is also important to remember that the patient may experience significant guilt feelings regarding the traumatic event.[130] Failure of the mother to wear her seat belt, for example, may provide considerable guilt feelings and even recrimination and hostility from other family members. The patient must be encouraged to verbalize her feelings on these matters and be provided with adequate support measures.

We have instituted a trauma support team to assist the patient and her family in the entire rehabilitative process. In case of fetal death, such support measures are even more important. The mother can be expected to undergo a normal grieving process and may request information as to the child's sex, weight, time of death, and even hair color.[126] Such requests are not unusual and must be anticipated by the team.

Even in cases where the infant survives, the injuries to the mother may result in enforce separation of the mother and child. The mother would again require reassurance regarding the status of her baby. On the other hand, if the mother is injured, "conversation should not always focus on the baby,"[126] and the mother is allowed time to deal with her fears and anxiety regarding her own injuries. Attention to these psychosocial impacts of trauma in the pregnant patient facilitate the rehabilitative process.[131]

Seat Belts

It has been clearly shown that the proper use of seat belts during pregnancy reduces the incidence and severity of maternal injuries and increases fetal survival.[132] Although it is true that women are more likely to wear their seat belts than men, women are also less likely to wear their belts during pregnancy.[133,134] Many pregnant women fear that the use of the belt (especially during late pregnancy) will increase the risk to the fetus. The opposite is in fact the truth. Seat belts reduce the risk of ejection from the vehicle, and ejection increases the chances of death 20-fold.[135] The seat belt must be properly worn and adjusted.[136] It is worn low across the pelvis with the shoulder strap in the normal position between the breasts, not against the neck.[126,131] High-riding belts have been implicated in uterine and fetal injuries,[137,138] but even the risk of these injuries does not justify nonuse of the belts.[139]

The use of seat belts during all stages of pregnancy has been advocated by the American College of Obstetricians and Gynecologists (ACOG), the United States Department of Transportation, the Nurses Association of ACOG, and the American Academy of Pediatrics.[20]

Conclusions

When managing a pregnant trauma patient the challenge of resuscitating and saving two lives must be met. Injuries to the mother take precedence, as maternal death is the most common cause of traumatic fetal death. The general principles of trauma management for nonpregnant patients are followed in the pregnant patient, but several important differences need to be remembered. Principal among them are the different response to blood loss in the pregnant patient and the vena caval obstruction syndrome. All needed diagnostic and therapeutic tests, including radiographs, CT scans, and diagnostic peritoneal lavage, are performed.

Specific types of injury to the uterus and fetus must be anticipated. Uterine rupture and abruptio placentae must be considered. The management of intrauterine fetal injuries depends on fetal maturity and degree of distress. Cesarean section is used only for specific indications.

Attention is paid to the prevention of trauma during pregnancy. Advice in this regard must be a routine part of prenatal care and counseling for all pregnant patients. Specific encouragement is given to the patient to wear her seat belt at all times and at all stages of pregnancy. Adherence to these principles results in a significant reduction in the morbidity and mortality associated with trauma in the pregnant patient.

References

1. Baker DP. Trauma in the pregnant patient. Surg Clin North Am 1982;62:275.
2. Peckman CH, King RW. A study of intercurrent conditions observed during pregnancy. Am J Obstet Gynecol 1963;87:609.
3. Buchsbaum HJ. Accidental injury complicating pregnancy. Am J Obstet Gynecol 1968;102:752.
4. Rothenberger D. Blunt maternal trauma: a review of 103 cases. J Trauma 1978;18:173.
5. Jackson FC. Accidental injury—the problem and the initiatives. In Buchsbaum HJ (ed): Trauma in Pregnancy. Philadelphia, Saunders, 1979.
6. National Center for Health Statistics: Advance Report of Final Natality Statistics 1983. Supplement Monthly Vital Statistics Report, 20 September 1985.
7. Crosby WM. Trauma during pregnancy: maternal and fetal injury. Obstet Gynecol Surv 1974;29:683.
8. National Safety Council: Accident Facts. Chicago, NSC, 1975.
9. Baker S, O'Neill B, Karpf R. The Injury Fact Book. Lexington, MA, D. C. Heath, 1984.
10. Buchsbaum HJ, Staples PP Jr. Self-inflicted gunshot wound to the pregnant uterus: report of two cases. Obstet Gynecol 1985; 65(suppl 3):328.
11. Hillard P. Physical abuse in pregnancy. Obstet Gynecol 1985;66:185.
12. Lefton D. Expectant mothers and unexpected mishaps. Family Safety Winter 1980/1981.
13. Fort A, Harlin R. Pregnancy outcome after noncatastrophic maternal trauma during pregnancy. Obstet Gynecol 1970;35:912.
14. Velano K, Novy MJ, Petersen EN, et al. Maternal cardiovascular dynamics. Am J Obstet Gynecol 1969;104:854.
15. Deswiet M. The cardiovascular system. In Hytten FE, Chamberlain, (eds): Clinical Physiology in Obstetrics. Oxford, Blackwell, 1980.
16. MacGillivray I, Rose GA, Rowe B. Blood pressure survey in pregnancy. Clin Sci 1969;37:395.
17. Walters WAW, MacGregor WG, Hills M. Cardiac output at rest during pregnancy and the puerperium. Clin Sci 1966;30:1.
18. Colditz RJ, Josey WE. Central venous pressure in the supine position during normal pregnancy. Obstet Gynecol 1970;36:769.
19. Kerr MG. The mechanical effects of the gravid uterus in the late pregnancy. J Obstet Gynecol Br Commonw 1965;72:513.
20. Barrett SM. Trauma during pregnancy. Prog Crit Care Med 1984;1:248–272.

21. Lees MM, Scott DB, Kerr MG, et al. The circulatory effects of recumbent postural change in late pregnancy. Clin Sci 1967; 32:453.
22. Hollander AG, Crawford JH. Roentgenologic and electrocardiographic changes in the normal heart during pregnancy. Am Heart J 1943;26:364.
23. Marriott HJL. Practical Electrocardiography, 6th ed. Baltimore, Williams & Wilkins, 1977.
24. Pritchard JA. Changes in the blood volume during pregnancy and delivery. Anesthesiology 1965;26:343.
25. Letsky E. The haematological system. In Hytten FE, (ed): Clinical Physiology in Obstetrics. Oxford, Blackwell, 1980.
26. Hytten FE, Lind T. Diagnostic Indices in Pregnancy. Basel, Ciba-Geigy, 1973.
27. Buchsbaum HJ. Traumatic injury pregnancy. In Barber G, (ed): Surgical Disease in Pregnancy. Philadelphia, Saunders, 1974.
28. Gibson A. On leucocyte changes during labour and puerperium. J Obstet Gynaecol Br Emp 1937;44:500.
29. Andrews WC, Bonsnes RW. The leucocytes during pregnancy. Am J Obstet Gynecol 1951;61:1129.
30. Mitchell GW, McRipley RJ, Selvaraj JR, et al. The role of phagocyte in host-parasite interactions. Am J Obstet Gynecol 1966; 96:687.
31. Pechet L, Alexander B. Increased clotting factors in pregnancy. N Engl J Med 1961;265:1093.
32. Deswiet M. The respiratory system. In Hytten FE, (ed): Clinical Physiology in Obstetrics. Oxford, Blackwell, 1980.
33. Davison JS, Davison MC, Hay DM. Gastric emptying time in pregnancy and labour. J Obstet Gynaecol Br Commonw 1970;77:37.
34. Hytten FE. The alimentary system. In Hytten FE, Chamberlain (eds): Clinical Physiology in Obstetrics. Oxford, Blackwell, 1980.
35. Parry E, Shields R, Turnbull AC. Transit time in the small intestine in pregnancy. J Obstet Gynaecol Br Commonw 1970;77:900.
36. Hytten F, Linda T. Diagnostic Indices in Pregnancy. Basel, Ciba-Geigy, 1973.
37. Albright J, Sprague B, El-Khoury G, Brand R. Fractures in pregnancy. In Buchsbaum HJ (ed): Trauma in Pregnancy. Philadelphia, Saunders, 1979.
38. Committee on Trauma American College of Surgeons. Advanced Trauma Life Support Course. Chicago, American College of Surgeons, 1984.
39. Patterson RM. Trauma in pregnancy. Clin Obstet Gynaecol 1984;27:32.
40. Maull K, Kozycki G. Injury to the female reproductive system. In Mattox K, Moore E, (eds): Trauma. East Norwalk, CN, Appleton & Lange, 1988.
41. Phillips TF, Goldstein AS. Airway management. In Mattox, Moore (eds): Trauma. East Norwalk, CN, Appleton & Lange, 1988.
42. Santos O. The obstetrical patient for nonobstetric surgery. Curr Rev Nurse Anesthetists 1984;9:66.
43. Kress TD, Balasubramaniam S. Cricothyroidotomy. Ann Emerg Med 1982;11:197.
44. Jackson C. High tracheotomy and other errors the chief causes of chronic laryngeal stenosis. Surg Gynecol Obstet 1921;32:392.
45. Jackson C. Tracheotomy. Laryngoscope 1909;19:285.
46. Brantigan CO, Grow JB. Cricothyroidotomy: elective use 46 respiratory problems requiring tracheotomy. J Thorac Cardiovasc Surg 1976;71:72.
47. Brantigan CO, Grow JB. Cricothyroidotomy revisited again. Ear Nose Throat J 1980;59:26.
48. Boyd AD, Romita MC, Conlan AA, et al. A clinical evaluation of cricothyroidotomy. Surg Gynecol Obstet 1979;149:365.
49. Habel DW. Cricothyroidotomy as a site for elective tracheotomy. Trans Pac Coast Otoophthalmol Soc Annu Meet 1977;58:181.
50. Carrio CJ, Ivey TD, Rusch VW. Injury to the lung and pleura. In Mattox K, Moore E (eds): Trauma. East Norwalk, CN, Appleton & Lange, 1988.
51. Suck JM, Snow NJ. Injury to the chest wall. In Mattox K, Moore E (eds): Trauma. East Norwalk, CN, Appleton & Lange, 1988.
52. Maloney JV, Schmotzer KJ, Raschke E. Paradoxical respiration and "pendelluft." J Thorac Cardiovasc Surg 1961;41:291.
53. Trinkle JK, Richardson JD, Franz JL, et al. Management of flail chest without mechanical ventilation. Ann Thorac Surg 1975; 19:355.
54. Richardson JD, Adams L, Flint LM. Selective management of flail chest and pulmonary contusion. Ann Surg 1982;196:481.
55. Garzon AA, Seltzer B, Karlson KE. Pathophysiology of crushed chest injury. Ann Surg 1968;168:128.

56. Kirsh MM, Sloan H. Blunt Chest Trauma. General Principles of Management. Boston, Little, Brown, 1977.
57. Saunders CR, Doty DB. Myocardial contusion. Surg Gynecol Obstet 1977;144:595.
58. Madoff IM, Desforges G. Cardiac injuries due to non-penetrating thoracic trauma. Ann Thorac Surg 1972;14:504.
59. Smith JM III, Grover FL, Marcos JJ, et al. Blunt traumatic rupture of the atria. J Thorac Cardiovasc Surg 1976;71:617.
60. Sutherland GR, Calvin JD, Driedger AA, et al. Anatomic and cardiopulmonary responses to trauma with associated blunt chest injuries. J Trauma 1981;21:1.
61. Mattox KL, Limacher MC, Feliciano DV, et al. Cardiac evaluation following heart injury. J Trauma 1985;25:758.
62. Stern T, Wolf RY, Reichart B, et al. Coronary artery occlusion resulting from blunt trauma. JAMA 1974;230:1308.
63. Fowler NO. Physiology of cardiac tamponade and pulsus paradoxus. 1. The mechanisms of pulsus paradoxus in cardiac tamponade. Mod Concepts Cardiovasc Dis 1978;47:109.
64. McGregor M. Pulsus paradoxus. N Engl J Med 1979;301:480.
65. Wechsler AA, Auerbach BJ, Graham TC, et al. Distribution of intramyocardial blood flow during pericardial tamponade: correlation with microscopic anatomy and intrinsic myocardial contractility. J Thorac Cardiovasc Surg 1974;68:847.
66. Romney SL, Gavel PV, Takeda V. Experimental hemorrhage in late pregnancy. Am J Obstet Gynecol 1963;87:636.
67. Reid DE. Shock in obstetrics. Am J Obstet Gynecol 1957;73:697.
68. Boba A, Linkie DM, Plotz EJ. Effects of vasopressor administration and fluid replacement on fetal bradycardia and hypoxia induced by maternal hemorrhage. Obstet Gynecol 1966;27:408.
69. Boba A, Plotz EJ, Linkie DM. Effect of atropine on fetal bradycardia and arterial oxygenation: experimental study in the dog during graded hemorrhage and following vasopressor administration. Surgery 1965; 58:267.
70. Baue AE, Tragus ET, Wolfson SK Jr, et al. Hemodynamic and metabolic effects of Ringer's lactate solution in hemorrhagic shock. Ann Surg 1967;166:29.
71. Carey LC, Lowery BD, Cloutier CT. Hemorrhagic shock. Curr Probl Surg 1971;8:26.
72. James PM, Bredenberg CE, Anderson RN, et al. Tolerance to long and short term lactate infusion in men with battle casualties subjected to hemorrhagic shock. Surg Forum 1969;20:543.
73. Trinkle JK, Rush BE, Eiseman B. Metabolism of lactate following major blood loss. Surgery 1968;63:782.
74. Moore FD. The effects of hemorrhage on body composition. N Engl J Med 1965; 273:567.
75. Aeder MI, Crowe JP, Rhodes RS, et al. Technical limitations in the rapid infusion of intravenous fluids. Ann Emerg Med 1985; 14:307.
76. Hogan MJ (ed): Xrays and the pregnant woman. Mayo Clin Update 1987;3:1.
77. Lazarus HM, Nelson JA. Peritoneal lavage with low morbidity. Clin Toxicol 1979;15:129.
78. Lazarus HM, Nelson JA. Refining the technique of diagnostic peritoneal lavage. ER Rep 1980;1:11.
79. Parvin S, Smith DE, Asher WM, Virgilio RW. Effectiveness of peritoneal lavage in blunt abdominal trauma. Ann Surg 1975;181:255.
80. Rothenberger DA, Quattlebaum FW, Zabel J, Fischer RP. Diagnostic peritoneal lavage for blunt trauma in pregnant women. Am J Obstet Gynecol 1977;129:479.
81. Griswold RA, Collier HS. Blunt abdominal trauma. Int Abstr Surg 1961;112:309.
82. Perry JF. A five-year survey of 152 acute abdominal injuries. J Trauma 1970;5:53.
83. Marx JA, Moore EE, Jordan RC, et al. Limitations of computed tomography in the evaluation of acute abdominal trauma: a prospective comparison with diagnostic peritoneal lavage. J Trauma 1985;25:933.
84. Moreno C, Moore EE, Rosenberger A, et al. Hemorrhage associated with major pelvic fracture: a multispecialty challenge. J Trauma 1986;26:987.
85. Hubbard SG, Bivins BA, Sachatello CR. Diagnostic errors with peritoneal lavage in patients with pelvic fractures. Arch Surg 1979;114:844.
86. MacCarty V, Risley DR. Traumatic rupture of the uterus in early pregnancy. J Int Coll Surg 1956;26:228.
87. Schrinsky DC, Benson RC. Rupture of the

pregnant uterus: a review. Obstet Gynecol Surg 1978;33:217.
88. Crosby WM. Traumatic injuries during pregnancy. Clin Obstet Gynecol 1983; 26:902.
89. Crosby WM, Costiloe JP. Safety of lap belt restraint for pregnant victims of automobile collisions. N Engl J Med 1977;284:632.
90. Mucha P Jr, Farnell MB. Analysis of pelvic fracture management. J Trauma 1984; 24:379.
91. Margolies MN, Ring EJ. Arteriography in the management of hemorrhage from pelvic fractures. N Engl J Med 1971;287:317.
92. Matalon TS, Athanasouilis CA. Hemorrhage with pelvic fractures: efficacy of transcatheter embolization. Am J Med 1979;133:859.
93. Penetta T, Selafani SJ, Goldstein AS, et al. Percutaneous transcatheter embolization for massive bleeding from pelvic fractures. J Trauma 1985;25:1021.
94. Eastman NJ. Editorial comment. Obstet Gynecol Surg 1958;13:69.
95. Delaney JJ. Obstetrical and gynecological injuries. In Zuidema GD, Rutherford RB, Ballinger WF (eds): The Management of the Trauma (3rd ed.). Philadelphia, Saunders, 1979.
96. Dyer I, Barclay DL. Accidental trauma complicating pregnancy and delivery. Am J Obstet Gynecol 1962;83:907.
97. Speer DP, Peltier LF. Pelvic fractures and pregnancy. J Trauma 1972;12:474.
98. Mulla N. Fracture of the pelvis in pregnancy. Am J Obstet Gynecol 1957;74:246.
99. Voegelin AW, McCall ML. Some acquired bony abnormalities influencing the conduct of labor. Am J Obstet Gynecol 1944;48:361.
100. Abright J, Sprague B, El-Khoury G, Brand R. Fractures in pregnancy. In Buchsbaum HJ (ed): Trauma in Pregnancy, Philadelphia, Saunders, 1979.
101. Lower J, Saletta JD, Read DR, et al. Should laparotomy be mandatory or selective in gunshot wounds of the abdomen? J Trauma 1977;17:903.
102. Merlotti GJ, Marcet E, Sheaff CM. Use of peritoneal lavage to evaluate abdominal penetration. J Trauma 1985;25:228.
103. Buchsbaum HJ. Penetrating injury of the abdomen. In Buchsbaum HJ (ed): Trauma in Pregnancy. Philadelphia, Saunders, 1979.
104. Fawzi AI, Samir NH, Buchsbaum HJ. Gunshot wounds of the pregnant uterus: report of two cases. J Trauma 1980;20:90.
105. Charters AC III, Charters AC: Wounding mechanisms of very high velocity projectiles. J Trauma 1976;16:464.
106. Fackler ML, Malinowski JA. The wound profile: a visual method for qualifying gunshot wound components. J Trauma 1985; 25:522.
107. Thompson JS, Moore EE, Van Duzer-Moore S, et al. The evolution of abdominal stab wound management. J Trauma 1980;20:478.
108. Thal ER. Evaluation of peritoneal lavage and local wound exploration in lower chest and abdominal stab wounds. J Trauma 1977;17:642.
109. Markovchick VJ, Moore EE, Moore JB, et al. Local wound exploration of anterior abdominal stab wounds. J Emerg Med 1985;2:287.
110. Mackersie RC, Lewis FR Jr. Local exploration and exploratory laparotomy. In Dailey RH, Callaham M, (eds): Clinics in Emergency Medicine, Vol. 6: Controversies in Trauma Management. New York, Churchill Livingstone, 1985.
111. Moore EE, Marx JA. Penetrating abdominal wounds: the rationale for exploratory laparotomy. JAMA 1985;10:253
112. Petersen FR, Sheldon GF. Morbidity of negative finding at laparotomy abdominal trauma. Surg Gynecol Obstet 1979;148:23.
113. Moss LK, Schmidt FE, Creech O. Analysis of 550 stab wounds of the abdomen. Am Surg 1962;28:483.
114. Merlotti GJ, Dillon BC, Lange DA, et al. Peritoneal lavage in penetrating thoraco-abdominal trauma. J Trauma 1988;28:17.
115. Conner E, Curran J. In utero traumatic intra-abdominal deceleration injury to the fetus—a case report. Am J Obstet Gynecol 1976;125:567.
116. Alexander E, Davis CH. Intra-uterine fracture of the infant's skull. J Neurosurg 1969;30:446.
117. Akhter MS, Savior JF, Akhter IS, Deacon JS. Fatal exsanguination associated with antepartum hemorrhage. Can Med Assoc J 1978;118:652.
118. Stuart GCE, Harding PGR, Davies EM. Blunt abdominal trauma in pregnancy. Can Med Assoc J 1980;122:901.
119. Dillon WP, Egan EA. Aggressive obstetric management in late second-trimester deliveries. Obstet Gynecol 1981;58:685.
120. Pritchard JA. Fetal death in utero. Obstet Gynecol 1959;14:573.
121. Depace NL, Betesh JS, Kotler MN. "Post-

mortem" cesarean section with recovery of both mother and offspring. JAMA 1982; 248:971.
122. Arthur RK. Postmortem cesarean section. Am J Obstet Gynecol 1978;132:175.
123. Fort AT, Harlin RS. Pregnancy outcome after noncatastrophic maternal trauma during pregnancy. Obstet Gynecol 1970;35:912.
124. Javert CT. Role of the patient's activities in the occurrence of spontaneous abortion. Fertil Steril 1960;11:550.
125. Hertig AT, Sheldon WH. Minimal criteria required to prove prima facie case of traumatic abortion or miscarriage. Ann Surg 1943;117:596.
126. Smith LG. The pregnant trauma patient. In Cardona VD (ed): Trauma Nursing. Philadelphia, Saunders, 1988.
127. McCormick RD. Seatbelt injury: case of complete transection of pregnant uterus. JAM Obstet Assoc 1968;67:1139.
128. Dunham CM, Cowley RA. Shock Trauma/ Critical Care Handbook. Rockville, MD, Aspen, 1986.
129. Auerbach P. Trauma in the pregnant patient. In Meislin H (ed): Priorities in Multiple Trauma. Germantown, MD, Aspen, 1980.
130. Smith LG. Abdominal trauma during pregnancy. In Strange J (ed): Shock Trauma Care Plans. Springhouse, Springhouse Corp., MD, 1987.
131. Benfield DG, Nichols J. Living with newborn death. In Almed P (ed): Pregnancy. Childhood and Parenthood. New York, Elsevier, 1981.
132. Pepperill R, Rubinstein E, MacIsaac I. Motor car accidents during pregnancy. Med J Aust 1977;1:203.
133. Crosby W, Costiloe J. Safety of lap belt restraint for pregnant victims of automobile collisions. N Engl J Med 1971;284: 632.
134. Automobile Passenger Restraints for Children and Pregnant Women. ACOG Technical Bulletin No. 74, 1983.
135. Tringa G. Medical aspects of seatbelt usage. J Traffic Med 1980;8:37.
136. Should I wear a seatbelt when I'm pregnant? Maryland Institute for Emergency Medical Systems and Services (MIEMSS) Brochure, 1984; Baltimore, MD.
137. Handel CK. Case report of uterine rupture after an automobile accident. J Reprod Med 1978;20:90.
138. Matthews CD. Incorrectly used seat belt associated with uterine rupture following vehicular collision. Am J Obstet Gynecol 1975;121:1115.
139. Crosby WM, Synder RG, Snow CC, Hanson PG. Impact injuries in pregnancy. I. Experimental studies. Am J Obstet Gynecol 1968;101:100.

11
Breast Cancer and Pregnancy

DONALD J. FERGUSON

Recognition of mammary tumors is made appreciably more difficult by the normal enlargement of the breast occurring with pregnancy and lactation, as well as by the restriction on the use of radiography and scans considered necessary to protect the fetus. To what extent treatment of breast cancer in pregnant women differs from that in other women has not been established by controlled studies, mainly because of the rarity of the combination but also because the policies for nonpregnant women are controversial. About 3 of 10,000 pregnancies are complicated by breast cancer; about 2% of breast cancers are found within 1 year of pregnancy.[1] We here review selected series of cases and our own experience in order to formulate reasonable policies.

Diagnosis

Obstetricians have a good opportunity to detect breast cancer at the routine prenatal visits.

History

Previous biopsies of the breast should be carefully checked for premalignant changes such as papillomas or sclerosing adenosis. Any recent mammograms should also be reviewed. Breast cancer in the patient's mother or sister warns that special risk is present. One or more abortions prior to a first live birth was reported to increase the incidence of breast cancer 4.5 times above that in women who have had only one full-term pregnancy.[2] Prior use of birth control hormones may affect the incidence or prognosis of breast cancer, but the data so far are inconsistent.[3,4] Any history of a breast lump, pain, nipple discharge, or skin change may be significant. In patients who present with a large tumor or palpable nodes, systemic symptoms such as bone pain, cough, weight loss, or altered mental state should be elicited or ruled out.

Examination

Simple inspection can reveal telltale findings: Paget's change of the nipple (a sort of eczema), spontaneous bloody nipple discharge, a visible mass or asymmetry, inflammatory change of the skin, and dimpling, which can be brought out by raising the arms or by compressing the tissue adjacent to a suspected mass. Palpation, including the neck and axillae, with the patient both sitting and lying requires much practice to do expertly, that is, to recognize what is recognizable and to remain completely aware of the pitfalls and limitations of the method, which are greater than usual in the pregnant woman. Changes that appear to be mastitis or cellulitis, manifested by edema and redness, should be suspect and usually biopsied if they last more than a few days because they may signify inflammatory carcinoma.

Needling

Palpable masses in young women frequently are simple cysts that may feel hard. The best aid to diagnosis of cysts is aspiration. Generous use of 0.5% lidocaine prevents serious discomfort. A size 18 needle should be used carefully to probe several times if fluid is not immediately encountered. In small women the breast may need to be lifted as the needle is inserted to guard against perforation of the pleura. All fluid should be drained to be sure no residual mass remains. Cytology of cyst fluid is not useful, but needle aspiration cytology of solid tumors can strongly suggest carcinoma when one is present. A needle biopsy can be definitive and apparently is harmless.[5]

Mammography and Ultrasound

Fibroadenomas may be tentatively identified by ultrasonography, which shows internal shadows in a smoothly lobulated mass. Mammography is of relatively little help in the diagnosis of a palpable mass in young women; in good hands 35% of cancers were not seen.[6] It is perhaps less helpful during pregnancy. Although the uterus can be shielded from stray radiation, it is general policy to avoid mammography during pregnancy, especially during the first trimester. Certainly pregnant women must not undergo screening mammography. However, if a significant mammographic lesion is recognized and no mass is palpable, needle localization, with shielding of the uterus, and biopsy are required.

Biopsy

Biopsy of indeterminate masses is the safest approach, and it must never be delayed because of the pregnancy. It can be made painless under local anesthesia by the liberal use of 0.5% lidocaine. Complete hemostasis is mandatory for good healing; it may require prolonged effort in the engorged breast. Use of bipolar cautery is the least traumatic method.

Systemic Metastases

When the pregnant woman presents at a later stage of local disease, the most important work-up for distant spread is the history and physical examination, a chest radiograph, and an alkaline phosphatase assay.[7] A bone or liver scan is contraindicated during pregnancy. Bone pain in the pelvis or hips requires radiographic study, also contraindicated. If there is systemic metastatic disease, the mother is almost certainly doomed. Chemotherapy, which is teratogenic, is unwise during pregnancy. Decisions in the light of these facts are likely to be difficult and cannot be standardized. Abortion may be appropriate during the first or second trimester. Early induction or cesarean section may be indicated after 35 weeks. By these means the mother can be prepared for adequate metastatic work-up and treatment. The fetus is not directly affected by spread of maternal cancer.

Treatment and Results

Experience at the University of Chicago

We have treated and followed 1958 patients with stage I or II breast cancer between 1927 and 1982. The oldest pregnant patient was 47 years. There were 620 patients, 32% of the total, under age 47. Of these women, 30 were pregnant within 1 year before mastectomy, constituting 4.8% of the younger women or 1.5% of the total. Cancers 1 cm in diameter are likely to have been growing at least 2 years[8]; hence the carcinomas were all probably present throughout the pregnancy. The 30 women were treated by mastectomy similar to the non-pregnant group; four in early pregnancy were aborted, and one underwent early induction. Survival was 63% at 5 years at stage I, 67% at stage II, and 44% at stage III. These proportions seem not remarkably different from survival of the nonpregnant patients, but the numbers are too small to analyze statistically. (The same difficulty pertains to other studies in the literature.) Only three women are known to have carried pregnancy to term following mastectomy; one died of breast cancer at 91 months, and two survive at 79 and 167 months.

Other Series

One of the first groups to find that pregnancy did not preclude adequate treatment of breast cancer was Byrd et al.[9] in 1962. They reviewed the interesting earlier opinions, including the one that declared pregnant patients to be inoperable for cure. In their own experience there were 29 who were pregnant among 1162 breast cancer patients. General anesthesia for biopsy and radical mastectomy was used for all 29 patients without maternal or fetal mortality. Five-year survival was 71%. Twelve patients were long-term survivors.

Holleb and Farrow[10] treated 193 operable mammary carcinomas in pregnant women. About one-half were aborted, and the other one-half were delivered. Survival was 33% and 17%, respectively, from which these authors concluded that interruption of pregnancy was not helpful for operable patients, though it was recommended during the first and second trimester for inoperable cancer. Among their stage I patients, the 5-year survival was 58%. They observed that carcinomas had no apparent effect on the fetus. Horsley et al.[11] reported that 10 of their 17 pregnant patients survived more than 10 years after mastectomy.

Experience at the Mayo Clinic was reported by King et al.[12] Between 1950 and 1980 there were 63 pregnant patients aged 22 to 44 who were found to have mammary carcinoma. Most were treated by radical mastectomy. Pregnancy was interrupted in 24, with survival of 43% at 5 years; among the 39 patients who delivered viable infants, survival was 59% at 5 years. Thus there was no apparent benefit from interruption in this nonrandom comparison. The 5-year survival of 22 patients at stage I was 82%. Nine patients had 17 pregnancies after mastectomy; 15 of the children did well, but 2 died of breast cancer in their teens.

Petrek et al.[13] found that 63 patients with pregnancy-associated breast cancer had essentially the same survival by stage as a comparable group of nonpregnant cancer patients. At stage I, survival was 76% and 75%, respectively, at 10 years.

In a French study 68 patients who became pregnant after mastectomy were followed.[14] In 27 the pregnancy was interrupted. In 41 there were one or more full-term deliveries. Analysis by stage revealed no significant difference in survival of the 68 patients compared to that of 136 controls who had no pregnancy after mastectomy. The 10-year survival rate of 68 patients was 71%. Interruption of pregnancy did not change survival.

In another French series[15] the patients who became pregnant after mastectomy had a higher survival rate than matched controls. A selection effect was postulated: Those who were doing well were more likely to become pregnant.

A British series[16] likewise showed a high survival rate (100%) at 5 years among women who became pregnant 2 to 5 years after mastectomy. There were 121 women with concomitant pregnancy and cancer treated by radical or total mastectomy or by local excision, often with adjuvant radiotherapy. Normal birth occurred in 79%. Nodes were involved in 72%. Survival at 5 years was 37%.

Surgical Recommendation

We have reviewed all randomized comparisons of the extent of surgery followed for at least 10 years.[17] Although these studies did not single out pregnant women in their results, they do constitute the best available evidence concerning the relation between surgical treatment and survival. There are just eight studies. All but one showed better survival with more complete operations; in two of the studies these differences are statistically significant. The conclusion we reached from these and our own ongoing research is that survival of breast cancer patients at clinical stages I and II is dependent on the nominal extent of the mastectomy and on the detailed care in its performance.[18] Our studies of adjuvant radiotherapy after radical mastectomy have demonstrated a high rate of serious long-term toxicity, with no improvement in survival.[19]

The application of these facts to treatment of pregnant women is reasonable so long as we have no equal or better evidence to the contrary. In particular, pregnant women are probably at no greater surgical risk than those in the nonpregnant group.[9]

Anesthesia

Local infiltration of anesthetic is strongly indicated for biopsies and segmental resections. Mastectomy under local anesthesia is possible but not recommended.

The principal concerns regarding general anesthesia in pregnant women are the safety of the mother, adequate oxygenation of the mother to prevent fetal hypoxia in utero, and avoidance of any drug that may harm the fetus.[20] Physiologic changes in pregnancy include increasing alveolar ventilation (25% above the nonpregnant level at 4 months and 70% above it at term), a corresponding fall in end-tidal pCO_2, and an increase in oxygen consumption. The requirement for halogenated anesthetics decreases as much as 40% at midpregnancy, and the hazard of overdose is correspondingly increased. Blood volume and cardiac output increase by about one-third at this time. The effective dose of spinal and epidural anesthetics is decreased about one-fourth, perhaps because nerves become more sensitive to anesthetics during pregnancy, although other factors may play a role.[21]

Lower esophageal sphincter tone is reduced during pregnancy, as intraabdominal pressure is increased, predisposing women to esophageal regurgitation. Acid secretion by the stomach may be increased, and esophagitis may occur. Treatment using oral nonsystemic antacids such as calcium carbonate and gastric intubation preoperatively to prevent aspiration are the appropriate remedies.

Potentially teratogenic drugs remain a possible hazard. Investigation is difficult because experimental animals differ from humans in terms of susceptibility, and they differ among species; moreover, adequate research on agents affecting the human fetus has not been completed. In general, therefore, all drugs are best avoided during pregnancy, especially during the first trimester. Narcotics inducing sleep may damage the fetus indirectly, as may anesthetics, by causing hypoxia and hypercarbia in the mother.

Studies of anesthetic risk to pregnancy in humans are retrospective surveys lacking in actual measurements of exposure to specific anesthetics. From these data the principal risk of general anesthetics of any kind appears to be an increased incidence of spontaneous abortion. The risk of an increase in teratogenicity is apparently small but cannot be ruled out by the scanty data available at present.[22,23] No particular anesthetic, on the other hand, is known to be teratogenic in humans.

Summary and Conclusions

Obstetricians need to be especially alert to recognize a possible breast cancer because mammary growth tends to hide findings during pregnancy, and radiologic studies are contraindicated. Needle aspiration and ultrasonography are useful for identifying cysts and fibroadenomas. No delay in biopsy of a suspicious breast lesion should be tolerated. Biopsy should be done under local anesthesia.

Survival of women treated surgically for localized breast cancer during pregnancy appears to be within the range of survivals found in nonpregnant women. General anesthesia during pregnancy requires special expertise; it is not hazardous to the mother, although the rate of spontaneous abortions is increased. No particular anesthetic is known to be teratogenic in the human.

Systemic spread of cancer cannot be safely treated during pregnancy. Pregnancies following mastectomy have been successful and have not apparently increased the risk of recurrent cancer.

References

1. White TT. Prognosis of breast cancer for pregnant and nursing women; analysis of 1,413 cases. Surg Gynecol Obstet 1955;100:661–666.
2. Hadjimichael OC, Boyle CA, Meigs JW. Abortion before first livebirth and risk of breast cancer. Br J Cancer 1986;53:281–284.
3. McPherson K. The pill and breast cancer: why the uncertainty? Br Med J 1986;293:709–710.
4. Paul C, Skegg DCG, Spears GFS, et al. Oral contraceptives and breast cancer; a national study. Br Med J 1986;293:723–726.
5. Berg JW, Robbins GF. A late look at the safety of aspiration biopsy. Cancer 1962; 15:826–827.
6. Cohen MI, Mintzer RA, Matthies HJ, et al. Mammography in women less than 40 years of age. Surg Gynecol Obstet 1985;160:220–222.
7. Coombes RC, Abbott M, Ford HT, et al. Physical test for distant metastases in patients with breast cancer. J R Soc Med 1980;73:617–623.
8. Gershon-Cohen J, Berger SM, Klickstein HS. Roentgenography of breast cancer moderating concept of "biologic predeterminism." Cancer 1963;16:961–964.
9. Byrd BF, Bayer DS, Robertson JC, et al. Treatment of breast tumors associated with pregnancy and lactation. Ann Surg 1962; 155:940–947.
10. Holleb AI, Farrow JH. The relation of carcinoma of the breast and pregnancy in 283 patients. Surg Gynecol Obstet 1962;115: 65–71.
11. Horsley JS, Alrich EM, Wright CB. Carcinoma of the breast in women 35 years of age or younger. Ann Surg 1969;169:839–843.
12. King RM, Welch JS, Martin JK, et al. Carcinoma of the breast associated with pregnancy. Surg Gynecol Obstet 1985;160:228–232.
13. Petrek JA, Dukoff R, Crowe JP, et al. Pregnancy-associated breast cancer. Surg Forum 1986;37:478–480.
14. Mignot L, Morvan F, Berdah J, et al. Pregnancy after treated breast cancer: results of a case-control study. Presse Med 1986;15:1961–1964 [French].
15. Querleu D, Laurent JC, Verhaeghe M. Pregnancy following surgery for cancer of the breast. J Gynecol Obstet Biol Reprod 1986;15:633–639 [French].
16. Ribeiro G, Jones DA, Jones M. Carcinoma of the breast associated with pregnancy. Br J Surg 1986;73:607–609.
17. Ferguson DJ. The actual extent of mastectomy: a key to survival. Perspect Biol Med 1987;30:311–323.
18. Meier P, Ferguson DJ, Karrison T. A controlled trial of extended radical mastectomy. Cancer 1985;55:880–891.
19. Ferguson DJ, Sutton HS, Dawson PJ. Late effects of adjuvant radiotherapy for breast cancer. Cancer 1984;54:2319–2323.
20. Levinson G, Schnider SM. Anesthesia for surgery during pregnancy. In Schnider SM, Levinson G (eds): Anesthesia for Obstetrics. Baltimore, Williams & Wilkins, 1987;188–205.
21. Datta S, Lambert DH, Gregus J, et al. Differential sensitivities of mammalian nerve fibers during pregnancy. Anesth Analg 1983;62:1070–1072.
22. Duncan PG, Pope WDB, Cohen MM, et al. Fetal risk of anesthesia and surgery during pregnancy. Anesthesiology 1986;64:790–794.
23. Kolata GB. Behavioral teratology: birth defects of the mind. Science 1978;202:732–734.

12
Carcinoma of the Cervix During Pregnancy

ELIZABETH K. SENEKJIAN

Historically, the association of cervical carcinoma with pregnancy has been the subject of controversy with respect to the growth of the tumor, the prognosis for the patient, and the risk of cancer dissemination accompanying vaginal delivery.[1–5] It has been postulated that factors such as the altered hormonal milieu, increased vascularity, and the immunologic tolerance of gestation influence tumor behavior. This chapter reviews methods of diagnosis, treatment options, and the prognosis of cervical carcinoma during pregnancy.

Incidence and Age

Cervical carcinoma is the eighth most common cancer affecting women in the United States (13,500 new cases were estimated for 1990). Malignant tumors of the cervix rank seventh as a cause of cancer death among women (6000 cancer deaths were estimated for 1990).[6] Carcinoma of the cervix is the most common gynecologic cancer in pregnant women.

The reported incidence of cervical carcinoma in association with pregnancy varies among individual studies owing to differences in the duration and interval of study, the birth rate, the race distribution and socioeconomic status of the population, the referral base of the institution, and the length of the postpartum period included. Hacker et al.[7] obtained data from 15 reports and found the average incidence of cervical carcinoma during pregnancy and the first 12 months postpartum to be approximately one case per 2205 pregnancies; nearly 3% of patients with cervical carcinoma were pregnant at the time of diagnosis. The average age of pregnant patients with cervical cancer (33.8 years)[1,7] is 15 years younger than the average age of nonpregnant patients.[2,8] This difference seems to be related to the lower pregnancy rate in older women.

Detection and Diagnosis

The most common initial symptom of cervical carcinoma is abnormal vaginal bleeding.[2–4] Abortion, ectopic pregnancy, placenta previa, and placental abruption are most often responsible for bleeding in the obstetric patient; however, the possibility of a cervical malignancy warrants timely, thorough, and judiciously performed inspection and palpation of the cervix. Undue physician delay in the diagnosis of cervical cancer in pregnant patients has been reported in 62% of cases with an average delay interval of 4.5 months.[9] In up to one-third of patients bleeding does not occur, and the tumor is detected at routine prenatal pelvic examination or on evaluation of an abnormal cytologic smear. A directed biopsy of a cervical lesion that is visible by gross inspection or colposcopic examination is indicated to establish the diagnosis.

TABLE 12-1. FIGO staging of cervical carcinoma.

Stage I	Carcinoma confined to the cervix
Stage Ia	Preclinical lesions diagnosed microscopically
Stage Ib	Lesions with a depth of invasion exceeding 5 mm from the base of the epithelium with a horizontal measurement exceeding 7 mm
Stage IIa	Carcinoma involving the upper two-thirds of the vagina with no parametrial infiltration
Stage IIb	Carcinoma extending into the parametrium but not to the pelvic wall
Stage III	Carcinoma involving the lower one-third of the vagina and/or extending to the pelvic wall and/or associated with hydronephrosis or nonfunctional kidney due to tumor
Stage IV	Carcinoma involving the bladder or rectal mucosa and/or extending beyond the true pelvis

Constructed with material from Annual Report on Gynecologic Cancer. FIGO, Vol. 19, 1985. Changes in definitions of clinical staging for carcinoma of the cervix and ovary: International Federation of Gynecology and Obstetrics. Am J Obstet Gynecol 1987; 156:263. (With permission of FIGO and C.V. Mosby Company).

Of 958 cases in a review by Hacker et al.,[7] squamous carcinomas accounted for more than 93% of cervical cancers associated with pregnancy, adenocarcinomas comprised 3% of cases, adenosquamous carcinomas approximately 1%, and miscellaneous subtypes 3%. Cherry and Glucksmann[10] reported on 16 cases of adenosquamous carcinoma in 26 (62%) pregnant patients, whereas only 19% of 548 premenopausal, nonpregnant patients had lesions of this histologic variety. Other studies focus on two less common histologic subtypes, namely, six antepartum cases of clear cell adenocarcinoma[11] and four cases of glassy cell carcinoma which were uncovered during pregnancy and up to 26 months postpartum.[12]

The FIGO staging classification for cervical carcinoma is presented in Table 12-1. The clinical extent of the lesion is assessed by the bimanual and rectovaginal examination. Cystoscopy and proctoscopy are indicated in symptomatic patients and in women with advanced disease.

The routine pretreatment evaluation of cervical cancer during pregnancy includes a chest radiograph with lead shielding of the abdomen and a modified intravenous pyelogram that consists of a single film obtained 20 minutes after the injection of contrast. With one diagnostic abdominal radiograph the conceptus is probably exposed to less than 1 cGy, which does not seem to increase the risk of malformation, childhood cancer, or genetic defect.[13] Therapeutic abortion has generally not been recommended if the fetal dose during the first 4 months of pregnancy is below 10 cGy, as the risk of genetic or structural damage to the fetus is small.[14]

The distribution by stage of cervical carcinoma in pregnant and nonpregnant women from four large studies is shown in Table 12-2. The proportion of stage I tumors in the pregnant group is greater than in the nonpregnant group. Advanced stages are more common when the diagnosis is made during the first year after delivery than when the lesion is uncovered antepartum (Table 12-3).

TABLE 12-2. Distribution by stage for pregnant and nonpregnant women with cervical carcinoma.

	Pregnant		Nonpregnant	
Stage	No.	%	No.	%
I	271	31	2,115	16
II	343	40	5,842	43
III	200	23	3,828	28
IV	47	6	1,834	14
Total	861		13,619	

Data from Gustaffson and Kottmeier,[3 b,c]; Waldrop and Palmer,[2 b,d]; Creasman et al.[4 a,d] Sablinska et al.[1 b,d]

[a,b]Disease was diagnosed during pregnancy or within 6[a] or 12[b] months postpartum.

[c]Stage was assigned according to the League of Nations classification.

[d]Staging criteria were not defined.

TABLE 12-3. Distribution of advanced stage cervical carcinoma by time of diagnosis relative to the gestation.

Source of data	Antepartum diagnosis		Postpartum diagnosis	
	Total cases (No.)	Stage III/IV (%)	Total cases (No.)	Stage III/IV (%)
Gustafsson & Kottmeier[3]	48	13	65	28
Waldrop & Palmer[2]	50	28	132	44
Creasman et al.[4]	146	12	181	44
Sablinska et al.[1]	82	16	157	27
Total	326	15	733	27

Management

The treatment of cervical carcinoma during pregnancy must be individualized, with consideration being given to the gestational age, the stage of the cancer, and the wishes of the mother. In general, during the first half of pregnancy, treatment is undertaken promptly, without regard to the fetus. When the diagnosis is made after 20 weeks' gestation, therapy can be instituted after delivery.

From the limited data available, a short delay in initiating treatment does not seem to adversely affect the outcome of therapy. In two studies, no evidence for tumor progression was reported despite treatment delay among 18 pregnant patients. Although therapy was postponed for 11 to 17 weeks during the second half of pregnancy, six patients with stage I cervical cancer survived disease free for 3 to 10 years.[5,15]

Mothers who wish to defer treatment until after the delivery of a viable fetus, particularly when the malignancy is diagnosed during the first half of pregnancy, therein necessitating a long treatment delay, are counseled with regard to the risks of interval tumor progression. Delivery usually follows soon after documentation of fetal lung maturity unless tumor growth warrants earlier intervention.

Patients who are anxious about their disease, who do not wish to continue the pregnancy, and who have immature amniotic fluid phospholipid levels may be given prophylactic corticosteroid therapy 24 to 48 hours prior to a planned delivery. Although several studies have demonstrated a decrease in the incidence and severity of respiratory distress syndrome with antenatal corticosteroid therapy the efficacy of these agents when used to treat the fetus before 26 weeks' gestation is unclear.

Mothers who are uncertain as to whether to elect a termination of pregnancy, particularly when the diagnosis is made before 24 weeks, can be provided with information regarding the predictability of fetal outcome to facilitate the decision-making process. Birth weight seems to be the most important determinant of neonatal survival. The data from the Chicago Lying-In Hospital, University of Chicago[16] are similar to those from other institutions. The survival rate for infants in the 500- to 899-g birth weight group is 28%, whereas for those in the 900- to 1199-g and 1200- to 1500-g groups are 79% and 86%, respectively. An accurate antenatal estimate of the fetal weight may be calculated using formulas based on ultrasonographically measured fetal dimensions. Ultrasonographic study may also influence management when a major congenital malformation associated with a poor fetal outcome is detected.

Survival expectations based on both birth weight and gestational age may be of additional value. For example, a fetus with a birth weight below 500 g and a gestational age of less than 26 weeks is unlikely to survive outside the uterus.

The stage of the disease is also an important factor when planning therapy. Radical hysterectomy with pelvic lymphadenectomy seems to be suitable for patients with small stage Ib and IIa tumors when there is minimal risk of lymphatic involvement, a desire to preserve ovarian and coital function, a history of diverticulitis or pelvic inflammatory disease, and an interest in a short treatment interval. Gestational edema facilitates the operative procedure. Morbidity associated with this form of therapy is acceptable. During the first half of pregnancy radical hysterectomy can be done with the fetus in situ, whereas after 20 weeks' gestation it is performed after a vertical fundal hysterotomy or classic cesarean section.

Radiation therapy is as effective as radical hysterectomy for small early stage cervical cancers and is the treatment of choice for locally advanced lesions. When the diagnosis is made during the first half of pregnancy, treatment is initiated with external beam irradiation without termination of the pregnancy. During the first trimester spontaneous abortion usually occurs before brachytherapy (within 4–7 weeks; mean 33 days).[15] The interval between the start of irradiation and abortion is longer during the second trimester (within 5–9 weeks; mean 44 days). If spontaneous abortion fails to take place, the uterus is evacuated prior to intracavitary therapy. Alternatively, a combination of surgery and radiotherapy may be chosen for early stage lesions that demonstrate an appropriate response to radiation therapy; an extrafascial hysterectomy may be undertaken after external beam irradiation and one intracavitary insertion.

When therapy is instituted during the third trimester, surgical staging including lymph node sampling can be done at the time of abdominal delivery with radiation therapy being administered postoperatively. Extended field radiotherapy is utilized in selected cases when pelvic, common iliac, and aortic node metastases are documented. For cancers detected during the postpartum interval, radiotherapy is instituted as in the nonpregnant patient. Major radiation complications[17–19] and interruption of therapy[3] have been reported in patients undergoing radiotherapy during pregnancy.

Radiation therapy is avoided during the waiting period when a decision has been made by the mother to continue the pregnancy until fetal viability. Most infants exposed to radiation doses above 250 cGy between 4 and 11 weeks' gestation have had severe malformations of major organs. A similar dose delivered between 11 and 20 weeks' gestation has been associated with growth retardation, mental retardation, and microcephaly.[20] After 20 weeks' gestation the risk to the fetus is essentially the same as postnatal exposure. Irradiation may increase the possibility of later carcinogenesis.[21]

Treatment may be withheld during pregnancy in patients with advanced disease when maternal life expectancy is limited, especially if the diagnosis is made during the third trimester. Only one case of maternal cervical carcinoma involving the placenta has been reported to date; there are no cases of metastasis to the fetus. Therefore, therapeutic abortion is not warranted for fetal indications. Gross and microscopic examina-

TABLE 12-4. Five-year survival of patients with cervical carcinoma associated with pregnancy.[a]

Stage	Pregnant: Total cases (No.)	Pregnant: % Alive	Nonpregnant: Total cases (No.)	Nonpregnant: % Alive
I	271	75	2,115	70
II	343	50	5,842	53
III,IV	223	16	5,228	20
Total	1008	41	13,185	43

Data from Gustafsson and Kottmeier,[3] Waldrop and Palmer,[2] Creasman et al.,[4] [b] and Sablinska et al.[1] [c]
[a]Disease was diagnosed during pregnancy or up to 12 months postpartum.
[b]Actuarial survival rates obtained from graph for stages I and II; stages III and IV were excluded; these data include cases diagnosed up to 6 months postpartum.
[c]About 16% of 97 pregnant patients versus 30% of 190 nonpregnant patients with stage III and IV lesions were alive at 5 years ($p < 0.02$).

TABLE 12-5. Five-year survival of patients with cervical carcinoma according to mode of delivery.

	Vaginal delivery		Abdominal delivery	
Source of data	Total cases (No.)	% Alive	Total cases (No.)	% Alive
Waldrop & Palmer[2]	148	43	34	27
Creasman et al.[a,4]	29	76	13	77
Hacker et al.[b,7]	210	63	55	33

[a]Actuarial survival; includes only stages I and II.
[b]Data may include cases from Waldrop and Palmer[2] and/or from Creasman et al.[4]

tion of the products of conception is advisable.[22]

Prognosis

The overall 5-year survival rates for pregnant and nonpregnant patients with cervical carcinoma are similar (Table 12-4). As expected, the clinical extent of disease correlates with survival. It has been suggested but unproved that pregnancy has an adverse effect on the prognosis of advanced disease.[1,7] In an evaluation of 896 cases Hacker et al.[7] observed a better prognosis when the diagnosis was made during the first trimester (5-year survival rate 69%) compared to postpartum (5-year survival rate 47%). The difference seems to be related to the greater proportion of more advanced lesions among cases of cervical cancer diagnosed postpartum. Among pregnant women with stage I tumors, the 5-year survival rate for 232 patients treated with radiotherapy was lower than for 48 who underwent radical surgery (74% versus 81%).[7] The more favorable outcome for the latter group is probably due to the selection of patients with smaller tumors for surgical therapy.

Mode of Delivery

Cesarean section is the recommended method of delivery because of the possibility of hemorrhage. Published data do not substantiate concerns regarding an increased risk of tumor dissemination or adverse prognosis with vaginal delivery. The 5-year survival rate after vaginal delivery seems to be as good as or better than after abdominal delivery (Table 12-5).

Microinvasive Carcinoma of the Cervix

The diagnosis of microinvasive cervical carcinoma is made by the pathologic examination of a cervical conization specimen in which the margins are free of disease. Hacker et al.[7] reviewed seven studies addressing microinvasive cervical carcinoma in association with pregnancy. No standard definition or treatment approach for microinvasion could be identified. Recurrence was not reported among the 60 patients who underwent definitive treatment. Twelve patients were allowed to deliver vaginally. Treatment has been delayed for up to 28 weeks after the diagnosis of microinvasion by conization.[17] In such cases, evidence of progression may be detected promptly by monthly assessment of the cervix by cytologic smear and colposcopic examination.

References

1. Sablinska B, Tarlowska L, Stelmachow J. Invasive carcinoma of the cervix associated with pregnancy: correlation between patient age, advancement of cancer and gestation, and result of treatment. Gynecol Oncol 1977;5:363–373.
2. Waldrop GM, Palmer JP. Carcinoma of the

cervix associated with pregnancy. Am J Obstet Gynecol 1963;86:202–212.
3. Gustafsson DC, Kottmeier HL. Carcinoma of the cervix associated with pregnancy. Acta Obstet Gynecol Scand 1962;41:1–21.
4. Creasman WT, Rutledge FN, Fletcher GH. Carcinoma of the cervix associated with pregnancy. Obstet Gynecol 1970;36:495–501.
5. Lee RB, Neglia W, Park RC. Cervical carcinoma in pregnancy. Obstet Gynecol 1981;58: 584–589.
6. American Cancer Society: Cancer facts and figures, 1990.
7. Hacker NF, Berek JS, Lagasse LD, et al. Carcinoma of the cervix associated with pregnancy. Obstet Gynecol 1982;59:735–746.
8. Williams TJ, Brack CB. Carcinoma of the cervix in pregnancy. Cancer 1964;17:1486–1491.
9. Stander RW, Lein JN. Carcinoma of the cervix and pregnancy. Am J Obstet Gynecol 1960;79:164–167.
10. Cherry CP, Glucksmann A. Histology of carcinomas of the uterine cervix and survival rates in pregnant and nonpregnant patients. Surg Gynecol Obstet 1961;113:763–776.
11. Senekjian EK, Hubby M, Bell DA, et al. Clear cell adenocarcinoma (CCA) of the vagina and cervix in association with pregnancy. Gynecol Oncol 1986;24:207–219.
12. Seltzer V, Sall S, Castadot MJ, et al. Glassy cell cervical carcinoma. Gynecol Oncol 1979; 8:141–151.
13. Wagner LK, Lester RG, Saldana LR. Prenatal risks from ionizing radiations, ultrasound, magnetic fields and radiofrequency waves. In Wagner LK, Lester RG, Saldana LR (eds): Exposure of the Pregnant Patient to Diagnostic Radiations: A Guide to Medical Management. Philadelphia, Lippincott, 1985;61–78.
14. Hammer-Jacobsen E. Therapeutic abortion on account of x-ray examination during pregnancy. Dan Med Bull 1959;6:113–122.
15. Prem KA, Makowski EL, McKelvey JL. Carcinoma of the cervix associated with pregnancy. Am J Obstet Gynecol 1966;95:99–108.
16. Hibbard JR, Moawad AH. Umbilical cord blood gas values and Apgar scores in very low birth weight infants. In CT Jones (ed): Ithaca, NY, Perinatology Press, 1988.
17. Thompson JD, Caputo TA, Franklin EW III, et al. The surgical management of invasive cancer of the cervix in pregnancy. Am J Obstet Gynecol 1975;121:853–863.
18. Dudan RC, Yon JL, Ford JH. Carcinoma of the cervix and pregnancy. Gynecol Oncol 1973;1:283–289.
19. Nisker JA, Shubat M. Stage IB cervical carcinoma and pregnancy: report of 49 cases. Am J Obstet Gynecol 1983;145:203–206.
20. Dekaban AS. Abnormalities in children exposed to x-radiation during various stages of gestation: tentative timetable of radiation injury to the human fetus, part 1. J Nucl Med 1968;9:471–477.
21. Doll R. Radiation hazards: 25 years of collaborative research. Br J Radiol 1981;54:179–186.
22. Dildy GA 3d, Moise KJ Jr, Carpenter RJ Jr, et al. Maternal malignancy metastatic to the products of conception: a review. Obstet Gynecol Surv 1989;44:535–540.

13
Ovarian Tumors During Pregnancy

KARL C. PODRATZ and CHARLES S. FIELD

The coexistence of an ovarian tumor and pregnancy with potential untoward sequelae can present arduous diagnostic and therapeutic decisions for the obstetrician if maximal fetal salvage and maternal well-being are to be realized. The increasing awareness by patients of the benefits of early prenatal care and the routine use of high-resolution echography presumably facilitate earlier and more frequent detection of adnexal abnormalities. Therefore familiarity with the incidence and clinical manifestations of various ovarian tumors aids in establishing a differential diagnosis and a therapeutic approach that in turn can be appropriately articulated to the patient and her family. In addition, perioperative management requires the concerted expertise of multiple health care providers to optimize pregnancy outcome. Considering that the number of pregnancy-associated ovarian tumors encountered at any single institution is limited, selected previous reported experiences were summarized in support of the management profiles detailed below.

Incidence

Although the presence of a clinically detected adnexal mass during pregnancy is considered an infrequent occurrence, the incidence rates generally cited in the literature range from one ovarian tumor per 81 deliveries, as reported by Grimes et al.,[1] to only one in several thousand deliveries.[2] This wide range is a reflection of the variability in the inclusion criteria used among investigators addressing this phenomenon. The composition of study populations and the facility with which ovarian lesions are identified are influenced by multiple factors, including gestational age at initial antenatal examination, use of routine ultrasonographic imaging, definition of significant tumor size, accepted management approaches, and the requirement for laparotomy for histologic documentation. When no restriction has been imposed on tumor size and surgical intervention for verification has been limited, a modest incidence (1 per 81) has been noted, but the incidence is approximately 70-fold less (1 per 6226) when the patient population forgoes antenatal care until late in pregnancy.[2]

Using sonographic evaluation of the adnexa during the initial 20 weeks of gestation, Lavery et al.[3] demonstrated that ovarian lesions exceeding 2 cm in greatest dimension were not uncommon through 10 weeks of pregnancy. Imaging identified significant adnexal masses in 8.8% of patients at 0 to 5 weeks of gestation and in 7.4% of patients at 6 to 10 weeks; these figures decreased to 1.3% at 11 to 15 weeks of gestation and to 0.35% at 16 to 20 weeks. These observations emphasize the frequent association of cystic adnexal enlargement with early pregnancy and the dramatic, natural attrition that occurs after the initial 10 weeks of gestation.

The incidence of ovarian tumors that attain a size sufficiently significant to be clinically detected and to warrant histologic definition has been debated. Table 13-1 illustrates the incidence rates of ovarian tumors and percentages of tumor-associated malignancies in nine major studies reported since 1973.[4–12] Collectively, 643 ovarian lesions were evaluated surgically during intervals when approximately 491,000 deliveries occurred at the corresponding institutions; thus the overall incidence was 1 per 764. Furthermore, only 25 (3.9%) of the evaluable tumors were identified as malignant; thus the frequency of ovarian cancer during pregnancy was 1 in 19,640 deliveries.

In contrast to the nonpregnant state, only a small percentage (4%) of persistent ovarian tumors during gestation are malignant. Hence the obstetrician must be familiar with the natural history of the more common pregnancy-associated benign ovarian lesions. Table 13-2 shows the frequency and distribution of several histologic tumor types that were detected and required treatment during pregnancy as reported by eight investigative groups.[5–12] Considering that the peak incidence for mature teratomas occurs during the latter half of the third decade of life,[13,14] dermoid tumors were encountered most frequently and accounted for 29% of all ovarian tumors. Because most of these lesions exceed a diameter of 6 cm and possess sufficiently unique echographic characteristics, preoperative diagnosis is facilitated. Overall, functional cysts accounted for 21% of the ovarian tumors surgically removed during pregnancy. Rather noticeable variations (ranging from 0 to 52%) were noted among the reporting institutions, presumably reflecting antenatal diagnosis and expeditious surgical intervention. When the "presence of symptoms and large tumor size" were used as indications for prompt exploration, operation was performed at an average gestational age of 9.1 weeks by Hill et al.[7] and yielded functional cysts in more than 50% of patients subjected to laparotomy.

Collectively, serous and mucinous simple cysts and cystadenomas occur at a rate approximating that of mature teratomas and frequently attain impressive dimensions. Finally, endometriotic cysts, paraovarian cysts, luteomas of pregnancy, and miscellaneous ovarian lesions, although they are less frequent, present diagnostic challenges and require individualized therapy. The histologic subtypes of benign ovarian tumors and their distribution during pregnancy (Table 13-2) are strikingly similar to those of such lesions in 265 nonpregnant patients described by Portuondo et al.[15]

TABLE 13-1. Incidence of clinically detectable ovarian tumors requiring surgical management during pregnancy.

		Ovarian tumors			
Reference	Year	No.	Incidence	No. malignant	%
Chung & Birnbaum[4]	1973	199	1:808	6	3.0
Buttery et al.[5]	1973	164	1:938	4	2.4
White[6]	1973	37	1:958	3	8.1
Hill et al.[7]	1975	47	1:536	3	5.3
Sivanesaratnam et al.[8]	1976	27	1:512	2	7.4
Karpathios et al.[9]	1977	16	1:351	0	0.0
Teoh et al.[10]	1984	40	1:649	2	5.0
Struyk & Treffers[11]	1984	90	1:640	3	3.3
Hopkins & Duchon[12]	1986	23	1:556	2	8.7
Total		643	1:764	25	3.9

TABLE 13-2. Frequency of histologic subtypes in ovarian tumors during pregnancy.

Reference	Serous	Mucinous	Teratoma	Functional	Other	Malignant	Total
	Ovarian tumor subtype (No.)						
Buttery et al.[5]	18	38	45	29	30	4	164
White[6]	8	4	8	10	4	3	37
Hill et al.[7]	4	1	7	30	12	3	57
Sivanesaratnam et al.[8]	5	5	12	1	2	2	27
Karpathios et al.[9]	13	0	3	0	0	0	16
Teoh et al.[10]	4	5	17	5	5	2	38
Struyk & Treffers[11]	17	8	25	5	11	3	69
Hopkins & Duchon[12]	1	0	8	9	3	2	23
Total	70	61	125	89	67	19	431
%	16	14	29	21	16	4	

Although ovarian tumors have been reported to be malignant in 18 to 22% of cases in nonpregnant populations, the overall incidence is influenced by the portion of sampled patients older than 40 years. Assessing 1391 ovarian tumors in nonpregnant patients, Buttery et al.[5] noted cancer in less than 2% of patients during the first two decades of life, in 10% during the fourth decade, and in approximately 50% during the sixth decade. The consolidated incidences of ovarian cancers in these nonpregnant patients before 40 years of age was 6.0%, a rate only slightly higher than the 3.9% observed in the pregnant population ($n = 643$) amassed from the literature (Table 13-1). However, the distribution among the histologic subtypes and the extent of metastatic disease at presentation are dramatically different in the pregnant and nonpregnant populations.

According to published studies, essentially one-half of the cancers encountered during pregnancy were either germ cell or gonadal stromal tumors (Table 13-3), in contrast to the preponderance of epithelial carcinomas associated with the typically nonpregnant older population.[4–8,10–12,16,17] Likewise, disease was limited to the ovaries in more than 70% of the cases diagnosed during pregnancy, presumably reflecting the natural history of the associated tumors as well as early clinical or imaging detection during antenatal examinations. Further-

TABLE 13-3. Distribution of histologic subtype in ovarian cancers associated with pregnancy.

Reference	Epithelial	Germ cell	Gonadal stromal	Other
	Ovarian tumor subtype (No.)			
Creasman et al.[16]	9	6	2	—
Chung & Birnbaum[4]	4	1	1	—
Buttery et al.[5]	2	2	—	—
White[6]	3	—	—	—
Hill et al.[7]	3	—	—	—
Novak et al.[17]	24	23	14	8
Sivanesaratnam et al.[8]	—	1	1	—
Teoh et al.[10]	2	—	—	—
Struyk & Treffers[11]	2	1	—	—
Hopkins & Duchon[12]	—	—	2	—
Total	49	34	20	8
%	44	31	18	7

more, the more favorable survival rates generally ascribed to pregnancy-related ovarian cancers correlate again with the higher frequency of early stages of disease and the nonepithelial tumor types observed during gestation. Therefore the obstetrician must be mindful of the low incidence (1 per 19,640), the more favorable histologic subtypes, and the low stage of ovarian malignancy when determining the management of ovarian tumors during gestation.

Clinical Presentation and Diagnosis

The routine use of continually evolving imaging technology during pregnancy and the increasing role of the obstetrician and gynecologist as primary health care providers for women in the reproductive age-group enhance the early detection and facilitate definition of both occult and symptomatic ovarian tumors. Initiation of antenatal care during early gestation aids the clinical identification of adnexal abnormalities through careful pelvic examination.[18] Subsequent serial bimanual and sonographic assessment of suspicious ovarian lesions dictates appropriate management. By using ultrasonographic evaluation of the adnexa during the first 20 weeks of pregnancy, Lavery et al.[3] detected significant ovarian enlargement in 7.5% of patients during the initial 10 weeks of gestation with rapid resolution during the following 10 weeks; surgical intervention was required in only 0.2% of monitored patients.

In obstetric practice settings where early antenatal care and liberal use of echography are the norm, ovarian tumors are detected in a finite number of asymptomatic patients and only an occasional patient develops symptoms or harbors a tumor that is sufficiently large or progressively expanding to prompt surgical removal. If one assumes that the indications for operation denote a significant lesion, most such adnexal masses become clinically identifiable during the first half of pregnancy. In a collected series of 382 patients requiring operative definition of ovarian abnormality during pregnancy (Table 13-4), 45% were diagnosed during the first trimester, 24% during the second trimester, and 31% during labor, delivery, or the puerperium.[5,6,8,10–12,16,19] As noted by Buttery et al.,[5] 30% of the 164 ovarian tumors associated with pregnancy were detected by routine pelvic examination in asymptomatic patients: 62% during the first

TABLE 13-4. Presentation and surgical definition of ovarian tumors according to gestational status.

	No. of patients, by trimester			
Reference	First	Second	Third[a]	Puerperium
Buttery et al.[5]	80	44	23	17
White[6]	12	15	10	1
Sivanesaratnam et al.[8]	9	3	9	6
Teoh et al.[10]	17	7	14	2
Struyk & Treffers[11]	49	13	18	10
Hopkins & Duchon[12]	5	8	10	—
Total	172	90	84	36
%	45	24	22	9
Jubb[b19]	6	12	9	6
Creasman et al.[b16]	1	3	7	6
Total	7	15	16	12
%	14	30	32	24

[a]Includes labor and delivery.
[b]Ovarian cancers only.

trimester and more than 90% during the initial two trimesters, figures that emphasize the value of early antenatal assessment.

Considering that the most common presentation for benign ovarian tumors in nonpregnant populations is abdominal pain,[13–15] it would rightfully be anticipated that pain is similarly the chief complaint when this process occurs during pregnancy.[5,7,11,17] Most commonly, the cause of such abdominal pain is ovarian capsular distention, tissue ischemia, and chemical or inflammatory irritation of the abdominal peritoneum, with the extent of the unnatural event determining acuity and thus dictating the need for emergency operation. During early gestation the combination of vaginal bleeding and lower abdominal pain warrants consideration of an ectopic pregnancy, which occasionally requires surgical assessment.[7] Regardless of the gestational status or site, an incidental distended ovarian cyst, invariably functional, is usually encountered.

Presumably any rapidly enlarging cystic lesion within the ovary, irrespective of pathophysiology, results in sufficient distention to cause abdominal discomfort. Furthermore, the pain associated with adnexal torsion reflects tissue ischemia and may be accompanied by a mild temperature elevation and leukocytosis. Although the question of whether torsion indeed occurs more frequently during pregnancy continues to be debated, no apparent correlation exists between the more common histologic types or tumor size and the frequency of torsion.[5,11,14,15,18] Nevertheless, the clinical declaration of ovarian torsion during pregnancy has a predilection for the time frame between 10 and 15 weeks of gestation, a suggestion that the size and mobility of both the gravid uterus and the ovaries at this point during pregnancy predispose to adnexal rotation.

Furthermore, pain may reflect irritation of the abdominal peritoneum from contents released from a ruptured ovarian cyst or infection of an intact adnexal mass. Infection invariably presents initially as a puerperal infection and subinvoluted uterus, and definition awaits careful clinical evaluation. In contrast, rupture of an ovarian cyst with spillage of blood or other contents into the abdomen generally causes acute pain with occasional nausea and vomiting. Trauma, particularly during labor and delivery, torsion, spontaneous intracystic hemorrhage, and infection during the puerperium are predisposing etiologic settings in which rupture occurs.[5,10,20] Luteal cyst rupture with accompanying hemoperitoneum seems to occur with some frequency and predominantly during early gestation, whereas mature teratomas, cystadenomas, and endometriotic and other cysts are more prone to rupture late in pregnancy, particularly during labor, delivery, and the puerperium.

Obstruction of labor, abdominal distention, palpable abdominal masses, and masculinization (from luteomas of pregnancy) are additional clinical presentations for pregnancy-associated ovarian tumors. It must also be recognized that a significant number of ovarian lesions elude detection during pregnancy only to be discovered incidentally at laparotomy (e.g., cesarean section) or at the time of the postpartum clinical evaluation.

Retrospective evaluation of patients presenting with ovarian cancer during pregnancy reveals that most cases were diagnosed late in pregnancy or during the puerperium, in contrast to the early diagnosis of benign ovarian tumors (Table 13-4). Consistent with these observations and the general insidious nature of ovarian cancer was the lower frequency of associated clinical symptoms and major complications noted during pregnancy.[5,16,17,19] Furthermore, these findings suggest a more favorable prognosis than is typically associated with the characteristically advanced stage of disease and histologic types of ovarian cancer in nonpregnant patients. Evaluation of the stage of disease at the time of diagnosis (Table 13-5) demonstrated that 28 (74%) of 38 patients presented with cancer localized to the ovary, a figure that is probably attributable to timely diagnoses related to the frequent antenatal and postpartum evalua-

TABLE 13-5. Ovarian cancer during pregnancy according to stage of disease.

Reference	No. of patients, by stage I	II	III	IV
Creasman et al.[16]	10	2	4	1
Chung & Birnbaum[4]	6	—	—	—
White[6]	2	—	1	—
Hill et al.[7]	3	—	—	—
Lutz et al.[21]	3	—	1	—
Struyk & Treffers[11]	2	—	1	—
Hopkins & Duchon[12]	2	—	—	—
Total	28	2	7	1
%	74	5	18	3

tions during and after pregnancy.[4,6,7,11,12,16,21] In addition, the nonepithelial histologic types, including six gonadal stromal tumors and nine germ cell tumors (predominantly dysgerminomas), were more frequently associated with women in the reproductive age-group and tended to remain localized for longer periods, thereby rendering a more favorable prognosis.

Preoperative Management Considerations

The management of adnexal tumors detected during pregnancy is similar to the approach used in nonpregnant patients of comparable age. Clinical evaluation is conducted in an expeditious manner, assimilating and sharing with the patient and her partner in a timely fashion the significance of symptoms, physical and sonographic findings, estimated relative risks of untoward sequelae (including malignancy), gestational age, and fetal viability. On the basis of the information collected from the clinical and laboratory assessment, previously reported experiences, and the patient's perspective, an acceptable disposition is arrived at by the physician and patient (Fig. 13-1).

In approximately 15% of pregnant patients harboring an adnexal tumor with a diameter exceeding 5 cm, significant sequelae ensue to require emergency surgical intervention.[10,11] The cause is generally torsion of the adnexa around the suspensory ligament of the ovary, which results in vascular compromise with progressive edema, infarction, necrosis, and pain. Less frequently encountered causes for an acute abdomen include intracystic hemorrhage, tumor rupture with accompanying hemoperitoneum or chemical peritonitis, and infection (invariably postpartum). Although the degree of acuity determines the urgency of surgical intervention, careful systematic sonographic imaging of the fetus, placenta, uterus, adnexae, and pelvis aids in rendering a differential diagnosis, a plan of management, and an informed patient. After identification of the acute adnexal process, intraoperative management proceeds as described below.

In the nonemergency setting, the management of a pregnancy-associated ovarian tumor is predominantly dictated by: (1) the gestational age of the fetus; (2) the risk to the patient of developing complications related to the ovarian tumor during the antepartum period, labor, delivery, or the puerperium; and (3) the possible risk of providing sanctuary for a malignant process. A properly abstracted medical history, including information addressing prior menses and pelvic examinations, and meticulous echographic assessment provide insight regarding gestational age and assessment of relative maternal and fetal risks.

Tumors considered at risk for developing untoward sequelae include lesions with a diameter exceeding 10 cm (specifically the risk of torsion during the early part of the second trimester), a thick capsule, vegetative margins, multiple internal septae, a solid or nodular consistency, an increasing size, and any sign of cul-de-sac infiltrates, adnexal fixation, or ascites. The presence of one or more risk determinants necessitates consideration of timely surgical definition of the ovarian abnormality. With rare exception, prompt surgical exploration is entertained during the second trimester after completion of the obstetric and preanes-

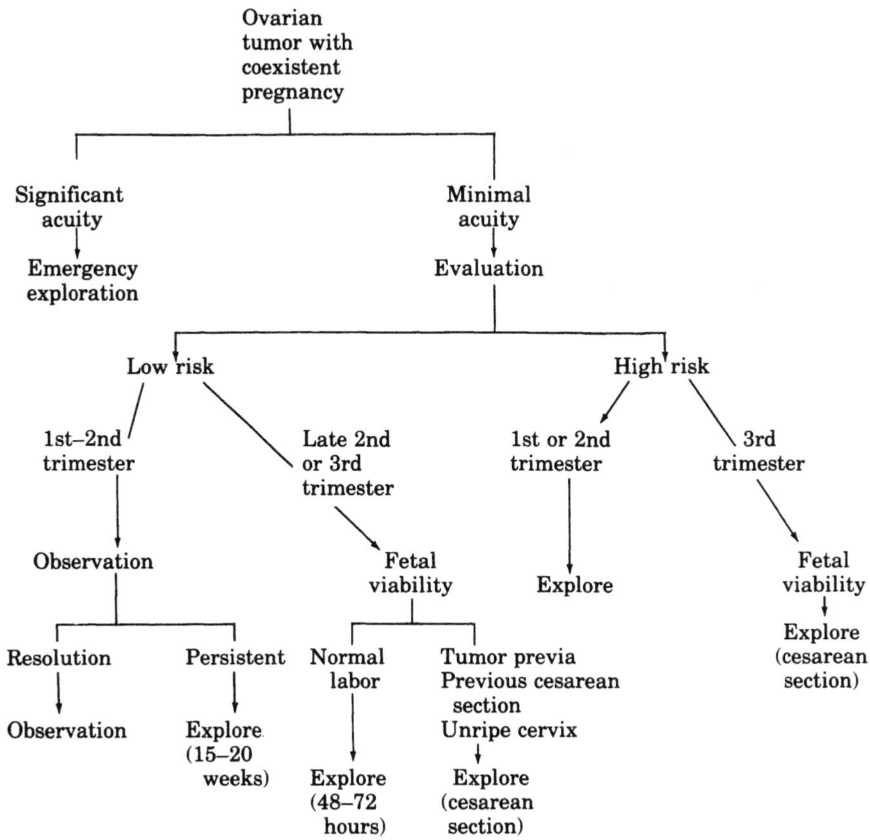

FIGURE 13-1. Management of ovarian tumor during pregnancy.

thetic medical evaluations. In contrast, when high risk lesions are detected during the latter portion of the first trimester or the early phases of the third trimester, brief delays in surgical treatment may be considered advisable to permit enhancement of fetal salvage. However, the patient must be counseled regarding these timing options and be apprised of her vulnerability. In addition, she and her family must be forewarned of potential interim complications to afford early recognition and medical intervention if such occur.

The therapeutic approach to ovarian tumors with a minimal risk for unfavorable sequelae is more expectant and thereby allows optimization of conditions for both patient and fetus should surgical intervention become mandatory. With increasing patient awareness of the benefits of early antepartum care and routine use of ultrasonography, the obstetrician can expect to encounter ovarian tumors more frequently and must be familiar with the management of such lesions, most of which are benign physiologic (functional) cysts. In support of this supposition, Lavery et al.[3] demonstrated cystic or solid-appearing ovarian lesions exceeding 2 cm (mean area 15.2 sq cm) in 7.6% of 964 patients examined echographically during the initial 10 weeks of gestation, whereas only 0.7% of 254 patients examined between weeks 11 and 20 had detectable ovarian tumors. Furthermore, only 8 of 94 patients (8.5%) with sonographic evidence of ovarian enlargement eventually required surgical definition. Hence, after detection of an enlarged ovary during the first and early second trimesters of pregnancy, the obstetrician should arrange for serial physical and sonographic assessments and anticipate spontaneous resolution. However, if the lesion size persists or increases, surgical excision is indicated and is preferably performed between the 15th and 20th weeks of gestation.

Likewise, the identification of low-risk adnexal abnormality during the latter phases of the second trimester or the third trimester permits cautious observation, which includes periodic imaging until fetal viability can be ascertained. Tumor previa, prior cesarean section, and a decision to deliver the patient with an unripe cervix are clinical settings warranting consideration for surgical excision of the adnexal process concurrently with transabdominal delivery of the mature fetus. However, if the patient remains asymptomatic and a normal, uneventful labor and delivery are anticipated and experienced, or an ovarian mass is initially diagnosed during a routine labor or delivery or the immediate postpartum period, surgical exploration should be contemplated 48 to 72 hours after delivery.

Operative Management

Once the decision has been made to define an ovarian tumor surgically and histologically, the technical approach should allow for adequate cytoreduction in the event that a malignant process is encountered. After appropriate bowel preparation, a generous vertical midline incision is used to facilitate exposure for appropriate resection while minimizing uterine manipulation. When the abdominal cavity is entered, either ascitic fluid is sampled or peritoneal washings are secured from both the cul-de-sac and the right upper quadrant for cytologic assessment. All viscera are systematically inspected and palpated, adhesions are lysed, and any suspicious adhesions or lesions are excised for histologic review. The ovaries are assessed to determine whether a partial resection, oophorectomy, or more extensive surgical manipulation is warranted.

Benign Ovarian Tumors

Considering that only 4% of surgically excised pregnancy-associated ovarian tumors are malignant (Tables 13-1 and 13-2), resection of the tumor from the ovary is the treatment of choice, provided its texture and appearance are consistent with a benign process. Immediate histologic examination dictates the need for additional surgical treatment. In the absence of neoplasia, the ovarian defect or pelvic peritoneum is repaired and the opposite ovary is carefully inspected. Unless an obvious lesion is appreciated, we prefer not to resect the contralateral ovary in that the yield of usable information is miniscule and resection may be a potential cause of secondary infertility.

In the presence of a hemoperitoneum or soiling from preoperative or intraoperative rupture of a benign cyst, the ruptured cyst is removed and the abdominal cavity is lavaged with copious quantities of normal saline. We prefer a moderately warm lavage to aid removal of the sebum from rupture or leakage of contents from teratomas and mucinous cysts. In the presence of intense chemical irritation and bowel distention, transnasal long-tube decompression, hydration, and meticulous electrolyte monitoring are advocated. Furthermore, if a superimposed, infectious process is suspected (particularly during the puerperium), specimens are submitted for aerobic and anaerobic culturing; and transperitoneal suction drainage and antimicrobial therapy are instituted.

Torsion, the most frequently encountered complication of pregnancy-associated adnexal masses, invariably produces acute symptoms proportional to the degree of resulting tissue ischemia. Therefore emergency operative intervention is generally required. When only partial torsion has occurred and inspection suggests minimal vascular compromise, the contorted adnexa is untwisted, conservative resection is performed cautiously, and the ovary is transfixed to avoid recurrent torsion. Conversely, complete torsion requires an oophorectomy (and invariably concomitant salpingectomy) with identification and sufficient mobilization of the ureter to enable a high ligation of the gonadal vessels, thereby minimizing the potential release of thrombotic emboli during the surgical procedure. It must be emphasized that malignant ovarian lesions rarely present in association with adnexal torsion.

Although rare, luteoma of pregnancy is a benign, frequently virilizing, smooth, yellow, encapsulated ovarian tumor that regresses spontaneously after parturition; it seldom recurs.[22,23] Approximately 30% of patients present with evidence of masculinization, and 50% of their female infants likewise demonstrate virilization. In addition, nearly one-half of the affected women have bilateral lesions of varying size. The condition is usually diagnosed during a pregnancy-related surgical procedure. The obstetrician must consider luteoma of pregnancy at cesarean section, tuboligation, or another operative procedure if a smooth, yellow, encapsulated tumor is encountered (particularly if it is bilateral) and the mother or female infant are masculinized. The luteoma of pregnancy should be observed after a biopsy specimen is obtained for histologic confirmation.

Malignant Lesions

The major prognostic and therapeutic determinants of ovarian cancer include histologic subtype, stage, and grade of the neoplasm. The distribution of tumor histologic types during pregnancy does not seem to differ from that in the corresponding nonpregnant, age-adjusted patient population; a higher frequency of germ cell and gonadal stromal tumors are detected in this age-group than is customarily observed in the typically older nonpregnant patients with ovarian cancer. Fortunately, the extent of tumor growth is more limited, reflecting, at least in part, the increased incidence of nonepithelial tumors and the earlier diagnosis of epithelial malignant lesions with a poorer prognosis. Survival analysis for 42 patients with pregnancy-associated ovarian malignancies (Table 13-6) revealed extended longevity in 83% of patients with gonadal stromal tumors, 73% with germ cell tumors, and 52% with epithelial tumors.[4–7,11,12,16,21]

The overall management of ovarian cancer during pregnancy differs minimally from standard treatment in the age-adjusted nonpregnant state. After the initial intraoperative evaluation and histologic confirmation

TABLE 13-6. Survival according to histologic subtype of ovarian cancer associated with pregnancy.

	No. DOD/total No., by histologic subtype			
Reference	Epithelial	Germ cell	Gonadal stromal	Total
Creasman et al.[16]	5/9	0/6	0/2	5/17
Chung & Birnbaum[4]	3/4	1/1	1/1	5/6
Buttery et al.[5]	2/2	1/2	—	3/4
White[6]	2/3	—	—	2/3
Hill et al.[7]	0/3	—	—	0/3
Lutz et al.[21]	0/2	0/1	0/1	0/4
Struyk & Treffers[11]	0/2	1/1	—	1/3
Hopkins & Duchon[12]	—	—	0/2	0/2
Total	12/25	3/11	1/6	16/42
%	48	27	17	38

DOD = dead of disease.

of ovarian cancer, definitive surgical staging and tumor reduction are expeditiously but methodically performed. If the neoplasm appears to be localized to the gonad(s), biopsy specimens of the peritoneum from the colic gutters, large and small bowel mesentery, pelvic sidewalls, cul-de-sac, bladder, serosa from the sigmoid and ileum, and right hemidiaphragm are submitted for histologic evaluation with the omentum and generous samples of pelvic and periaortic lymph nodes. The requirement for additional extirpation and adjunctive therapy is dictated by the specific histologic type (epithelial, germ cell, and stromal) and stage at diagnosis, and it is influenced by the degree of tissue differentiation and the patient's age and desire to preserve reproductive function.

Epithelial Carcinomas

As noted by Novak et al.,[17] most epithelial lesions diagnosed during pregnancy are of low malignant potential. Furthermore, early antepartum physical and imaging examinations favor timely detection of asymptomatic ovarian lesions before carcinomatous spread beyond the gonad. Unilateral salpingo-oophorectomy is sufficient treatment if the patient desires preservation of reproductive function and proper staging has confirmed a localized intracystic (stage IA) well differentiated carcinoma. We recommend that such patients be monitored closely until childbearing has been completed, at which time reassessment operation is performed in conjunction with a hysterectomy and removal of the remaining adnexa. For more advanced primary or poorly differentiated disease, definitive operation, including hysterectomy, bilateral salpingo-oophorectomy, omentectomy, and tumor cytoreduction, is performed, sacrificing the pregnancy if fetal maturity is not obtainable with a short delay in treatment. After optimal surgical reduction, postoperative adjuvant therapy that includes cisplatin-based chemotherapy commences during convalescence.

Germ Cell Tumors

Although germ cell tumors characteristically attain a large size, they are generally localized and unilateral (except for dysgerminomas, which are bilateral in 10 to 15% of cases) but have a greater propensity for torsion, rupture, and lymphatic dissemination than other ovarian malignant tumors. After intraoperative documentation of the extent of disease, unilateral salpingo-oophorectomy and ipsilateral pelvic and periaortic node sampling are performed together with excision of extraovarian disease if such exists. Dysgerminoma is the most common germ cell tumor encountered during pregnancy[16,17] and is distinguished by its radiosensitivity and occasional bilaterality.[24,25] Because of the potential for bilateral dysgerminomas, biopsy of the contralateral ovary is recommended and oophorectomy is performed if involvement is documented. After parturition, patients with localized disease should be examined periodically and treated with radiotherapy if recurrence develops.[26] However, patients with lymphatic involvement or intraperitoneal dissemination require adjuvant irradiation, preferably after hysterectomy and tumor cytoreduction.

In contrast, the nondysgerminomatous (endodermal sinus, embryonal, choriocarcinoma, immature teratomas, and mixed tumors) germ cell tumors are less common, more virulent, and radioresistant. Again, unilateral adnexal resection for disease involving a single ovary is recommended and is followed by combination chemotherapy. Dramatic improvements in survival rates have been witnessed with the use of VAC (vincristine, actinomycin D, cyclophosphamide), VBP (vinblastine, bleomycin, cisplatin), or BEP (bleomycin, etoposide, cisplatin), alone or sequentially. Because these chemotherapy regimens may provide a cure for a generally fatal disease, delays in initiation should be avoided, patients should be consulted regarding the effects on fetal development, and pregnancy termination should be extended as an option during early gestation. However, successful pregnancy outcomes can be anticipated through coordinated efforts by the obstetrician, oncologist, and pediatrician.

Gonadal Stromal Tumors

Granulosa cell and Sertoli-Leydig cell tumors are infrequently associated with pregnancy and seldom present with significant manifestations of hyperestrogenism or excess androgens, as are frequently appreciated in the nonpregnant state.[27] Of 36 gonadal stromal tumors retrospectively reviewed by Young et al.,[27] all tumors were localized to the ovaries (stage I) and only a single patient had bilateral involvement. Although more than one-third of the tumors ruptured, no significant sequela was recorded, and the need for postoperative treatment was not realized. Considering the excellent prognosis, unilateral salpingo-oophorectomy after definitive staging constitutes appropriate primary management.

Outcome of Pregnancy

One must recognize that any major operative procedure has associated risks and that these risks are possibly compounded during gestation; moreover, they are potentially further confounded if adnexal resection or excision is required. Although the stresses attributable to the surgical process as well as the procedure, anesthesia, and postoperative complications (particularly in the presence of preexisting medical problems) can have a negative impact on pregnancy outcome (as judged by fetal wastage), adnexal surgery is considered to superimpose further uncertainties that can compromise fetal salvage.

A review of selected studies[5–9,11,12] since 1973 addressing fetal wastage after operation for ovarian tumors during pregnancy demonstrated an overall associated fetal loss rate of 14% (range 0 to 21%) (Table 13-7). Furthermore, several reports[5–7,17] suggested a fetal loss rate of approximately 30% or more when surgery was performed during the first trimester of pregnancy. However, most of these investigations included study intervals initiated more than two decades ago. Assessment of three reports benefiting from more advanced technology and pharmaceuticals showed overall fetal loss rates of 7.6%, 3.0%, and 3.0% after adjustment for losses not attributable to operation.[8,9,12] In addition, the practice of delaying surgical intervention when possible until the second trimester resulted in only 8 of 66 procedures being performed during the first trimester, with only one observed spontaneous abortion occurring within this small subset.

In addition, optimal use of preoperative imaging, surgical planning and expertise, intraoperative support, and postoperative

TABLE 13-7. Fetal wastage with operation for ovarian tumors during pregnancy.

Reference	Study interval	Tumors excised (No.)	Fetal loss	
			No.	%
Buttery et al.[5]	1947–1969	119	22	18
Struyk & Treffers[11]	1952–1980	90	10	11
Hill et al.[7]	1955–1971	51	8	16
White[6]	1961–1970	34	7	21
Sivanesaratnam et al.[8]	1968–1973	27	4	15
Karpathios et al.[9]	1969–1975	16	1	6
Hopkins & Duchon[12]	1982–1984	23	0	0
Total		360	52	14

monitoring and treatment when indicated minimize fetal wastage. Prompt echographic evaluation provides valuable information regarding gestational dates and tumor characteristics and thereby facilitates management decisions. When possible, and without jeopardizing maternal short- or long-term well-being, operation is preferably delayed either beyond the first trimester or until fetal maturity is established if the tumor is diagnosed during the latter half of gestation. The operative procedure is performed expeditiously through a generous vertical incision, uterine manipulation is minimized, and available oncologic expertise is used to render appropriate surgical therapy if a malignant process is encountered. Equally important is the avoidance of hypotension and hypoxia through judicious attention to hydration, oxygenation, intubation, maternal positioning, and intraoperative monitoring. Likewise, continuous postoperative monitoring for uterine irritability after 20 weeks of gestation is routine, and tocolytic agents are used liberally.

The efficacy of using progestational agents before 20 weeks of gestation to reduce pregnancy loss continues to be debated. However, complete or partial removal of the corpus luteum before a sufficient placental source of progesterone is established has the potential of leading to spontaneous uterine evacuation. Csapo et al.[28] demonstrated that uterine activity correlated with decreasing steroid levels after luteectomy but that progesterone replacement prevented the anticipated abortions. Hill et al.[7] conducted a retrospective assessment and observed no decrease in pregnancy loss when progestational agents were administered before or after operation. In contrast, Karpathios et al.[9] administered isoxsuprine and progestational agents simultaneously before and after operation in patients at 20 weeks or less of gestation and reported acceptable salvage rates. More recently, Hopkins and Duchon[12] administered progesterone alone before and after operation during the first and early second trimesters with no associated pregnancy losses. Unfortunately, no prospectively controlled trial has been conducted to address the value of progestational agents after adnexal operation, and hence their efficacy remains unknown.

References

1. Grimes WH Jr, Bartholomew RA, Colvin ED, et al. Ovarian cyst complicating pregnancy. Am J Obstet Gynecol 1954;68:594–605.
2. Sinnathuray TA. Ovarian tumors in pregnancy: a clinicopathologic study of 19 surgically proven cases in a Southeast Asian hospital. Int Surg 1971;55:422–430.
3. Lavery JP, Koontz WL, Layman L, et al. Sonographic evaluation of the adnexa during early pregnancy. Surg Gynecol Obstet 1986; 163:319–323.
4. Chung A, Birnbaum SJ. Ovarian cancer associated with pregnancy. Obstet Gynecol 1973; 41:211–214.
5. Buttery BW, Beischer NA, Fortune DW, et al. Ovarian tumours in pregnancy. Med J Aust 1973;1:1345–1349.
6. White KC. Ovarian tumors in pregnancy: a private hospital 10 year survey. Am J Obstet Gynecol 1973;116:544–550.
7. Hill LM, Johnson CE, Lee RA. Ovarian surgery in pregnancy. Am J Obstet Gynecol 1975;122:565–569.
8. Sivanesaratnam V, Ang LT, Sinnathuray TA. Ovarian tumours complicating pregnancy in a Malaysian study. Med J Malaysia 1976; 30:291–295.
9. Karpathios S, Lolis D, Tzigounis B, et al. Ovarian neoplasms and pregnancy. Int Surg 1977;62:80–81.
10. Teoh SH, Lim E, Vengadasalam D. Ovarian tumours in pregnancy: a review. Singapore Med J 1984;25:165–167.
11. Struyk APHB, Treffers PE. Ovarian tumors in pregnancy. Acta Obstet Gynecol Scand 1984;63:421–424.
12. Hopkins MP, Duchon MA. Adnexal surgery in pregnancy. J Reprod Med 1986;11:1035–1037.
13. Lakkis WG, Martin MC, Gelfand MM. Benign cystic teratoma of the ovary: a 6-year review. Can J Surg 1985;28:444–446.
14. Pepe F, Panella M, Pepe G, et al. Dermoid cysts of the ovary. Eur J Gynaecol Oncol 1986;7:186–191.
15. Portuondo JA, Gimenez B, Rivera JM, et al. Clinical and pathologic evaluation of 342 be-

nign ovarian tumors. Int J Gynaecol Obstet 1984;22:263–267.
16. Creasman WT, Rutledge F, Smith JP. Carcinoma of the ovary associated with pregnancy. Obstet Gynecol 1971;38:111–116.
17. Novak ER, Lambrou CD, Woodruff JD. Ovarian tumors in pregnancy: an ovarian tumor registry review. Obstet Gynecol 1975;46:401–406.
18. Ballard CA. Ovarian tumors associated with pregnancy termination patients. Am J Obstet Gynecol 1984;149:384–387.
19. Jubb ED. Primary ovarian carcinoma in pregnancy. Am J Obstet Gynecol 1963;85:345–354.
20. Mehra U, O'Connor T, Ostapowicz F, et al. Pregnancy with bilateral benign cystic teratomas. Am J Obstet Gynecol 1976;124:361–366.
21. Lutz MH, Underwood PB Jr, Rozier JC, et al. Genital malignancy in pregnancy. Am J Obstet Gynecol 1977;129:536–542.
22. Garcia-Bunuel R, Berek JS, Woodruff JD. Luteomas of pregnancy. Obstet Gynecol 1975; 45:407–414.
23. Cronjé HS. Luteoma of pregnancy. S Afr Med J 1984;66:59–60.
24. Karlen JR, Akbari A, Cook WA. Dysgerminoma associated with pregnancy. Obstet Gynecol 1979;53:330–335.
25. Gershenson DM. Malignant germ-cell tumors of the ovary. Clin Obstet Gynecol 1985;28: 824–838.
26. Buskirk SJ, Schray MF, Podratz KC, et al. Ovarian dysgerminoma: a retrospective analysis of results of treatment, sites of treatment failure, and radiosensitivity. Mayo Clin Proc 1987;62:1149–1157.
27. Young RH, Dudley AG, Scully RE. Granulosa cell, Sertoli-Leydig cell, and unclassified sex cord-stromal tumors associated with pregnancy: a clinicopathological analysis of thirty-six cases. Gynecol Oncol 1984;18:181–205.
28. Csapo AI, Pulkkinen MO, Wiest WG. Effects of luteectomy and progesterone replacement therapy in early pregnant patients. Am J Obstet Gynecol 1973;115:759–765.

14
Pregnancy in the Kidney Transplant Recipient

LUIS A. CIBILS, JOAN K. STUART, and FRANK P. STUART

The pituitary-ovarian function appears to be seriously altered in the patient with severe renal insufficiency requiring a renal transplant. The reasons for this dysfunction are not clearly understood. However, it is not difficult to imagine that minor disturbances may occur in the many marvelously coordinated physiologic processes preparing the woman for conception when an organ of such importance as the kidney does not fulfill its function adequately. One manifestation of that dysfunction is the relatively common occurrence of amenorrhea often rapidly corrected after a successful transplant operation.[1]

Complications of Transplant

Implantation of a tissue-compatible kidney in an individual is a process carrying potential significant problems that go far beyond the surgical risk itself. Mortality within a few years after the transplant has been reported to be in the order of 20 to 30%.[2,3] The most frequent causes have been infections, cardiovascular complications, and suicide. As the inherent dangers of the operation have diminished with the enlarged experience of organ transplants, so have those appearing after complete recovery from surgery. Nevertheless, chronic problems closely related to either the preexisting renovascular dysfunction or the effect of immunosuppressive medications are inevitable. Among the most frequent in the female patient are persistent hypertension,[4] glucose metabolism abnormalities,[5,6] bone decalcification or aseptic necrosis, and gastrointestinal disorders.[4] Of course, these problems are continuously potential complications in the life of an allograft recipient and perhaps more so in the patient who, by planning or by accident, becomes pregnant. The omnipresent possibility of rejection is a heavy burden for these individuals and may be the reason underlying the high incidence of psychological impairment observed by those who have studied their personalities.[7]

Effect of Transplant

The benefits of the new kidney are quickly manifested in the female patient, despite the problems just briefly reviewed. Outstanding among them is the return to normal ovarian function and, with it, menstrual periods.[1,8] This change occurs within 6 months, in general, but often ovulation takes place shortly after transplant and the sexually active patient may conceive during that cycle. It is not uncommon to examine some of these women for postoperative follow-up and find them to be pregnant without having menstruated after surgery (three patients of the series here reported presented in that manner). This fact underscores the importance of knowing and discussing this point with the patient before surgery. This aspect becomes critical when the recipient may have some other important problems contraindicating pregnancy. The careful planning of contraceptive means is thus part of the preoperative evaluation. It appears that the fertility of the recipient of a successful allograft kidney is not different from that of the general population. It follows that physicians and patients must be aware of this situation in order to prevent unplanned pregnancies.

With more experience in the management of the pregnant transplant patient, and the popular dissemination of this information, more recipients are willing to conceive and attempt to successfully carry a pregnancy to viability. Unfortunately, not all of them have the discipline and determination required to follow the tedious steps necessary to continuously evaluate the conceptus as well as the effect of pregnancy on the patient. The consequences of this lack of "compliance" have been a gamut of complications ranging from loss of pregnancy to rejection of the transplant.

Effect of Pregnancy

In an elegant prospective evaluation of renal function in the prepregnant and pregnant state, Davison demonstrated that the changes known to occur in normal patients, such as increased glomerular filtration rate (GFR) and creatinine clearance, also occur in the transplant recipient, although at lesser levels.[9] Along the same line Absy and co-workers[5] showed that these changes are contemporaneous with an increase in the volume of the graft, and that this volume change parallels the slow decrease in these functions seen later during the third trimester of gestation. As seen in normal gestation, this slight decrease in function is not indicative of deterioration of the graft. Small increases in proteinuria (above the generally accepted 300 mg in 24 hours) are frequently seen in the transplant recipient, and the proteinuria may become heavier during the third trimester of gestation, even among those who have normal function before conception. This change has been observed in some of the patients in this series. On the other hand, preexisting hypertension or borderline graft function are clear predictors of further deterioration during gestation manifested by progressive hypertension, increased serum creatinine and blood urea nitrogen, and proteinuria.[10] It has been reported that between 40 and 60% of recipients have significant proteinuria as pregnancy advances into the midsecond to third trimester.[9] Several of our patients had that response to pregnancy.

Favorable factors in our series appeared to be *family-related donors,* as none of that group had a complicated gestation (with one exception to be discussed later). Conversely, among the recipients of *cadaver grafts* were grouped the patients with moderate impaired renal function and those whose condition deteriorated during the evolution of pregnancy. These patients probably constitute the bulk of the estimated 10 to 20% whose function steadily deteriorates with gestation when one surveys combined series.[11,12] A number of these patients may continue the unfavorable change during the postpartum period, terminating in rejection,[10,13,14] whereas others revert to better function.[5,10,12,15] The age of the donor kidney does not seem to influence its function during gestation, the extreme case being the patient reported by Coulam et al.[16] who conceived when the kidney was 75 years old and 10 years after transplant.

Graft rejection has been reported to occur in around 9% of patients, ostensibly not significantly different from rejections in the nonpregnant state.[11] There are no known predictable factors for this accident. The clinical manifestations are the same and sometimes mimic pregnancy complications, such as pyelonephritis and developing preeclampsia. Ancillary tests may help to evaluate the situation because it has been shown that *acute* increases in size assessed by sonogram simultaneous with function deterioration herald a rejection episode.[5,17] High resolution ultrasonography may further confirm the suspicion, as there have been reported changes in the echogenic characteristics of the parenchyma when rejection occurs.[18,19] Confirmatory evidence may be obtained by needle biopsy, as in the nonpregnant state. However, most authors prefer to first attempt a medical antirejection treatment trial. One of the patients in our series had a rejection episode during her gestation. All received corticosteroids and azathioprine therapy. Warfarin, used by some[12] as antirejection therapy, is not part of the standard medication in our institution.

Pregnancy Losses

As fertility is quickly restored after a successful allograft, one might also expect that pregnancy losses would be comparable to those in the nontransplant population. It is not the case, however, due perhaps to a combination of factors. Since the publication of the first successful pregnancy in a transplant recipient over 30 years ago[20] probably several thousand recipients have become pregnant, many more than once, and have delivered viable infants. Their pregnancy evolution has varied from being almost completely uneventful to stormy, some requiring therapeutic intervention at various stages in an attempt to preserve the graft or, in more critical situations, the patient's life.[10] As was mentioned before, the number of unplanned or unexpected gestations is relatively high.

The effect of pregnancy on the transplanted organ or the general health of the patient is not well understood by a large number of patients or the physicians who assess them during early gestation. This misunderstanding is a strong element in the decision of the patients to request a therapeutic termination. The usual spontaneous miscarriage rate of 10 to 15% in the general population has been reported based on data from large series[11,13]; this fact combined with the situations just discussed, (that is the large population of voluntary-therapeutic terminations, reaching about 30%) add up to an impressive 40% rate of pregnancy losses during the first trimester. A comparable number of voluntary abortions has been observed in our own series of transplants, but in no case was severe deterioration of renal function the reason for terminating the pregnancy.

Pregnancy Complications

If one disregards the first trimester pregnancy losses just discussed, the remainder of gestations may be expected to terminate in a viable birth. However, the continuation of pregnancy does not imply a trouble-free evolution. On the contrary, few of these patients continue their gestation without one or more of the serious complications that affect "high risk" gestations. Among the more important complications are hypertension, infections of various organ systems, intrauterine growth retardation, and prematurity.

Hypertension

Moderate to severe degrees of hypertension affect a large proportion of allograft recipients.[2,10,11,21,22] Despite the close follow-up given to these patients, preeclampsia (or "pregnancy-induced hypertension") alone has been diagnosed in about 30% of all pregnancies.[10,22] Even a case of fulminant eclampsia and intrauterine fetal death has been reported,[23,24] as has another with delivery of a premature infant who died.[21] One must add to these cases all those of progressive increase of blood pressure among the numerous cases with borderline high readings, as well as those undergoing a rejection episode. Of course a precise diagnosis of which one of these syndromes affects a given patient can be made only by means of a kidney biopsy, not frequently practiced even though its clear usefulness (and low risk) have been demonstrated. It is for this reason that these hypertensive conditions are discussed together. No single clinical or laboratory sign (except biopsy) is pathognomonic.

Most of the disorders produce moderate to severe proteinuria (up to 10–12 g/day), decreases in creatinine clearance, or increases in blood urea nitrogen and serum creatinine, none of which confirms the diagnosis of preeclampsia.[2,9–11,25] Likewise, thrombocytopenia and increased liver function tests may be observed in all graft recipients taking immunosuppressants.

Some of our patients presented with a syndrome suggesting preeclampsia, whereas others appeared to clearly have an exacerbation of an underlying hypertension. Only six exhibited no hypertension throughout their gestation. Among the more severely affected, five necessitated premature termination of the pregnancy because of either uncontrollable hypertension or the severe risk to the fetus if the needed medications had to be increased.

Infections

The state of induced and sustained immunosuppression predisposes all graft recipients to infections. As previously mentioned, this complication is the immediate cause in nearly one-half of all transplant deaths.[3] This danger is not eliminated by the state of gestation, even though in some instances it is possible to diminish or even discontinue the immunosuppressants.[22,26] A high incidence of infections was observed in almost all reported small series[8,27–29] with few exceptions.[2] The most frequently observed infection is probably that of the urinary tract—cystitis or pyelonephritis[13,21,27]—which must be vigorously treated with specific antibiotics and for sustained periods.

Viral infections have shown to be highly prevalent among transplant recipients[30] as well as among candidates for transplant (dialysis patients).[13] Those more frequently identified are herpes and hepatitis B, with fewer cases of cytomegalovirus infection.[30] The current availability of vaccines has markedly ameliorated the prognosis for the neonates of hepatitis B carriers. Septic infections may be due to a variety of organisms, from *Pseudomonas* and *Staphylococcus* to *Pneumocystis carinii*.[3] They are particularly dangerous in patients with altered glucose metabolism.

The patients in our small series have been fortunate because none has developed a severe infection, although two had demonstrated a urinary tract infection. Each responded promptly to antibiotic therapy.

Intrauterine Growth Retardation

The steady growth of the conceptus, commensurate with the chronologic age of gestation, reflects good general maternal health and intervillous space circulation as well as an absence of interfering factors, such as malnutrition, viral infections, and genetic abnormalities. The allograft recipient is, in principle, in a disadvantageous position because in many cases she had had pretransplant hypertension and regularly takes immunosuppressive medications, which predispose to a variety of chronic and acute infections. It is an accepted fact that chronic infections may cause intrauterine (fetal) growth retardation (IUGR). What is not well known is whether the ingestion of immunosuppressive substances may predictably result in insufficient fetal growth. The observation of human gestations indicates that a significant number of graft recipients carry fetuses small for gestational age at delivery. The incidence has been reported to be 16 to 40%.[11,15,22,28,31] The mechanism whereby the fetus is affected has not been clarified, but it could be mediated through alteration of several organ systems. The general depression of the immune system could affect the growth of the fetus. Although not unequivocally demonstrated in humans, such an affect has been shown in animals.[31] It has not been ruled out that a direct effect of the drugs utilized may have minor toxic action on the growing fetus, in particular related to cell division and growth. The general vascular abnormality that is part of the hypertensive syndrome must affect the intervillous space circulation and thereby fetal nutrition. Perhaps the combination of these factors underlies the insufficient fetal growth.

In our small series, four newborns were clearly below the expected weight for their gestational age, and each was a product of a complicated gestation. Conversely, the fetuses carried in uncomplicated gestations demonstrated somatic growth well within (or even slightly above) the expected average.

Prematurity

Premature termination of pregnancy is one of the most frequent complications encountered by the allograft-recipient patient. General surveys of published material reveal an incidence ranging from 25 to 60%.[12,13,15,22,28,31] The conditions leading to the preterm birth are varied. Among the obstetric problems, the most frequently observed is premature rupture of membranes (PRM), occurring in a higher proportion than in the general population.[11] The reasons for this phenomenon are not well understood, although the weakening effect of steroids on the connective tissues of the membranes have been implicated. Of course, the determination to induce labor, albeit premature, is a natural decision under the circumstances in view of the danger of infection as discussed above. PRM occurred in four patients in our series. Idiopathic premature labor occurs with an unusual frequency among these patients, but again there is no good explanation for its happening.[2,11] The hypertensive conditions uncontrolled with standard therapy are often a strong reason to terminate the pregnancy, regardless of age of gestation. Probably premature births due to hypertension are as numerous as those due to PRM. In our series 8 of 19 patients delivered prematurely because of several of the reasons just discussed.

Miscellaneous Complications

The prolonged use of steroids has been implicated in the *development of diabetes* among renal allograft recipients.[5,6,21] In addition to hyperglycemia and glucosuria, the grafts of these patients undergo a significant increase in volume, similar in size to that seen with early rejection; unlike these cases, however, the increase in volume is accompanied by a steady improvement in function.[5] When *diabetes antedates* the transplant operation, pregnancy is still attainable even when the disease is poorly controlled.[25] However, under this circumstance the incidence of the complications so far discussed is exceedingly high and includes fetal and maternal deaths. This somber prognosis is understandable with the overlap of two severe systemic diseases (peripheral vascular disease and abnormal glucose metabolism on the one hand and possible hypertension and chronic immunosuppression on the other) upon which a pregnancy is added with its normal alteration of glucose metabolism and propensity to induce hypertensive alterations. There was one case of abnormal glucose metabolism in our series, probably induced by steroid therapy.

Other complications of gestation seen in nontransplant patients have been reported among allograft recipients. *Ectopic gestation* has occurred among these patients,[10,29] although the incidence may be misleadingly low. Among more than 2000 reported cases of pregnancy, only three have been reported in the English literature; the current incidence for nontransplant patients is in the order of 1 to 3% of live births in tertiary centers of large metropolitan hospitals. Either most cases are not reported among graft recipients, or they have intrinsic factors that prevent the ectopic implantation of the conceptus.

Likewise, benign *trophoblastic disease* has been reported to occur following gestation,[32] as has flare-up of inactive choriocarcinoma after transplant.[33] Of course these conditions create special management problems because they may require aggressive chemotherapy in already somewhat debilitated and anemic subjects.

The association of *sickle cell anemia* and renal transplant[34,35] has also been reported.

A case of spontaneous rupture of the uterus at term was suspected to be due to chronic steroid effect on the connective tissues.[36]

Fetal Surveillance

From the foregoing it follows that in order to have a pregnancy carried to viability at or near term the physician in charge must strictly monitor the maternal response to the pregnant state as well as the fetal response to the pharmacologic manipulations required to protect the allograft from rejection. Unfortunately, unplanned gestations have occurred with amazing frequency, and the ideal planning (see below) has seldom been practiced.

During the early part of gestation a sonogram should establish the accurate date of gestation and expected date of delivery. Repeated at regular intervals of 4 to 5 weeks, sonography is the most precise means to diagnose inappropriate fetal growth and amniotic fluid volume. When pregnancy has reached about 28 weeks, fetal heart rate and uterine contractions recordings may be started, specially in hypertensive patients and those having signs of IUGR. A *reactive* nonstress test is an excellent indicator that the fetus is in satisfactory condition (Fig. 14-1). Conversely, a *nonreactive nonstress test* is probably evidence of fetal jeopardy; this result is strongly suggestive of fetal compromise after the 32nd week of pregnancy in patients not receiving medication that affects the sympathetic nervous system (either mimetics or blockers). In these cases a contraction stress test should be performed to assess fetal oxygen reserve; it is also an indirect indicator of intervillous space blood flow. A *positive contraction stress test* suggests deficient oxygen supply and thus the necessity to correct its causes or, more realistically, to terminate the pregnancy (Fig. 14-2). These tests should be carried out weekly in the patients with complicated gestations. In uncomplicated pregnancies they are started at about the 33rd week and continued weekly. Immediate reevaluation of the case is mandatory when the tests indicate fetal compromise. Another means of evaluating arterial blood circulation has recently been incorporated into our diagnostic

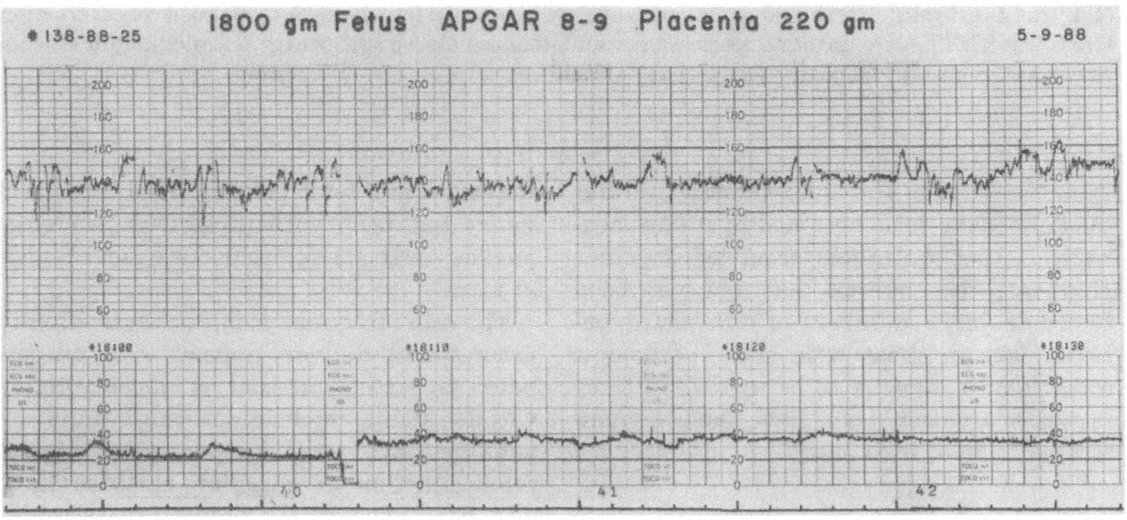

FIGURE 14-1. Thirty-five minutes of fetal heart rate (FHR) and uterine contractions (UC) recordings of a 35 weeks gestation with an allograft kidney. There was minimal uterine activity. The FHR accelerates with movements, and the baseline has good oscillations. Reactive nonstress test. Labor was induced because of premature rupture of the membrane and terminated in spontaneous delivery.

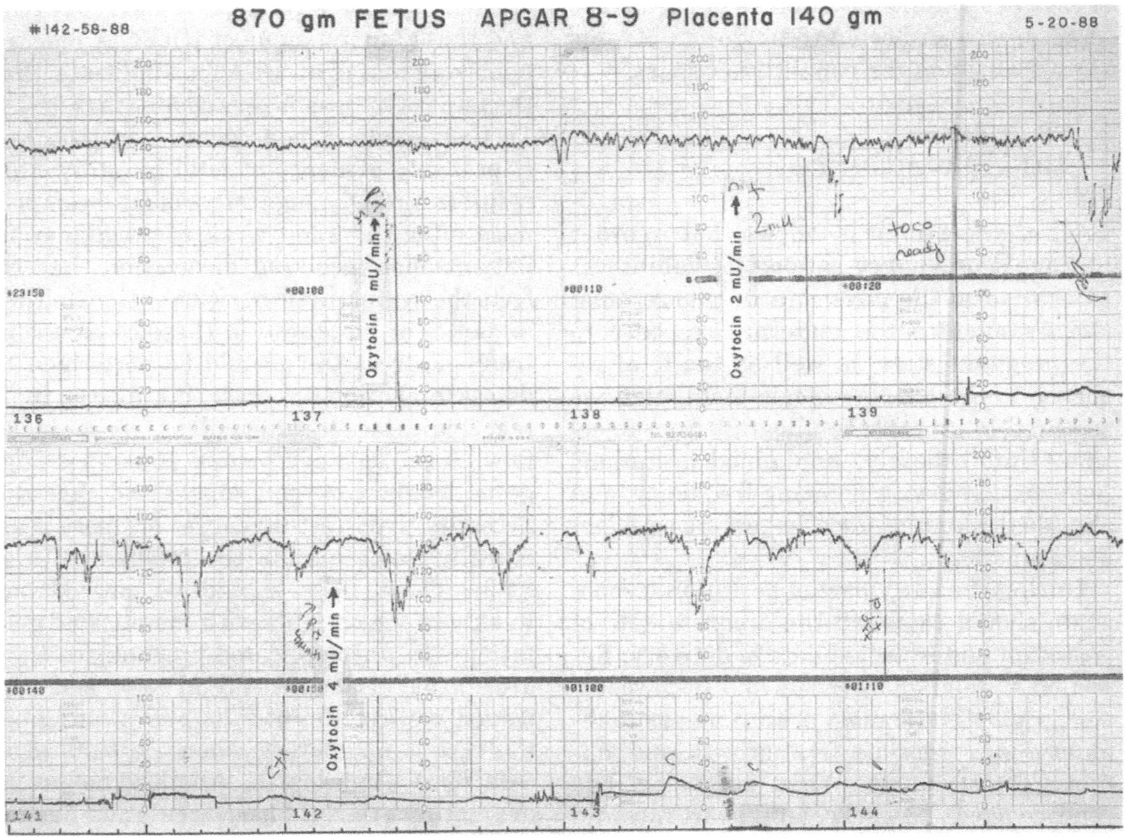

FIGURE 14-2. Continuous 80 minutes of FHR and UC recordings of a 31 weeks gestation with an allograft kidney. Severe IUGR, nonreactive nonstress test (NST; top left) followed by oxytocin infusion (1, 2, and 4 mU/min). Uterine contractions triggered repetitive late decelerations: positive contraction stress test (CST). An immediate cesarean section produced a severely growth retarded infant (with a very small placenta) who survived without difficulties.

armamentarium: the blood velocity systolic/diastolic wave ratio obtained by ultrasonography (either continuous or pulsed Doppler). However, this method has not yet been tested in large numbers of transplant patients, but in those with other pathologic conditions it appears to be useful.[37,38] Hormones evaluation tests have been demonstrated to be of little use.[39]

Mode of Delivery

The decision on when and how to terminate the pregnancy should be made—as for normal gestations—after careful evaluation of maternal and fetal factors. When the pregnancy has been without serious incidents, one may wait for spontaneous onset of labor at term.

The most frequent *maternal indication* to proceed with delivery appears to be exacerbated hypertension that is uncontrollable without using medications that would severely jeopardize the fetus. Deterioriation of renal function or rejection episodes have also been reasons for premature termination. Premature rupture of membranes, a relatively frequent complication (up to 25% in many series), often prompts the obstetrician to take action.

Fetal indications for interruption of pregnancy are characteristically associated with maternal hypertension. In cases of acute exacerbation the fetus is generally of normal size for gestational age; however it may not be able to tolerate further medications to be given to the mother (in addition to immunosuppressants, antihypertensives, diuretics, sympathetic blockers), or there may already be evidence of intrauterine stress or distress. Another important group is that exhibiting IUGR, some with evidence of intrauterine distress in addition to the defective growth.

The route of choice is *vaginal delivery,* which may be anticipated in the majority of patients. Regardless of whether the pregnancy progresses to term or interruption is necessary before that, a meticulous plan must be followed. No pregnancy should be allowed to go beyond well documented term. Steroids must be administered during labor and early puerperium in doses used for stress cases. Continuous fetal heart rate and uterine contractions recordings are mandatory to assess the contractility pattern, the oxytocin dose when used, and the fetal response to labor. When in doubt how to interpret abnormal fetal heart rate tracings, scalp blood sampling should be used without hesitation. Prophylactic antibiotics must be administered in cases of PRM and a history of genital tract infection in consideration of the state of immunosuppression of these patients. In the present series 10 of the 19 deliveries were vaginal and, of them, four were prematurely induced for PRM.

Among the important special aspects to consider in immunosuppressed patients is the general laxity of their connective tissues (due to the prolonged use of corticosteroids). It causes a relative friability of the perineum and danger either of lacerations or severe extensions of the episiotomy. Caution must likewise be observed when performing cesarean section where lateral extensions of transverse incisions could cause serious complications. This undesirable side effect on the connective tissues has been blamed for a reported spontaneous rupture of the uterus.[36]

A *cesarean section* delivery is done only for the usual maternal or fetal indications. In other words, the operation should be performed when facing uncontrollable maternal deterioration that does not permit the time to obtain the cervical changes needed for effacement and dilatation. Likewise, fetal deterioration during pregnancy or fetal distress during labor are indications for abdominal delivery. The obstetric indications, such as real or relative cephalopelvic disproportion, abnormal fetal lie, unsatisfactory progress of labor, or obstructed labor by an allograft transplant placed too low in the false pelvis,[40] are all good reasons to perform a cesarean section.

The type of uterine incision should be chosen, as in any other case, according to the fetal size, the extent of the lower segment, and the type of anesthesia available. Consideration must be taken of the relative friability of the tissues when handling them and when choosing the suture material. Retention sutures have been systematically applied in the nine cases of our series in consideration to the much slower healing ability of the immunosuppressed steroid-recipient patients. For the same reason, prophylactic antibiotics were used because septic complications could be very serious.[10] With these guidelines for delivery there have not been postoperative complications. With two vaginal deliveries there were extensions of episiotomy and excessive bleeding, but they did not require blood replacement. The incidence of cesarean sections varies widely among the series published, ranging from 9%,[21] to 50%,[13] to 75%,[15] to even 100%.[14,25]

Maternal Morbidity and Mortality

Severe rejection episodes might require that the pregnancy be terminated immediately. When they occur in early gestation therapeutic abortion is the choice. A decision may be more problematic for cases that develop during the second half of pregnancy. Pregnancies not responding to antirejection medications may further deteriorate, either losing the conceptus or necessitating intervention without regard to its viability. Cases requiring nephrectomy and even hysterectomy have been reported,[10,12] and maternal deaths have occurred.[25] However, by far the most frequent severe complications, aside from or in conjunction with hypertension, are rejection episodes successfully managed during pregnancy.[10,12,15] Nevertheless, during the postpartum period many of these patients successfully managed during pregnancy continue to deteriorate and lose the allograft function within months or a few years.[10,12,14,15] Implantation of a new graft is usually required with all the potential complications of this procedure, which could be as severe as death of the patient.[10] Only one patient in our series had a rejection episode that prompted early termination of pregnancy with a fetus who survived.

The usual intrapartum, postpartum, or postcesarean complications may be observed. Of particular importance are hemorrhage,[12] infectious episodes such as chorioamnionitis (following PRM), endomyometritis, pelvic thrombophlebitis, or cystitis-pyelitis.[10,28] It is advisable to be generous with prophylactic antibiotics.

There seems to be no significant long-term effect of pregnancy on the allograft function or life expectancy of the recipient when compared to patients who did not become pregnant.[11,13] The reported proportion of rejections and deaths in long-term follow-ups have been the same as for the overall transplant population.[3,13] In our series there was one maternal death due to a cerebrovascular accident: a patient who neglected her controls and died 2 years after delivery. Two postpartum graft losses occurred, 2 and 5 years after delivery. One of these patients is the recipient of a sister's kidney, but it was later found that this family has a genetic condition causing membranous glomerulonephritis.

Neonatal Mortality and Morbidity

The risk of neonatal death is proportionally increased because of the high incidence of *prematurity* and IUGR. Despite that somber perspective, most series report an excellent survival rate, ranging from 90% to 100%.[7,10,15,25,28] The more common causes of death were respiratory distress syndrome (RDS),[7,10,12] congenital malformations,[12,13] or infections.[10] No perinatal deaths were observed in our series. The morbidity is markedly higher owing to several factors. The already mentioned prematurity is a frequent cause of RDS with all its potential problems. Congenital *viral infections* (hepatitis B and cytomegalovirus) and *septic infections* are real threats to these neonates and require immediate attention. All infants born to mothers who are hepatitis B surface antigen (HBSAg)-positive should receive the standard immunoglobulin and vaccine within 48 hours of birth.

The risk of *congenital malformations,* in theory, is high because the more commonly used immunosuppressant azathioprine is markedly teratogenic in laboratory animals.[41] However, the incidence of reported malformations in neonates is low—not much higher than in the general population—when the medications taken are azathioprine and prednisone at the moderate recommended doses.[11,26] This fortunate circumstance is probably due to the fact that the degradation of azathioprine by the human embryo does not produce its teratogenic derivative.[41] Nevertheless, when warfarin has been given, the number of malformations is high.[12] None of the neonates in the present series had anatomic malformations. The overall experience suggests that the use of azathioprine and prednisone, at the standard moderate doses usually prescribed, do not predictably induce congenital malformations.

However, other effects of azathioprine on the neonate have been observed. It has been shown that neonatal thrombocytopenia and leukopenia are related to the maternal leukocyte count at 32 weeks' gestation[26] and that they may be prevented if the dose of azathioprine is adjusted to avoid the mild leukopenia observed in some pregnant patients. The thrombocytopenia may have been the cause of reported coagulopathies.[13] In the usual association with prednisone it has been shown to induce transient immunosuppression in neonatal life.[26,42] Other neonatal problems observed have been adrenocortical insufficiency[10] and transient chromosome abnormalities.[43] A less obvious but insidious aftereffect seen in the neonates is hypocalcemia secondary to parathyroid suppression, which is the consequence of the hypercalcemia frequently seen in transplant recipients.[13,44,45] Many delayed effects of azathioprine have not yet been reported. However, some caution must be observed because animal studies have demonstrated fertility alterations in the generation subjected to intrauterine exposure to metabolites of the drug.[46]

There is still limited experience with the newer immunosuppressant cyclosporine A and pregnancy. Its good effect in nonpregnant individuals appears to be clearly accepted[47]; however, only sporadic cases of its use during pregnancy have been reported, although the cases reported have had consistently good fetal outcomes.[34,48–50] Nevertheless, unpublished data (cited in ref. 34) have noted neonatal complications associated with the use of cyclosporine and even a severe congenital malformation. More experience is required before definitive statements can be made on this point.

Risk of Rejection During and Shortly After Pregnancy

The immunologic balance between recipient and graft is less precarious than frequently thought. Doses of immunosuppressive drugs are gradually reduced during the first year after transplantation. Although rejection episodes may be provoked by reducing immunosuppression, they can usually be reversed, and the likelihood of rejection diminishes as the interval from transplantation increases. The nature of the adaptation or homeostasis that develops, as months pass, depends on changes in both the graft and the host; the process has both passive and active components. The donated organ becomes less immunogenic with time, and the recipient develops feedback mechanisms that hold back immunologic damage to the graft.[51]

Stable renal allograft function has been observed for as long as 30 years, but its maintenance requires continued small doses of immunosuppressive drugs such as prednisone (10–15 mg daily) and azathioprine (100–150 mg daily). Only a few recipients can discontinue these drugs without losing the graft, but it is still not possible to predict who they are. Severe viral infections such as influenza can precipitate rejection as the viral episode begins to improve. Renal function must be monitored weekly for 3 to 4 weeks after severe viral illnesses (the common cold does not precipitate rejection).

If a kidney transplant is functioning normally at the end of the first or second year, it is unlikely that it will be rejected unless the recipient discontinues immunosuppressive drugs. Of the 19 women in our series, 18 experienced no rejection during pregnancy and maintained normal renal function more than a year after delivery. Thirteen of these women continue to have normal renal function as long as 15 years after delivery.

The only rejection episode that occurred during pregnancy in our experienced was in a woman who was known to be noncompliant with the immunosuppression regimen and was suffering from chronic rejection before she became pregnant. Her serum creatinine was more than 2.0 mg/dl prior to pregnancy, and she was hypertensive. The rejection during midpregnancy was reversed temporarily, but the graft failed several months after delivery.

Two other women have lost their grafts. One graft failed 5 years after delivery. Renal function was known to be normal until 4 years after delivery; at that time she stopped taking her immunosuppressive drugs and stopped visiting her physician. The other woman died 2 years after delivery from a cerebrovascular accident. Her renal function was normal for the first year after delivery, but during the second year she stopped seeing her physician. Her native kidneys had failed because of malignant hypertension, and it is likely that uncontrolled hypertension led to her stroke and death.

We strongly support Davison's main criteria for predicting the likelihood of successful pregnancy after kidney transplantation: good health for at least 2 years after transplantation, absence of hypertension, and serum creatinine of 2 mg/dl or less.[11] When these criteria are fulfilled the possibility of chronic rejection is significantly diminished; moreover, it is ensured that the reserve of renal function required for a successful pregnancy is present in the allograft.

Cyclosporine and Pregnancy

Of the 19 pregnancies here reported, 18 occurred in women whose maintenance immunosuppression consisted of prednisone (5–15 mg daily) and azathioprine (50–150 mg daily). Only the most recent delivery was from a woman who was taking cyclosporine, which the U.S. Food and Drug Administration approved in late 1983. Most of the reported experience with pregnancy after renal transplantation has involved azathioprine and prednisone. The limited medical literature on cyclosporine and pregnancy suggests that cyclosporine presents neither more nor less risk to mother and baby than azathioprine so long as renal function is good and blood pressure is normal.[34,48–50] The major problem posed by cyclosporine is that it is nephrotoxic to many recipients and frequently causes or aggravates hypertension.[52] These side effects tend to diminish as the dose decreases and clearly affect some recipients more than others at the same dose. Nevertheless, it is common for the serum creatinine of a patient on cyclosporine to be greater than 2.0 mg/dl a year or two after transplantation; many of these patients require antihypertensive drugs. Although some are normotensive and have a serum creatinine level of 2 mg/dl or less, it is likely that a larger proportion of women on cyclosporine will fail to meet the stated criteria for proceeding with pregnancy. These women would be forced to choose among foregoing pregnancy, proceeding despite compromised renal function, or switching from cyclosporine to azathioprine in hope of improving renal function prior to pregnancy. Unfortunately, the switch from cyclosporine to azathioprine triggers rejection in about 15% of recipients.[53] Rejection can almost always be detected and reversed when the recipient is monitored closely, but many are unwilling to take the risk.

Because cyclosporine, in therapeutic doses, is nephrotoxic to many renal transplant recipients and because it is expensive, a number of transplant specialists recommend accepting risk, 1 to 2 years after transplantation, of reducing the dose of cyclosporine or discontinuing it altogether in favor of azathioprine unless the recipient's renal function is excellent and purchase of cyclosporine does not cause financial hardship. We favor this approach, and it probably explains why we have had the occasion to carry only one woman through pregnancy on cyclosporine.

Treatment of Rejection During Pregnancy and Gynecologic Surgery

There are no effective special precautions that can be taken to avoid or prevent rejection other than to maintain a stable immunosuppressive regimen and to be cautious when using drugs that alter the metabolism of immunosuppressive drugs via the cytochrome P-450 system or some other pathway. Examples include phenytoin (Dilantin), cimetidine, and oral antifungal agents. It was common practice several years ago to avoid transfusions that contained platelets or leukocytes for fear that alloantigens present on their surfaces might trigger rejection. Current practice is to administer whole blood and platelets whenever they are indicated; there is no evidence that they precipitate rejection.

If rejection does occur during pregnancy, it is treated in the same way as for any other transplant recipient, which usually means two or three daily intravenous injections of methylprednisolone (250–1000 mg). If renal function does not improve by the third day, we move directly to a 10- to 14-day course of the pan-T-lymphocyte monoclonal antibody OKT-3.[54] Doses of azathioprine or cyclosporine are usually not increased during rejection episodes.

Management of Immunosuppressive Drugs During Pregnancy and Surgery

Maintenance immunosuppressive drugs are usually used in one of the following three combinations.

Azathioprine + prednisone
Cyclosporine + prednisone
Azathioprine + cyclosporine + prednisone

All three drugs are taken by mouth once or twice daily. The form, frequency, and common dosage of each are presented in Table 14-1. If the recipient cannot take medicine by mouth, each of the three drugs is available for intravenous injection. The most commonly used parenteral steroids are hydrocortisone or methylprednisolone. They are usually given every 8 or 12 hours, and the doses are increased appropriately to cover the stress needs of a person undergoing surgery. The intravenous salt of azathioprine is given only once a day and at the same dose as the oral form of the drug. The intravenous dose of cyclosporine is only one-third the corresponding oral dose and is usually divided into two injections 12 hours apart.

TABLE 14-1. Use of immunosuppressive drugs during pregnancy and surgery.

Drug	Form	Total daily dose (mg/kg)	Frequency
Azathioprine	Tablet	2.0	Daily
Prednisone	Tablet	0.2	Daily
Cyclosporine	Liquid	4.0	Divided q12h

General Considerations

It is not uncommon to examine an allograft recipient for amenorrhea and to find her to be in an advanced stage of gestation. In fact, many of the published series include such patients, and three in our own series presented between the 16th and 25th weeks. As previously pointed out, this situation is probably due to the patient's ignorance of this possibility and the lack of warning or advice by medical personnel. Ideally, pregnancy in allograft recipients must be planned, and conception should take place only if the woman fulfills a minimum set of guidelines intended to minimize the chances of pregnancy failure or deterioration of the transplant. Davison and Lindheimer[11] listed the following criteria: good health for at least 2 years after transplantation; absence of hypertension and proteinuria; no evidence of rejection and calyceal distention on urography; plasma creatinine of 2 mg/dl or less; therapy with prednisone 15 mg/day or less and azathioprine 2 mg/kg/day or less. Of course this list is neither complete nor inflexible.

To follow the stated guidelines the patients should be counseled and prescribed contraceptive measures that are known to be safe and effective. In principle, intrauterine contraceptive devices (IUDs) may be contraindicated because of the associated small increase in the chance of genital infections, which can become very serious in an immunosuppressed individual. It has also been postulated that the protection of IUDs is lessened because of the prednisone and azathioprine effects.[55]

Oral contraceptives may be the ideal means, but the possibility of causing thromboembolic accidents raises questions about their safety. Nevertheless, the very low dose cyclic preparations maybe safer from this standpoint and can be appropriately used. Perhaps the safest methods are single or combination barrier devices, such as diaphragm and condoms.

Conclusion

Pregnancy is possible for the allograft recipient, although about 30% of pregnancies are lost before reaching the second trimester. Of the pregnancies that reach that stage, a large number develop serious problems, and there is a high incidence of perinatal complications for the fetus. Most, however, survive without major handicaps. These good results are achieved only when pregnancies are planned, they are followed closely by an obstetrician and a transplant specialist, and the fetus is handled at delivery by an experienced neonatologist.

As with all immunosuppressed subjects, the renal allograft recipient is at high risk to develop malignancies. Gynecologic malignancies have been shown to occur in a much higher incidence than in the general population.[2,11,56] Frequent evaluations should help to discover these processes while they are still in their early stages.

References

1. Emmanouel DS, Lindheimer MD, Katz AI. Endocrine function in uremia. In Eknoyan E, Kouchel JP, (eds): The Systemic Consequences of Renal Failure. New York, Grune & Stratton, 1984;177–232.
2. Meier PR, Makowski EL. Pregnancy in the patient with a renal transplant. Clin Obstet Gynecol 1985;27:902.
3. Washer GF, Schroter GPS, Starzl TE. Causes of death after kidney transplantation. JAMA 1983;250:49.
4. Guttman RD. Renal transplantation. N Engl J Med 1979;301:1038.
5. Absy M, Metreweli C, Matthews C, et al. Changes in transplanted kidney volume measured by ultrasound. Br J Radiol 1987;60:525.
6. Stuart FP. Selection, preparation, and management of kidney transplant recipients. Med Clin North Am 1978;62:1381.
7. Kalman TP, Wilson PG, Kalman CM. Psychiatric morbidity in long term renal transplant recipients undergoing hemodialysis. JAMA 1983;250:55.
8. Waltzer WC, Coulam CB, Zincke H, et al. Pregnancy in renal transplantation. Transplant Proc 1980;12:221.
9. Davison JM. The effect of pregnancy on kidney function in renal allografts recipients. Kidney Int. 1985;27:74.
10. Penn I, Makowski EL, Harris P. Parenthood following renal transplantation. Kidney Int 1980;18:221.
11. Davison JM, Lindheimer MD. Pregnancy in women with renal allografts. Semin Nephrol 1984;4:240.
12. O'Donnell D, Sevitz H, Seggie JL, et al. Pregnancy after renal transplantation. Aust NZ J Med 1985;15:320.
13. Registration Committee of the European Dialysis and Transplant Association. Successful pregnancies in women treated by dialysis and kidney transplantation. Br J Obstet Gynaecol 1980;87:839.
14. Whetham JCG, Cardella C, Harding M. Effect of pregnancy on graft function and graft survival in renal cadaver transplant patients. Am J Obstet Gynecol 1983;145:193.
15. Marushak A, Weber T, Boek J, et al. Pregnancy following kidney transplantation. Acta Obstet Gynecol Scand 1986;65:557.
16. Coulam CB, Zincke H, Sterioff S. Relationship between donor age and outcome of pregnancy in a renal allograft population. Transplantation 1982;33:97.
17. Frick MP, Feinberg SB, Sibley R, et al. Ultrasound in acute renal rejection. Radiology 1981;138:657.
18. Levzow BL. The appearance of renal transplant rejection with ultrasound. Med Ultrasound 1982;6:43.
19. Maklad NF, Wright CH, Rosenthal SJ. Gray scale ultrasonic appearances of renal transplant rejection. Radiology 1979;131:711.

20. Murray JE, Reid DE, Harrison JH, et al. Successful pregnancies after human renal transplantation. N Engl J Med 1963;269:341.
21. Hadi HA, Stafford CR, Williamson JR, et al. Pregnancy outcome in renal transplant recipients. South Med J 1986;79:959.
22. Rifle G, Traeger J. Pregnancy after renal transplantation: an international survey. Transplant Proc 1975;7:723.
23. Williams PF, Jelen J. Eclampsia in a patient who had had a renal transplant. Br J Med 1979;2:972.
24. Williams PF, Johnstone M. Normal pregnancy in renal transplant recipient with history of eclampsia and intrauterine death. Br Med J 1982;285:1535.
25. Ogburn PL, Kitzmiller JL, Hare JW, et al. Pregnancy following renal transplantation in class T diabetes mellitus. JAMA 1986;255: 911.
26. Davison JM, Dellagrammatikas H, Parkin JM. Maternal azathioprine therapy and depressed hemopoiesis in the babies of renal allograft patients. Br J Obstet Gynaecol 1985;92:233.
27. Leigh DA. The outcome of urinary tract infections in patients after human cadaveric renal transplantation. Br J Urol 1969;45:406.
28. Sciarra JJ, Toledo-Pereyra LH, Bendel RP, et al. Pregnancy following renal transplantation. Am J Obstet Gynecol 1975;123:411.
29. Scott JR, Cruikshank DP, Corry RJ. Ectopic pregnancy in kidney transplant patients. Obstet Gynecol 1978;51:56-S.
30. Spencer ES, Anderson HK. Viral infections in renal allograft recipients treated with long-term immunosuppression. Br Med J 1979; 2:829.
31. Scott JR. Fetal growth retardation associated with maternal administration of immunosuppressive drugs. Am J Obstet Gynecol 1977; 128:668.
32. Manifold IH, Champion AG, Goepel JR, et al. Pregnancy complicated by gestational trophoblastic disease in a renal transplant recipient. Br J Med 1983;287:1025.
33. Lelievre G, Ribet M, Gosselin B, et al. Choriocarcinoma apres transplantation. J Urol Nephrol 1978;84:345.
34. Mallat SG, Brensilver JM. Successful pregnancy in a cyclosporine-treated renal transplant recipient with sickle cell disease. Transplantation 1988;45:660.
35. Westney LS, Callender CO, Stevens J, et al. Successful pregnancy with sickle cell disease and renal transplantation. Obstet Gynecol 1984;63:752.
36. Merrill LK, Board JA, Lee HM. Complications of pregnancy after renal transplantation including a report of spontaneous uterine rupture. Obstet Gynecol 1973;41:270.
37. Trudinger BJ, Cook CM, Giles WB, et al. Umbilical artery flow velocity waveforms—a randomized controlled trial. Lancet 1987;1: 188.
38. Trudinger BJ, Cook CM, Jones L, et al. A comparison of fetal heart rate monitoring and umbilical artery waveforms in the recognition of fetal compromise. Br J Obstet Gynecol 1986;93:171.
39. Farber M, Kennison RD, Jackson HT, et al. Successful pregnancy after renal transplantation. Obstet Gynecol 1976;48:2-S.
40. Nolan GH, Sweet RL, Laros RK, et al. Renal cadaver transplantation followed by successful pregnancies. Obstet Gynecol 1974;43:732.
41. Williamson RA, Karp LE. Azathioprine toxicity: review of the literature and case report. Obstet Gynecol 1981;58:247.
42. Cotte CJ, Meuwissen HJ, Pickernig RJ. Effects on the neonate of prednisone and azathioprine administered to the mother during pregnancy. J Pediatr 1974;85:324.
43. Lower GD, Stevens LE, Najarian JS, et al. Problems from immunosuppressives during pregnancy. Am J Obstet Gynecol 1971;111: 1120.
44. David DS, Sakai S, Brennan BL, et al. Hypercalcemia after renal transplantation: long term follow-up data. N Engl J Med 1973;289:398.
45. Schoenike SL, Kaldenbaugh HH, Kaplan AM, et al. Transient hypoparathyroidism in an infant of a mother with a renal transplant. Am J Dis Child 1978;132:530.
46. Reimers TJ, Sluss PM. 6-Mercaptopurine treatment of pregnant mice: effects on second and third generations. Science 1978;201:65.
47. Merion RM, White DJG, Thiru S, et al. Cyclosporine: five years' experience in cadaveric renal transplantation. N Engl J Med 1984;310:148.
48. Burrows DA, O'Neil TJ, Sorrells T. Successful twin pregnancy after renal transplant maintained on cyclosporine A immunosuppression. Obstet Gynecol 1988;72:459.
49. Klintmalm G. Renal function in a newborn baby delivered of a renal transplant patient

taking cyclosporine. Transplantation 1984; 38:198.
50. Lewis GJ, Lamont CAR, Lee HA, et al. Successful pregnancy in a renal transplant recipient taking cyclosporine A. Br Med J 1983; 286:603.
51. Stuart FP, McKearn TJ, Weiss A, Fitch FW. Suppression of rat renal allograft rejection by antigen and antibody. Immunol Rev 1980; 49:127.
52. Myers BD, Ross J, Newton L, et al. Cyclosporine associated chronic nephropathy. N Engl J Med 1984;311:699.
53. Stuart FP, Thistlethwaite JR, Haag BW, et al. Conversion from cyclosporine to azathioprine six months after kidney transplantation. Transplant Proc 1985;17: 2681.
54. Thistlethwaite JR, Gaber AO, Haag BW, et al. OKT3 treatment of steroid resistant renal allograft rejection. Transplantation 1987; 43:176.
55. Buhler M, Papiernik E. Successive pregnancies in women fitted with intrauterine devices who take anti-inflammatory drugs. Lancet 1983;1:483.
56. Porreco R, Penn I, Droegemueller W, et al. Gynecologic malignancies in immunosuppressed organ homograft recipients. Obstet Gynecol 1975;45:359.

15
Pregnancy and Cardiac Prosthetic Valves

JOSE NIZ-RAMOS, CARLOS IBARRA-PEREZ, and ALBERTO ALVARADO-DURAN

Female patients with significant abnormalities of one or more of the heart valves had, until few years ago, a poor life expectancy, and their possibility of reproduction was severely curtailed. Currently, those women may have a relatively normal life if they undergo replacement of the diseased valve(s) with implantation of a prosthetic cardiac valve, and they may recover the possibility of reproducing. However, a number of problems pertinent to implantation and maintenance of prosthetic valves constitute a serious risk of increased morbidity and mortality for the patient or her conceptus. It may well be stated that this circumstance encompasses new pathologic entities that must be confronted and solved by the obstetrician, anesthesiologist, and neonatologist.

There seems to be general agreement that valve implantation is done in class III and IV patients (New York Heart Association classification), although other factors such as cardiomegaly, chronic atrial fibrillation, and the hemodynamic state may also be taken into consideration.

In 1975 approximately 40 prosthetic valves per year were implanted at the Cardiology Hospital of the Centro Medico Nacional I.M.S.S.[1] Currently about 30 per year are implanted,[2] and about one-half of the recipients may later require obstetric care.

Prosthetic valves may be classified into two distinct types: mechanical (e.g., cage and ball, cage and dish, tilted disk); and biologic, constructed with material of animal origin such as fascia lata, dura mater, or valves obtained from animal hearts. Generally biologic valves are preferred because they have a better hemodynamic function and require no anticoagulation, and complications such as thrombosis, thromboembolic episodes, hemolysis regurgitation, fractures, and deformation are seldom observed. On the other hand, the mechanical valves require anticoagulation to prevent thrombosis and thromboembolic episodes. Nevertheless, the biologic implants have the disadvantage of a limited average life (7 years) particularly because of calcification, with its corollary of valvular stenosis and malfunction necessitating a new replacement procedure with its increased mortality, reported at about 9%.[2] There may also be an indication for anticoagulation for biologic valve carriers in cases of either the presence of a large left atrium, a history of thromboembolism, or atrial fibrillation.[3]

The more commonly used anticoagulants are the coumarin derivatives with their property of interfering with synthesis of vitamin-K-dependent clotting factors (factors II, VII, IX, and X). On the other hand, their low molecular weight (about 1000) facilitates their passage through the placental barrier and the anticoagulant effect on the fetus, with reported fetal and neonatal bleeding episodes and congenital malformations. The warfarin embryopathy (so called because it was the first coumarin derivative associated with the syndrome) has clearly been shown to be caused by the maternal ingestion of these preparations.[4–7] It is characterized by hypoplasia of the nose bones and chondrodysplasia punctata. The mechanism of this alteration is unknown; it was originally postulated to be the outcome of microhemorrhages. However, Hall et al.[7] showed that the critical exposure period to the drug to produce the embryopathy is between gestational weeks 6 and 9, but at that stage the vitamin-K-dependent clotting factors are not yet synthesized[8]; this hypothesis can thus be ruled out. The mechanism proposed by Hall and colleagues involves an enzymatic inhibition during a critical period of osteogenesis.[7] The warfarin syndrome is a phenocopy of the rhizomelic form of chondrodysplasia punctata, different from the Conradi Hunerman type in terms of its distribution of lesions.[6,7] The other morphologic abnormalities (of the central nervous or ocular system) do not have a recognized constant pattern. Raivo et al.[6] found that they are related to the ingestion of the drugs during the second or third trimester of pregnancy with excessive anticoagulation, and that they seem to correspond to the healing processes that follow local hemorrhage.[9] Another problem observed among patients receiving coumarin derivatives during the first trimester is the high incidence of spontaneous abortions, ranging from 28[10] to 44%.[11] Some authors have reported that the anticoagulant effect disappears about 3 weeks after discontinuing the medication, although the duration of the anticoagulant effect is not known with precision.[5,9,12] On the other hand, it is well known that strict control of anticoagulation is needed to prevent maternal and fetal complications,[5] achieved by maintaining a prothrombin time around two times above control.

The other anticoagulant medication available for use in these patients is heparin, which because of its high molecular weight (around 8,000–15,000) does not cross the placental barrier and therefore has no anticoagulant effect on the fetus. However, its prolonged use could constitute a risk for the mother and even the fetus, as there have been observations of fetal losses among patients receiving heparin for long periods of time, not different from the figures reported among those taking coumarin derivatives. This observation suggests that prolonged use of heparin is no better than the oral medication for those pregnant patients who need anticoagulation for protracted periods (more than 100 days).[4,7,13–18] It has been suggested that the fetal mortality observed among patients receiving heparin is probably due to the secondary chelating side effect of heparin.[17] Furthermore, other undesirable side effects have been reported associated with the prolonged use of heparin: thrombocytopenia, alopecia, and osteoporosis causing spontaneous fractures.[18,19]

The platelet-antiaggregatory properties of acetylsalicylic acid and dipyridamole do not protect the mother from thrombosis and furthermore have been implicated in the increased risk of maternal hemorrhage, neonatal death, and intrauterine fetal death.[15] These substances have been suggested as the treatment of choice for the pregnant patients,[20–22] but some catastrophic maternal and fetal effects have been reported (sudden maternal deaths due to prosthetic dysfunction) following their substitution for coumarin derivatives.[3,14,16,23]

The ideal anticoagulant has not yet been found, but several treatment protocols have been devised aimed at avoiding the warfarin syndrome and the hemorrhagic problems of the fetus and neonate, either during labor or the early neonatal period. Heparin, because it does not cross the placental barrier, possesses a property that may be useful, but its prolonged use has been implicated in fetal losses. It is therefore recommended that it be used only during the first trimester of pregnancy at the time of the critical period, as reported by Hall et al.[7] and confirmed by Iturbe-Alesio et al.,[24] particularly between the 6th and 12th weeks. From the 13th week on, the use of coumarin derivatives is resumed. Thus the danger of warfarin embryopathy can be avoided, and heparin is used for fewer than 100 days (the critical length).

The substitution of heparin for coumarin

derivatives two to several days before delivery has been recommended,[3,4,9,14,24] but the results have not been satisfactory. Some authors have reported significant bleeding during the peripartum period in up to 13% of patients receiving low doses of heparin.[10,24,25] On the other hand, when switching from heparin to coumarin and vice versa, the patient goes through critical periods with increased chances of valvular thrombosis or thromboembolic episodes.[24] For these reasons we have proposed other protocols, such as rapid conversion of the anticoagulant effect of coumarin derivatives with fresh frozen plasma. Although theoretically this procedure does not correct the anticoagulation effect on the fetus because the vitamin-K-dependent factors do not cross the placenta in significant amounts,[26] our observations have shown the passage of factors II and IX. These factors, according to our results, are sufficient to prevent hemorrhagic problems in the fetus and neonate, without maternal risk of either thrombosis or excess bleeding. An important point to consider when using anticoagulants during pregnancy is the need to avoid standardized schemes but, instead, to insist on the *individualized* response, aiming at a heparin dose sufficient to maintain the partial thromboplastin time between 1.5 and 2.0 times the control figures. When staying within these values no problems have been observed.[27]

Material and Methods

From 1970 to 1985 there were 139 patients with heart valve prostheses who had 166 pregnancies and were cared for at the Gynecology-Obstetric and Cardiology Hospitals of the Centro Medico Nacional, I.M.S.S. A total of 108 patients had been fitted with one or two mechanical prostheses, 28 had biologic prostheses only, and 3 had one of each. The 108 patients with a mechanical prostheses had 131 pregnancies, the 28 with biologic prostheses had 32 pregnancies, and the 3 with combined prostheses had 3 pregnancies. Of the 166 pregnancies, in 143 cases coumarin derivatives were given: the 131 with mechanical prostheses, 9 with biologic prostheses who had other indications for anticoagulation, and the 3 with combined prostheses. The Therapeutic Abortion Committee of the hospital authorized termination during the first trimester for 12 of these pregnancies; one patient had an illegal abortion, and two had ectopic gestations. From the total group, 105 patients received coumarin derivatives throughout the whole pregnancy, aiming at maintaining the prothrombin time between 2.0 and 2.5 times above control; they constitute group I.

Prospectively we recruited 23 patients (group II) who planned their pregnancies and substituted intravenous heparin for coumarinics, from week 5–6 to 12–13 of amenorrhea (age of gestation confirmed by sonogram). Heparin was given every 8 hours with the purpose of maintaining a partial thromboplastin time between 1.5 and 2.0 times above control. From the 13th week until delivery they were switched back to coumarinics, aiming at the same prolongation of prothrombin time as during early gestation. In both groups I and II, the prothrombin time was checked once or twice per week. The 23 pregnancies with biologic prostheses that did not require anticoagulation constitute group III.

All of these patients were evaluated weekly in the obstetrics and cardiology outpatient clinics. Laboratory test were ordered as needed, serial ultrasonograms were done to evaluate the growth of the pregnancy, and weekly nonstress tests (NST) were done from the 33rd week on. Medications known to interfere with the effect of coumarin derivatives were avoided among the patients receiving it.

The following guidelines were applied during labor, delivery, and the early puerperium.

1. The type of delivery was decided by agreement among cardiologist, obstetrician, anesthesiologist, and neonatologist after thorough review of obstetric and fetal, as

well as cardiovascular and hemodynamic conditions.

2. In the majority of cases the coumarin derivatives were continued until delivery in the patients who were taking them.
3. The prothrombin time was checked at the onset of labor or at the preoperative evaluation. Fresh frozen plasma was given during labor or surgery (or both). When the prothrombin time was 1.75 times or more above control, it was rechecked after completion of the delivery. (This point of the protocol was started in 1978.)
4. There was continuous monitoring of maternal vital signs, fetal heart rate, and uterine contractions, with administration of the medications and analgesia-anesthesia necessary to maintain a stable cardiac function.
5. Prophylaxis for subacute bacterial endocarditis (SBE) was accomplished with aqueous penicillin 5 million units every 6 hours and streptomycin 1 g every 12 hours; this regimen was given for 5 days.
6. Most patients underwent tubal ligation for sterilization.
7. Coumarinics were resumed the same day (the following day at the latest) at the same dose as before and subsequently adjusted to the appropriate prothrombin time.

The management of the neonate was, in general, as follows.

1. Determination of prothrombin time in umbilical cord blood, complete blood clotting factors, and complete blood count
2. Administration of vitamin K_1 in the usual newborn dose
3. Complete physical examination with special emphasis on discovering congenital anomalies
4. In-depth neurologic evaluation and electroencephalograms.
5. Radiographs of head, spine, and extremities to evaluate for chondrodysplasia punctata
6. Psychometric evaluation
7. Genetic studies

The clinical characteristics, complications of pregnancy, cardiovascular complications, and perinatal results including the presence or absence of congenital anomalies were compared among the three groups. Especially important were the prothrombin times before labor, intrapartum, and postpartum in maternal and umbilical blood in the patients who received fresh frozen plasma, as stated above, and those who did not receive it because of rapid labor or late arrival in labor and immediate delivery.

Results

Table 15-1 summarizes the clinical characteristics of the three groups of patients. Age, gravity, and parity were similar in all groups. Cardiac alterations were also similar in groups I and II. The location of the prostheses and other surgical procedures were also distributed uniformly.

Complications of pregnancy in the three groups (Table 15-2) occurred with the same frequency in all groups and had an incidence equivalent to that in our general obstetric population. Heart failure occurred more often in group I than in groups II and III ($p < 0.05$), as seen in Table 15-3. Excessive bleeding due to overanticoagulation was obvious in group I when compared with group II. Of course, it did not occur in group III. Cerebral and pulmonary thromboembolic accidents registered in group I (four cerebral and three pulmonary) occurred among the cases managed early in the program, and all followed discontinuation of coumarin therapy. One patient from group III with a Hanckoc-type biologic prosthesis in the mitral valve had an episode of cerebral embolism in the immediate postpartum period following an uncomplicated delivery. She had sequelae of hemiplagia and subsequent hemiparesis. Another patient at the end of the first trimester, when switching from heparin to coumadin, had a syndrome suggestive of prosthetic dysfunction, confirmed and diagnosed as prosthetic thrombosis. She was reoper-

TABLE 15-1. Clinical data of patients in three groups.[a]

Parameter	Group I	Group II	Group III
No. of pregnancies	105 (69.5%)	23 (15.2%)	23 (15.2%)
Average age (± SD)	26.3 ± 4.8	26.4 ± 5.5	26.2 ± 5.6
Average parity (± SD)	2.2 ± 1.1	2.1 ± 1.2	2.4 ± 1.3
Average live children (± SD)	1.8 ± 0.9	1.7 ± 0.9	1.6 ± 0.8
Atrial fibrillation	53 (50.5%)	10 (43.5%)	0
Previous cerebral embolism	5 (4.8%)	1 (4.3%)	0
Pts. with prostheses			
Small prosthesis	2 (1.9%)	1 (4.3%)	0
Mitral prosthesis	82 (78.1%)	17 (73.9%)	16 (69.6%)
Aortic prosthesis	5 (4.8%)	1 (4.3%)	2 (8.7%)
Mitral + aortic prostheses	5 (4.8%)	2 (8.7%)	1 (4.3%)
Mitral + tricuspid prostheses	4 (3.8%)	1 (4.3%)	1 (4.3%)
Mitral + aortic + tricuspid prostheses	1 (1%)	0	0
Aortic + mitral commissurotomy	4 (3.8%)	1 (4.3%)	1 (4.3%)

[a]See text for definition of the groups.

ated under extracorporeal circulation and given a new biologic valve (of dura mater) with satisfactory postoperative evolution to term; she was eventually delivered of a healthy infant.

The perinatal results, as seen in Table 15-4, revealed that in group I (taking coumarin throughout the whole pregnancy) there were 49 pregnancy losses (47%): miscarriages, fetal deaths, and neonatal deaths. There were 48 healthy infants delivered (46%) and eight neonates with malformations (7%), five of whom had the typical warfarin syndrome (two severe), one with patent ductus arteriosus and pulmonary stenosis, one with genus equinovarus, and the last one with corneal leukoma. In group II there were only 9% total pregnancy losses with 91% healthy infants delivered. Likewise in group III there were 96% healthy infants delivered. There was statistical difference in results between group I and each of the other two groups.

The incidence of cesarean section, blood loss of more than 500 ml during the operation or postpartum, and wound hematomas (laparotomy or episiotomy) were the same in all groups (Table 15-5). In all cases cesarean sections were performed for obstetric indications.

The effects of administering fresh frozen plasma, given to correct the anticoagulant effect, are shown in Figure 15-1. The prothrombin time fell in the maternal blood and more so in the cord blood, to near-control values. No changes in prothrombin time were observed among those who had no chance of receiving fresh frozen plasma, as they had the same values during labor and

TABLE 15-2. Pregnancy complications.

Complication	Group I	Group II	Group III
No. of pregnancies	105	23	23
Threatened abortions	15 (14.3%)	4 (17.4%)	3 (13%)
Threatened premature labor	13 (12.4%)	3 (13%)	3 (13%)
Premature rupture of membranes	11 (10.5%)	3 (13%)	2 (8.7%)
Cervicovaginitis	6 (5.7%)	1 (4.3%)	1 (4.3%)
Preeclampsia	4 (3.8%)	1 (4.3%)	1 (4.3%)
Abruptio placentae	3 (2.9%)	0	0
Polyhydramnios	2 (1.9%)	0	0

TABLE 15-3. Complications during pregnancy.

Diagnosis	Group I	Group II	Group III
Heart failure	12 (11.4%)	2 (8.7%)	2 (8.7%)
Overanticoagulation	9 (8.6%)	1 (4.3%)[a]	0 (NA)
Cerebral embolism[b]	4 (3.8%)	0	1 (4.3%)
Pulmonary embolism[c]	3 (2.9%)	0	0
Prosthetic thrombosis and replacement	0	1 (4.3%)[d]	0
Myocardial infarction	1 (1%)	0	0
Maternal mortality[e]	2 (1.9%)	0	0

[a]Following indomethacin administration to treat premature labor (fetal death).
[b]In all four patients of group I the coumarinic was discontinued, and in two patients platelet antiaggregatory medication was substituted. The patient in group III was immediately postpartum.
[c]In all three cases platelet antiaggregatory medication was substituted for coumarinics.
[d]See text.
[e]Both were caused by cerebral embolism following discontinuation of the coumarinic.

the postpartum period. Nevertheless, none of the infants had bleeding complications, and all received vitamin K_1 during the immediate neonatal period.

Comments

Pregnant patients with a prosthetic heart valve constitute a high-risk obstetric group—even those with a biologic prosthesis and not requiring anticoagulation. In general, the perinatal results are good, but the risk of thromboembolic accident or valvular calcification and stenosis, although infrequent, is a real threat for the mother as well as the fetus.

Furthermore, the need to use anticoagulant medications increases the danger and potential complications, the prevention of which requires close, uninterrupted follow-up. For some patients the pregnancy represents their only opportunity for reproduction, whether it is because of hemodynamic instability or the patient's lack of desire to face new risks.

The best protocol to follow when caring for heart valve prosthesis-carrying pregnant patient appears to be the one applied to our group II. It affords the lowest chance of maternal complications and fetal loss by individualizing the heparin doses during the critical period to maintain the partial thromboplastin time at 1.5 to 2.0 times the control. The dose ranges between 6000 and 8000 units every 6 to 8 hours, similar to that reported by Lee and co-workers.[27]

On the other hand, *not switching* to hepa-

TABLE 15-4. Perinatal results.

Group	Pregnancies (No.)	Pregnancy losses[a]				Live births	
		A-M	FD	NND	Total	Healthy	Malformed
I (warfarin)	105	41	6	2	49 (46.7%)	48 (45.7%)	8 (7.6%)
II (warfarin, heparin)	23	1	1	0	2 (8.7%)	21 (91.3%)	0
III (no anticoagulant)	23	1	0	0	1 (4.3%)	22 (95.6%)	0

Statistical calculations by χ^2. Between groups I and II-III, $p = < 0.001$. Between groups I and II, $p = < 0.01$. Between groups I and III, $p = < 0.005$. Between groups II and III, p = NS.
[a]A-M = abortion-miscarriage; FD = fetal death; NND = neonatal death.

TABLE 15-5. Delivery and puerperal complications.

Complication	Group I	Group II	Group III
Cesarean section	48 (45.7%)	10 (43.5%)	8 (34.8%)
Blood loss > 500 ml	7 (6.7%)	1 (4.3%)	0
Wound hematoma	9 (8.6%)	1 (4.3%)	0

rin at the end of pregnancy after taking coumarinics from the 13th week on avoids the difficult adjustments of the substitution, which often cause thromboembolic or prosthetic dysfunction complications.[24] Furthermore, timing of delivery is problematic if one tries to time the switch for labor and delivery.

These observations suggest that administration of fresh frozen plasma during labor or immediately prior to cesarean section is a satisfactory method that can be utilized in most cases with excellent results. It does not increase the maternal risk of valvular thrombosis or thromboembolic complications, and it diminishes the chances of peripartum maternal bleeding as well as fetal complication.

Conclusions

1. Pregnant patients with a prosthetic heart valve are at high risk of fetal and maternal morbidity and mortality.
2. The highest risk is observed among patients with mechanical prostheses and anticoagulation. The highest rate of pregnancy losses is recorded among them when pregnancy is not planned prior to conception.
3. All patients with a prosthetic heart valve requiring anticoagulation should plan their pregnancies and receive heparin between weeks 6 and 12 of pregnancy to maintain the partial thromboplastin time at 1.5 to 2.0 times above the control valve.
4. Coumarin derivatives should be taken prior to the 6th week and after the 13th week to the end of pregnancy, to maintain the prothrombin time at 2.0 to 2.5 times the control value. All medications that alter their effect must be avoided.
5. For delivery (vaginal or abdominal) the patient should receive fresh frozen plasma to maintain the prothrombin time up to 1.5 times above the control value. The fetus must receive vitamin K_1. Coumarinics should be resumed the same day or the following day at the latest adjusting the

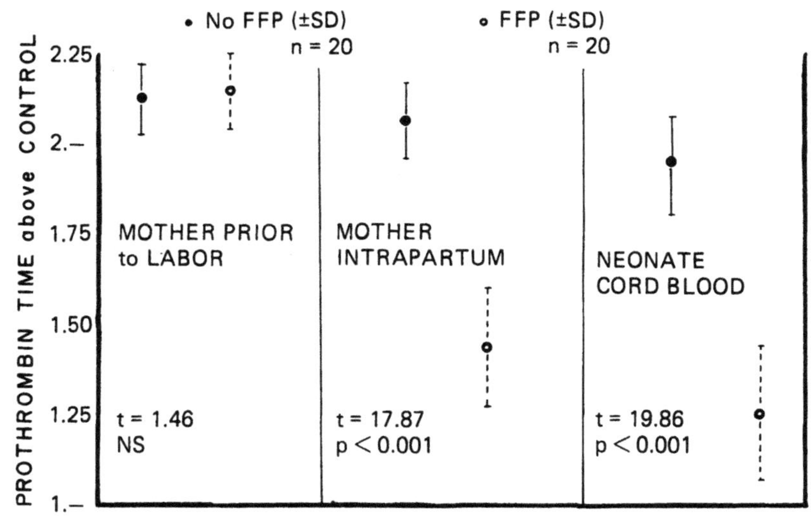

FIGURE 15-1. Corrections of prothrombin time with administration of fresh frozen plasma (FFP). The closed circles represent the group control, and the open circles the patients who received FFP. Findings are from mothers before labor (or surgery) and intrapartum, and from cord blood. Significant decreases are seen after FFP administration.

dose to prepartum prothrombin time values.
6. The type of delivery is determined by obstetric and hemodynamic conditions.
7. Subacute bacterial endocarditis prophylaxis is indicated using the appropriate antibiotics.
8. Fertility control is effected by either tubal ligation or contraceptive hormones at low dose, taking into consideration that estrogens decrease the anticoagulant action of coumarin derivatives. Therefore their dose should be adjusted accordingly to maintain the desired level of anticoagulation.

References

1. Memorias de las Jornadas del V aniversario del Hospital de Enfermedades del Torax. Centro Medico Nacional IMSS, 1976.
2. Palacios-Macedo J. Personal communication, 1987.
3. Harrison E, Roschke J. Pregnancy in patients with cardiac valve prostheses. Clin Obstet Gynaecol 1975;18:107–123.
4. Hirsch J, Cade JF, O'Sullivan E. Clinical experience with anticoagulant therapy during pregnancy. Br Med J 1970;1:270–273.
5. Hirsch J, Cade J, Gallus A. Anticoagulants in pregnancy: a review of indications and complications. Am Heart J 1972;83:301–305.
6. Raivo K, Ikonen E, Saarikoski A. Fetal risk due to warfarin therapy during pregnancy. Acta Pediatr Scand 1977;66:735–739.
7. Hall J, Pauli R, Wilson K. Maternal and fetal sequelae of anticoagulation during pregnancy. Am J Med 1980;68:122–140.
8. Bleyer W, Hakami N, Shepard T. The development of hemostasis in the human fetus and newborn infant. J Pediatr 1971;79:838–853.
9. Bloomfield D. Fetal deaths and malformations associated with the use of coumarin derivatives in pregnancy. Am J Obstet Gynecol 1970;107:888–892.
10. Salazar E, Zacarias A, Gutierrez N, Iturbe I. The problem of cardiac valve prostheses, anticoagulants, and pregnancy. Circulation 1984;70(suppl I):169–177.
11. Lutz DJ, Spittell JA Jr, Danielson GK, Fish CR. Pregnancy and its complications following cardiac valve prostheses. Am J Obstet Gynecol 1978;131:234–237.
12. Primidor B, Murray K, McAllen P. The management of anticoagulant therapy during and after pregnancy. Br J Obstet Gynaecol 1975; 82:740–743.
13. O'Neill H, Blake S, Sugrue D. Problems in the management of a patient with artificial valves during pregnancy. Br J Obstet Gynaecol 1982;89:940–943.
14. Matorras R, Reque J, Uzandizaga J. Prosthetic heart valve and pregnancy: a study of 59 cases. Gynecol Obstet Invest 1985;19: 21–26.
15. Tamari I, Eldear M, Rabinowitz B. Medical treatment of cardiovascular disorders during pregnancy. Am Heart J 1982;104:1357–1361.
16. Otterson W, McGranahan G, Fremman M. Successful pregnancy with McGovern aortic prostheses and long term heparin therapy. Obstet Gynecol 1968;31:273–278.
17. Muller M, Lebherz T. Antepartum thrombophlebitis: morbidity with long term heparin and proposed regimen therapy. Obstet Gynecol 1969;34:874–878.
18. Gervin A. Complications of heparin therapy. Surg Gynecol Obstet 1975;140:789–792.
19. Letsky EA, de Swiet M. Thromboembolism in pregnancy and its management. Br J Haematol 1984;57:543–552.
20. Casanegra P, Aviles G, Maturana G. Cardiovascular management of pregnant women with a heart valve prostheses. Am J Cardiol 1975;36:802–804.
21. Ahmad R, Rajah SM, Mearhs AJ. Dipyridamole in successful management of pregnant woman with prosthetic heart valve. Lancet 1976;2:1414–1416.
22. Sangines A, Escalante M, Cervantes F. Protesis cardiaca y embarazo: revision de 10 casos. Ginecol Obstet Mex 1981;49:109–112.
23. Ibarra-Perez C, Arevalo-Toledo N, Alvarez de la Cadena O, Noriega-Guerra L. The course of pregnancy in patients with artificial heart valves. Am J Med 1976;61:504–512.
24. Iturbe-Alessio I, Fonseca MC, Mutchinik O, et al. Risks of anticoagulant therapy in pregnant women with artificial heart valves. N Engl Med 1986;315:1390–1393.
25. Buxbaum A, Aygen MM, Shahin W, et al. Pregnancy in patients with prosthetic heart valves. Chest 1971;59:639–642.
26. Cade J, Hirsch J, Martin M. Placental barrier to coagulation factors, its relevance to the coagulation defect at birth and haemorrahage in the newborn. Br Med J 1969;2:281–286.
27. Lee PK, Wang RCY, Chow JSF, et al. Adjusted subcutaneous heparin in pregnant women with mechanical heart valves. J Am Coll Cardiol 1985;5:543–552.

Index

Italicized page numbers indicate figures or tables on page.

GPSR Compliance

The European Union's (EU) General Product Safety Regulation (GPSR) is a set of rules that requires consumer products to be safe and our obligations to ensure this.

If you have any concerns about our products, you can contact us on ProductSafety@springernature.com

In case Publisher is established outside the EU, the EU authorized representative is:

Springer Nature Customer Service Center GmbH
Europaplatz 3
69115 Heidelberg, Germany

Batch number: 10615115

Printed by Printforce, the Netherlands